IET ENERGY ENGINEERING 147

Distribution Systems Analysis and Automation

Other volumes in this series:

Volume 1	**Power Circuit Breaker Theory and Design** C.H. Flurscheim (Editor)
Volume 4	**Industrial Microwave Heating** A.C. Metaxas and R.J. Meredith
Volume 7	**Insulators for High Voltages** J.S.T. Looms
Volume 8	**Variable Frequency AC Motor Drive Systems** D. Finney
Volume 10	**SF_6 Switchgear** H.M. Ryan and G.R. Jones
Volume 11	**Conduction and Induction Heating** E.J. Davies
Volume 13	**Statistical Techniques for High Voltage Engineering** W. Hauschild and W. Mosch
Volume 14	**Uninterruptible Power Supplies** J. Platts and J.D. St Aubyn (Editors)
Volume 15	**Digital Protection for Power Systems** A.T. Johns and S.K. Salman
Volume 16	**Electricity Economics and Planning** T.W. Berrie
Volume 18	**Vacuum Switchgear** A. Greenwood
Volume 19	**Electrical Safety: A guide to causes and prevention of hazards** J. Maxwell Adams
Volume 21	**Electricity Distribution Network Design, 2nd Edition** E. Lakervi and E.J. Holmes
Volume 22	**Artificial Intelligence Techniques in Power Systems** K. Warwick, A.O. Ekwue and R. Aggarwal (Editors)
Volume 24	**Power System Commissioning and Maintenance Practice** K. Harker
Volume 25	**Engineers' Handbook of Industrial Microwave Heating** R.J. Meredith
Volume 26	**Small Electric Motors** H. Moczala *et al.*
Volume 27	**AC-DC Power System Analysis** J. Arrillaga and B.C. Smith
Volume 29	**High Voltage Direct Current Transmission, 2nd Edition** J. Arrillaga
Volume 30	**Flexible AC Transmission Systems (FACTS)** Y.-H. Song (Editor)
Volume 31	**Embedded Generation** N. Jenkins *et al.*
Volume 32	**High Voltage Engineering and Testing, 2nd Edition** H.M. Ryan (Editor)
Volume 33	**Overvoltage Protection of Low-Voltage Systems, Revised Edition** P. Hasse
Volume 36	**Voltage Quality in Electrical Power Systems** J. Schlabbach *et al.*
Volume 37	**Electrical Steels for Rotating Machines** P. Beckley
Volume 38	**The Electric Car: Development and future of battery, hybrid and fuel-cell cars** M. Westbrook
Volume 39	**Power Systems Electromagnetic Transients Simulation** J. Arrillaga and N. Watson
Volume 40	**Advances in High Voltage Engineering** M. Haddad and D. Warne
Volume 41	**Electrical Operation of Electrostatic Precipitators** K. Parker
Volume 43	**Thermal Power Plant Simulation and Control** D. Flynn
Volume 44	**Economic Evaluation of Projects in the Electricity Supply Industry** H. Khatib
Volume 45	**Propulsion Systems for Hybrid Vehicles** J. Miller
Volume 46	**Distribution Switchgear** S. Stewart
Volume 47	**Protection of Electricity Distribution Networks, 2nd Edition** J. Gers and E. Holmes
Volume 48	**Wood Pole Overhead Lines** B. Wareing
Volume 49	**Electric Fuses, 3rd Edition** A. Wright and G. Newbery
Volume 50	**Wind Power Integration: Connection and system operational aspects** B. Fox *et al.*
Volume 51	**Short Circuit Currents** J. Schlabbach
Volume 52	**Nuclear Power** J. Wood
Volume 53	**Condition Assessment of High Voltage Insulation in Power System Equipment** R.E. James and Q. Su
Volume 55	**Local Energy: Distributed generation of heat and power** J. Wood
Volume 56	**Condition Monitoring of Rotating Electrical Machines** P. Tavner, L. Ran, J. Penman and H. Sedding
Volume 57	**The Control Techniques Drives and Controls Handbook, 2nd Edition** B. Drury
Volume 58	**Lightning Protection** V. Cooray (Editor)
Volume 59	**Ultracapacitor Applications** J.M. Miller

Volume 62 **Lightning Electromagnetics** V. Cooray
Volume 63 **Energy Storage for Power Systems, 2nd Edition** A. Ter-Gazarian
Volume 65 **Protection of Electricity Distribution Networks, 3rd Edition** J. Gers
Volume 66 **High Voltage Engineering Testing, 3rd Edition** H. Ryan (Editor)
Volume 67 **Multicore Simulation of Power System Transients** F.M. Uriate
Volume 68 **Distribution System Analysis and Automation** J. Gers
Volume 69 **The Lightening Flash, 2nd Edition** V. Cooray (Editor)
Volume 70 **Economic Evaluation of Projects in the Electricity Supply Industry, 3rd Edition** H. Khatib
Volume 72 **Control Circuits in Power Electronics: Practical issues in design and implementation** M. Castilla (Editor)
Volume 73 **Wide Area Monitoring, Protection and Control Systems: The enabler for smarter grids** A. Vaccaro and A. Zobaa (Editors)
Volume 74 **Power Electronic Converters and Systems: Frontiers and applications** A.M. Trzynadlowski (Editor)
Volume 75 **Power Distribution Automation** B. Das (Editor)
Volume 76 **Power System Stability: Modelling, analysis and control** Abdelhay A. Sallam and Om P. Malik
Volume 78 **Numerical Analysis of Power System Transients and Dynamics** A. Ametani (Editor)
Volume 79 **Vehicle-to-Grid: Linking electric vehicles to the smart grid** J. Lu and J. Hossain (Editors)
Volume 81 **Cyber-Physical-Social Systems and Constructs in Electric Power Engineering** S. Suryanarayanan, R. Roche and T.M. Hansen (Editors)
Volume 82 **Periodic Control of Power Electronic Converters** F. Blaabjerg, K. Zhou, D. Wang and Y. Yang
Volume 86 **Advances in Power System Modelling, Control and Stability Analysis** F. Milano (Editor)
Volume 87 **Cogeneration: Technologies, optimisation and implementation** C.A. Frangopoulos (Editor)
Volume 88 **Smarter Energy: From smart metering to the smart grid** H. Sun, N. Hatziargyriou, H.V. Poor, L. Carpanini and M.A. Sánchez Fornié (Editors)
Volume 89 **Hydrogen Production, Separation and Purification for Energy** A. Basile, F. Dalena, J. Tong and T.N. Veziroğlu (Editors)
Volume 90 **Clean Energy Microgrids** S. Obara and J. Morel (Editors)
Volume 91 **Fuzzy Logic Control in Energy Systems with Design Applications in MATLAB®/Simulink®** İ.H. Altaş
Volume 92 **Power Quality in Future Electrical Power Systems** A.F. Zobaa and S.H.E.A. Aleem (Editors)
Volume 93 **Cogeneration and District Energy Systems: Modelling, analysis and optimization** M.A. Rosen and S. Koohi-Fayegh
Volume 94 **Introduction to the Smart Grid: Concepts, technologies and evolution** S.K. Salman
Volume 95 **Communication, Control and Security Challenges for the Smart Grid** S.M. Muyeen and S. Rahman (Editors)
Volume 96 **Industrial Power Systems with Distributed and Embedded Generation** R Belu
Volume 97 **Synchronized Phasor Measurements for Smart Grids** M.J.B. Reddy and D.K. Mohanta (Editors)
Volume 98 **Large Scale Grid Integration of Renewable Energy Sources** A. Moreno-Munoz (Editor)
Volume 100 **Modeling and Dynamic Behaviour of Hydropower Plants** N. Kishor and J. Fraile-Ardanuy (Editors)
Volume 101 **Methane and Hydrogen for Energy Storage** R. Carriveau and D.S.-K. Ting
Volume 104 **Power Transformer Condition Monitoring and Diagnosis** A. Abu-Siada (Editor)
Volume 106 **Surface Passivation of Industrial Crystalline Silicon Solar Cells** J. John (Editor)

Volume 107 **Bifacial Photovoltaics: Technology, applications and economics** J. Libal and R. Kopecek (Editors)
Volume 108 **Fault Diagnosis of Induction Motors** J. Faiz, V. Ghorbanian, and G. Joksimović
Volume 110 **High Voltage Power Network Construction** K. Harker
Volume 111 **Energy Storage at Different Voltage Levels: Technology, integration, and market aspects** A.F. Zobaa, P.F. Ribeiro, S.H.A. Aleem, and S.N. Afifi (Editors)
Volume 112 **Wireless Power Transfer: Theory, technology and application** N. Shinohara
Volume 115 **DC Distribution Systems and Microgrids** T. Dragičević, F. Blaabjerg, and P. Wheeler
Volume 117 **Structural Control and Fault Detection of Wind Turbine Systems** H. R. Karimi
Volume 119 **Thermal Power Plant Control and Instrumentation: The control of boilers and HRSGs, 2nd Edition** D. Lindsley, J. Grist, and D. Parker
Volume 120 **Fault Diagnosis for Robust Inverter Power Drives** A. Ginart (Editor)
Volume 123 **Power Systems Electromagnetic Transients Simulation, 2nd Edition** N. Watson and J. Arrillaga
Volume 124 **Power Market Transformation** B. Murray
Volume 125 **Wind Energy Modeling and Simulation Volume 1: Atmosphere and plant** P. Veers (Editor)
Volume 126 **Diagnosis and Fault Tolerance of Electrical Machines, Power Electronics and Drives** A.J.M. Cardoso
Volume 128 **Characterization of Wide Bandgap Power Semiconductor Devices** F. Wang, Z. Zhang, and E.A. Jones
Volume 129 **Renewable Energy from the Oceans: From wave, tidal and gradient systems to offshore wind and solar** D. Coiro and T. Sant (Editors)
Volume 130 **Wind and Solar Based Energy Systems for Communities** R. Carriveau and D.S.-K. Ting (Editors)
Volume 131 **Metaheuristic Optimization in Power Engineering** J. Radosavljević
Volume 132 **Power Line Communication Systems for Smart Grids** I.R.S Casella and A. Anpalagan
Volume 139 **Variability, Scalability and Stability of Microgrids** S.M. Muyeen, S.M. Islam, and F. Blaabjerg (Editors)
Volume 145 **Condition Monitoring of Rotating Electrical Machines** P. Tavner, L. Ran, and C. Crabtree
Volume 146 **Energy Storage for Power Systems, 3rd Edition** A.G. Ter-Gazarian
Volume 155 **Energy Generation and Efficiency Technologies for Green Residential Buildings** D. Ting and R. Carriveau (Editors)
Volume 157 **Electrical Steels, 2 Volumes** A. Moses, K. Jenkins, P. Anderson, and H. Stanbury
Volume 172 **Lighting interaction with Power Systems, 2 Volumes** A. Piantini (Editor)
Volume 905 **Power System Protection, 4 Volumes**

Distribution Systems Analysis and Automation
2nd Edition

Juan M. Gers

The Institution of Engineering and Technology

Published by The Institution of Engineering and Technology, London, United Kingdom

The Institution of Engineering and Technology is registered as a Charity in England & Wales (no. 211014) and Scotland (no. SC038698).

First published 2013

Second Edition published 2020

The Institution of Engineering and Technology
Michael Faraday House
Six Hills Way, Stevenage
Herts, SG1 2AY, United Kingdom

www.theiet.org

British Library Cataloguing in Publication Data
A catalogue record for this product is available from the British Library

ISBN 978-1-78561-871-0 (hardback)
ISBN 978-1-78561-872-7 (PDF)

Typeset in India by MPS Limited

To the loving memory of my late father Jose Gers and my brothers Jose Alejandro and Carlos Mauricio.

Contents

List of figures

List of tables

About the author

Juan M. Gers obtained his undergraduate degree as Electrical Engineer from the Universidad del Valle in Cali, Colombia in 1977. In 1981, he finished a master's degree in Power Systems Studies at the University of Salford in England, and his doctorate with research in distribution systems and automatization at the University of Strathclyde in Scotland in 1998. He was a professor at the Universidad del Valle in Colombia for more than 20 years and has been working as adjunct instructor at the Gonzaga University for more than 10 years. Dr. Gers served as the Vice Minister of Mines and Energy of Colombia in 2002. He is the author of *Protection of Electricity Distribution Networks*, is a Chartered Engineer of the IET and participates in several groups of the Power System Relaying and Control Committee of the IEEE.

Preface

There has been a great progress on Distribution Systems technology during the last years which encouraged the update of the first edition of this book that was published seven years ago. Undoubtedly, the reliability of the electrical system and the efficiency of the energy supply, depend largely on the Distribution Systems.

New developments have been achieved in different topics and in particular in switching and protective equipment, communication and automation. The evolution of Distributed Generation has advanced dramatically as well as the incorporation of Microgrids due to the growing concern for energy availability and cost of kWh, in some portions of electrical systems. Undoubtedly, the overall concept of the energy supply has experienced huge changes motivated mainly by the Distribution Systems.

The book keeps its original idea where the material has been prepared based on discussions on many international meetings, real applications on utility systems and personal research. The topics are elaborated first presenting the fundamental concepts but also the current state of the art as far as it was practical. The structure of this edition, although it is similar to that of the first edition, includes new material and exercises. Two new chapters were added to complement the content of the first edition. Chapter 10 covers Distributed Generation and Energy Storage Systems and Chapter 11 presents the foundations of Microgrids Technology.

All examples were reviewed and complemented and new ones were added with illustrations in MATLAB®. The scripts are included for readers and students to follow the corresponding solution.

I am indebted to several colleagues for their generous help but in particular to Carlo Viggiano and Juan Guardiola for their contribution to compile the material and test the exercises. Thanks are also due to Henry Smit for his contribution on reviewing several chapters. Other colleagues have contributed greatly with ideas and supply of reference sources. Thanks are given in particular to Leinyker Palacios, Andrej Souvent, Hugo Monterrubio, Jose Munoz, and Luis Aragon.

I express my sincere gratitude to Dr. Luigi Busarello for allowing me to use the NEPLAN Software for different exercises, which helped to illustrate the overall material. The examples and exercises that required a PSAT were all actually done using NEPLAN. My recognition will go always to Prof. K L Lo from Strathclyde University in the UK who introduced me to Distribution Automation concepts and gave me good guidance in my first year working on the field. I am also indebted to N Srinivasan of MPS Limited and Christoph von Friedeburg,

Olivia Wilkins, and other associates of IET for their help and assistance during the progress of the book.

My deepest thanks are always with my wife Pilar, our daughter Angela María, and our sons Juan Felipe and Juan Jose who have been a source of support and enthusiasm with the many hours devoted to the preparation of this edition.

Juan M Gers
Weston, April 2020

Chapter 1
Smart Grids overview

Smart Grid is a rather new concept that includes aspects of energy generation, transmission, and distribution and aims for a more reliable service, higher efficiency, more security, two-way utility-user communications, and promotion of green energy among other goals.

Smart Grid has been mainly associated with remote metering, which was later called AMR (automatic meter reading). The activities of AMR were encompassed within those of a broader field that was eventually called AMI (advanced metering infrastructure). Clearly, the metering system is one of the major elements of the Smart Grid but certainly is not the only one. Many elements are included in the overall field of Smart Grids. Those pertaining to distribution systems and in particular to the automation of distribution systems (or distribution automation) will be considered in this book.

1.1 Smart Grid for distribution systems

Distribution systems have been operated, with only occasional manual setting changes and a rather primitive automation that is known today as local intelligence. Automation in fact was first implemented on generation and transmission systems and gradually became popular also on distribution systems. Recently, in response to the growing demand to improve reliability and efficiency of the power system, more automation was introduced to distribution systems.

A good example of local intelligence is that used in reclosers operating in coordination with sectionalizers. After a local fault, reclosers start a set of reclosing operations, before locking out. Other good example is the operation of capacitor bank switches that rely on local signals, such as voltage level, power factor, or even time.

The Smart Grid policy requirements as outlined in the Energy Independence and Security Act (EISA) of December 2007 give a better understanding of the benefits and challenges of distribution automation for all of its stakeholders.

The ideas behind Smart Grids evolved from greater attention that Distribution Systems started to receive in the 1980s. Until then most of the attention was given to Generation and Transmission systems. Figure 1.1, taken from the paper "Automated power distribution" by Arthur C. M. Chen, published by IEEE spectrum in April 1982, shows a representation of a projected distribution system. The

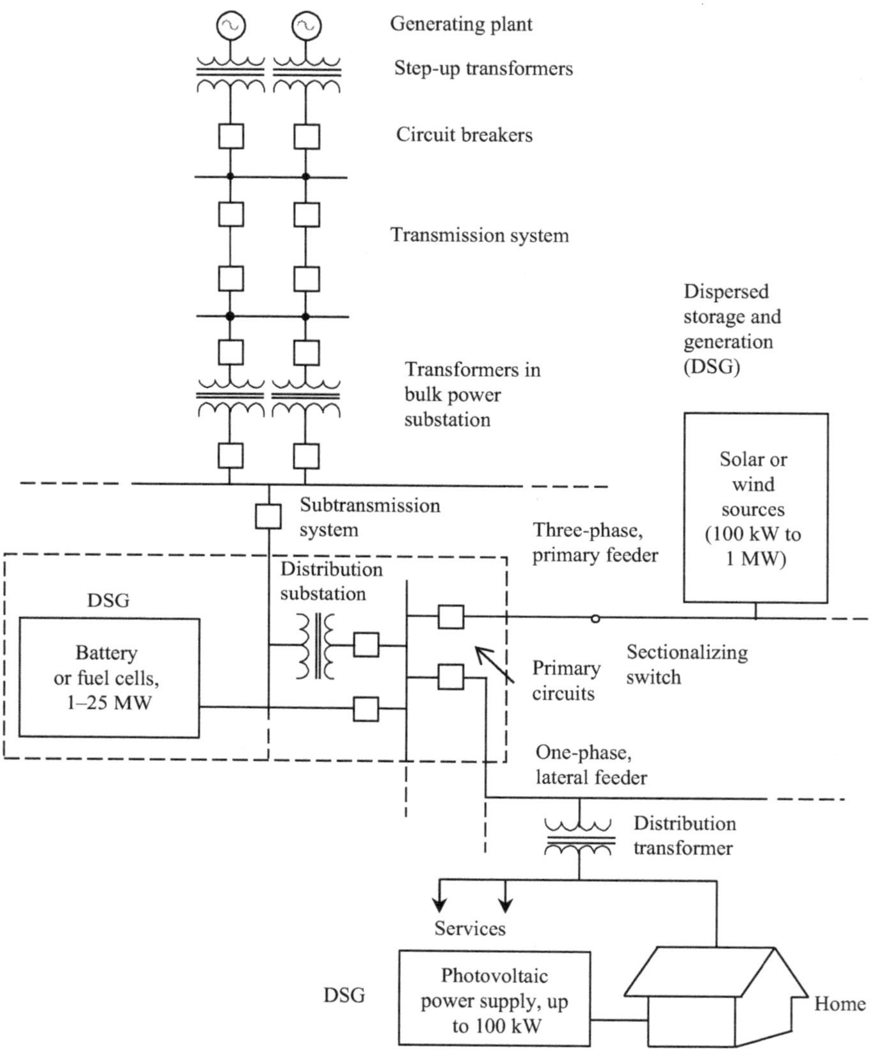

Figure 1.1 Power system as envisaged in 1982 (from "Automated power distribution," published by IEEE Spectrum, April 1982)

paper anticipated immediate detection and isolation of a faulted feeder and a reduction of the amount of time a line crew should spend to locate and fix it. It also referred to the increase of more distributed generation, called by then as dispersed storage and generation systems. Finally, the paper made a point on the importance of distribution automation for maintaining a reliable supply and to reduce operating costs. It is really interesting to see how all those provisions are now a reality.

Governments and utilities funding the development and modernization of grids have defined the functions required for Smart Grids. According to the United States Department of Energy's Modern Grid Initiative Report, a modern Smart Grid must satisfy the following requirements:

- motivate consumers to actively participate in operations of the grid,
- be able to heal itself,
- resist attack,
- provide a higher quality power that will save money wasted during outages,
- accommodate all generation and storage options,
- enable electricity markets to flourish,
- run more efficiently,
- enable higher penetration of intermittent power generation sources.

In order to achieve the goals of Smart Grid mentioned above and in particular the improvement in reliability, security, and efficiency, it is essential to have a well-developed digital technology. Among the significant challenges facing development of an Smart Grid are the cost of implementing it, and the new standards that regulatory bodies have to enact. Interoperability standards certainly will allow the operation of highly interconnected systems that include distributed generation plants.

Another challenge that the implementation of Smart Grid and distribution automation faces is the huge variety of technologies produced by multiple vendors. Establishing a proper development path is highly recommended to any utility before embarking on a comprehensive project. Maturity Models that are discussed in a latter chapter help one to establish this plan.

The implementation of the new technologies of Smart Grid will bring about changes that need to be addressed. For example, the connection of small generation plants on distribution feeders brings the possibility of having short circuit currents in two ways, and therefore the feeders should not be regarded as radial any longer. Problems with reclosing features should be closely examined to avoid out of synchronism closing. Likewise, the connection of charging stations for electrical vehicles will change the normal operation of the feeders.

Smart Grid has the great advantage of allowing two-way communication, i.e., utility–user and user–utility. This will allow a better and more effective relationship between the user and the utility. The latter will be able to monitor and control small appliances of each user. The user will have in turn the great advantage of getting information regarding the consumption level, new rates available, and load management schemes. This of course would require powerful communication systems that need to be flexible and reliable.

1.2 Definitions of Smart Grid

Many definitions have been written to describe Smart Grids. Every utility might have its own definition. Some possible meaningful definitions are the following.

EPRI: the intelligent grid

"A Smart Grid is one that incorporates information and communications technology into every aspect of electricity generation, delivery and consumption in order to minimize environmental impact, enhance markets, improve reliability and service, and reduce costs and improve efficiency."

Xcel energy: the Smart Grid

"While details vary greatly, the general definition of a smart grid is an intelligent, auto balancing, self-monitoring power grid that accepts any source of fuel (coal, sun, wind) and transforms it into a consumer's end use (heat, light, warm water) with minimal human intervention.

It is a system that will allow society to optimize the use of renewable energy sources and minimize our collective environmental footprint.

It is a grid that has the ability to sense when a part of its system is overloaded and reroute power to reduce that overload and prevent a potential outage situation; a grid that enables real-time communication between the consumer and utility allowing us to optimize a consumer's energy usage based on environmental and/or price preferences."

DOE (department of energy) definition

"An automated, widely distributed energy delivery network, the Smart Grid will be characterized by a two-way flow of electricity and information and will be capable of monitoring everything from power plants to customer preferences to individual appliances. It incorporates into the grid the benefits of distributed computing and communications to deliver real-time information and enable the near-instantaneous balance of supply and demand at the device level.

People are often confused by the terms Smart Grid and smart meters. Are they not the same thing? Not exactly. Metering is just one of hundreds of possible applications that constitute the Smart Grid; a smart meter is a good example of an enabling technology that makes it possible to extract value from two-way communication in support of distributed technologies and consumer participation."

The BC hydro definition of Smart Grid

"Smart Grid refers to a modern, intelligent electricity transmission and distribution system that incorporates traditional and advanced power engineering to enhance grid performance and support a wide array of functionality for customers and the economy. In other words: modernization and automation of the current power delivery system."

In summary, Smart Grid refers to a sustainable modernization of the electricity grid, integrating information and communication technologies to intelligently manage and operate generation, transmission, distribution, consumption, or even the electric energy market. This concept is illustrated in Figure 1.2.

The Smart Grid components comprise many or almost all of the elements of the utility and the relationships among the elements that constitute them. Figure 1.3 includes some of these elements, like smart meters, generation, transmission, substation, and feeders.

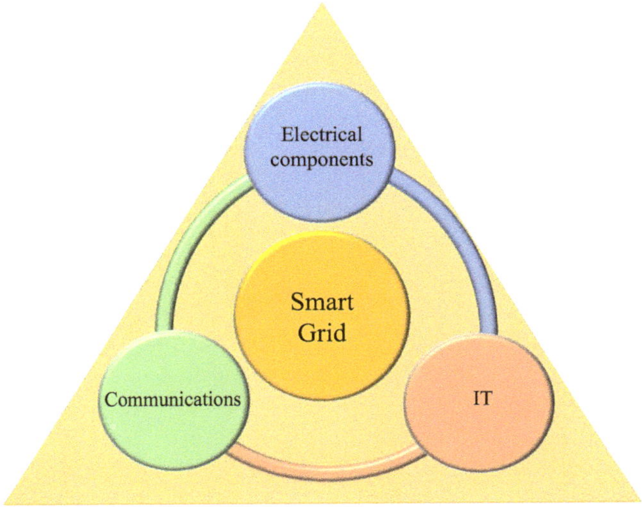

Figure 1.2 Smart Grid concept

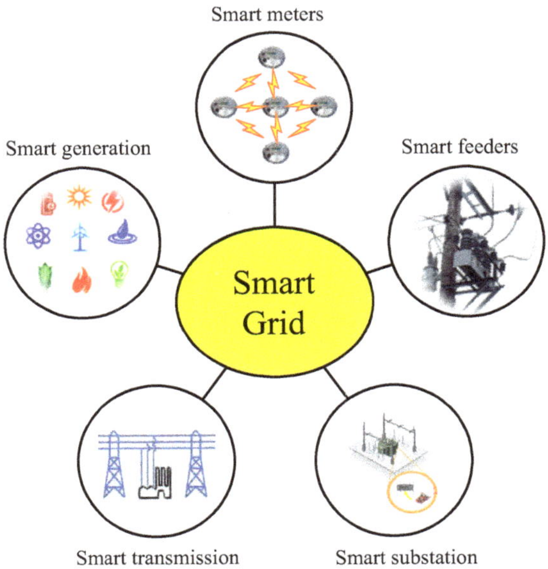

Figure 1.3 Smart Grid components

1.3 Benefits of the Smart Grid on distribution systems

The benefits of implementing an Smart Grid are many. They can be summarized in the following categories.

1.3.1 Enhancing reliability

The Smart Grid dramatically reduces the cost of power disturbances. This can be achieved by means of system reconfiguration using switches placed along the feeders. Communications and control technologies greatly help one to isolate faults and allow a faster restoration of service.

1.3.2 Improving system efficiency

The reduction of losses in electrical systems, both technical and nontechnical, is a purpose of every utility in the world. It not only reduces the demanded power but also contributes to the environment. The reduction of system losses results in capital deferral that gives an attractive payback of the investments. Power capacitors, voltage regulators, and proper design criteria are required to achieve this.

1.3.3 Distributed energy resources

Construction of generation plants at the user level is more frequent every day. These sources are referred to as distributed generation or distributed energy resources and are getting much attention from government authorities and environmental institutions as they reduce the pollution levels that some plants have, in particular those burning coal and oil. Distributed energy resources also contribute to better operating conditions of distribution systems as they are sources connected directly to the users' loads, increasing control of the voltage.

1.3.4 Optimizing asset utilization and efficient operation

Real-time data makes it possible to more effectively utilize assets during both normal and adverse conditions and to reduce the costs of outages. This results in a longer service life of the assets.

1.4 Maturity Models for Smart Grid applications

Utilities are aware of the need to implement Distribution Automation and Smart Grid programs. It is easy to appreciate the various individual efforts being made, given the number of existing applications. However, there is a lack of definite procedures and recommended practices to help utility organizations establish an order of priority in their implementation and development of these technologies.

Developing a Maturity Model is important for modern grid management and for selecting the best way of creating a sustainable path toward integration. This involves tools such as having a common strategy and vision, which not only helps in the development of an organized work plan but also allows for the implementation of profitable projects as is needed.

The Maturity Model helps utilities to implement Smart Grid application by prioritizing tasks and measuring the progress achieved. It also helps one to identify the characteristics of the organization by designing a roadmap and by promoting the exchange of common terms among internal and external actors. All share experiences with the community and prepare the organization to undertake the required changes.

Maturity Models, very common in information technology (IT) organizations, help an organization assess its methods and processes according to management criteria. The key to achieving a Maturity Model is a good strategy and a good vision in the Smart Grid context. The Maturity Model covers three major elements: communications, IT, and electrical components.

This chapter discusses the joint efforts among different entities that have come together to produce interesting procedures and schemes to aid organizations in using the most appropriate solutions for their requirements.

1.4.1 Smart Grid Maturity Model

The Smart Grid Maturity Model (SGMM) is a tool that provides the basis to help an organization guide, evaluate, and improve its efforts to best select applications of Smart Grid in order to achieve a proper transformation and modernization. From a methodological standpoint, the model allows one to create a map defining the task and technologies, to identify gaps in the strategy and execution, to support business opportunities that promote Smart Grid projects, to delineate the organization vision and strategy, and to evaluate alternative solutions and future goals that will help guide the future of the electric network.

1.4.2 Benefits of using a Smart Grid Maturity Model

In today's competitive world, the industry demands that organizations strive to achieve sustainable improvements and repeatable and scalable procedures and to promote improvements within an organization. Maturity Models were initially intended to be applied in the software development industry, driven by the necessity of evaluating various organizations under identical parameters. These models gave the possibility of developing improved planning, engineering, and governing practices to guarantee higher quality levels in both processes and results.

Many organizations approved the use of the Maturity Models in electrical utilities, considering the successful experience in the software industry. It allowed then one to determine the development level of networks and to visualize the gap between the current and the future situation. From this the best solutions can be proposed.

Maturity Models must be combined with an entire work methodology to identify the standards and technical solutions that will be considered in developing the Smart Grid roadmap. Maturity Models also support the implementation of applications.

The methodology has three goals: first, to identify the entity's current developmental state from an Smart Grid perspective and the desired state expressed as a maturity level. Gap analysis is then employed to obtain a simplified list of required steps; second, a cost–benefit analysis to determine which Smart Grid solutions are financially feasible; third, to arrange and describe the user requirements with use-cases based on the financial evaluations previously approved by the executives.

A robust model needs to recognize not only management activities being carried out at the individual project level, but also those activities within an organization that build and maintain a framework of effective project approaches and management practices.

By undertaking a maturity assessment against an industry standard model, an organization will be able to verify what they have achieved, where their strengths and weaknesses are, and then to identify a prioritized action plan to take them to an improved level of capability.

1.4.3 Genesis and components of an SGMM

The best known examples of Maturity Models are those developed by Carnegie Mellon University and IBM. In this chapter, the SGMM, which is one of the most accepted globally, will be explained.

The SGMM was originally proposed by four electrical utilities of the USA (Center Point Energy, Progress Energy, Pepco Holdings and Sempra Energy); DONG Energy from Denmark, NDPL from India; Country Energy from Australia; and IBM and APQC. In 2009 with the support of the Department of Energy (DOE), the development of the SGMM was given to Carnegie Mellon. That University still is responsible for its administration and update. The SGMM is maintained by the Carnegie Mellon Software Engineering Institute (SEI) as a resource for utility industry transformation. Carnegie Mellon encourages utilities to leverage SGMM to ensure that all aspects of transformation planning are considered, their options are prioritized, and their progress measured as the utilities implement an Smart Grid structure.

The main organizations using Maturity Models are depicted in Figure 1.4.

1.4.4 Development process of an SGMM

The Maturity Model development process can be carried out with the following steps:

1. *Information gathering in the eight domains of the related organization*: In this stage, the consulting engineer must explore all organization characteristics from a strategic point of view. Since the Navigator assists the organization in comprehending the questions included in the evaluation, it is vital that he has a full understanding of the model.
2. *Smart Grid and SGMM concept awareness*: The Navigator must be familiar with the organization; likewise the organization itself must understand the aspects to be evaluated. The Navigator must prepare the organization to understand the concepts of the *SGMM*. Next, the criteria must be adjusted and the evaluation process must be established.
3. *SGMM application*: In this stage, the Navigator researches the questions included in the Maturity Model. The idea is to evaluate the current condition of the organization as well as the desired future condition. It is vitally important that the organization responds to questions in a critical, objective, and honest manner. Also, it is important to involve individuals from the various areas of the organization, since the model covers all aspects of the organization and requires that answers are given with an exhaustive view of the problem. If this

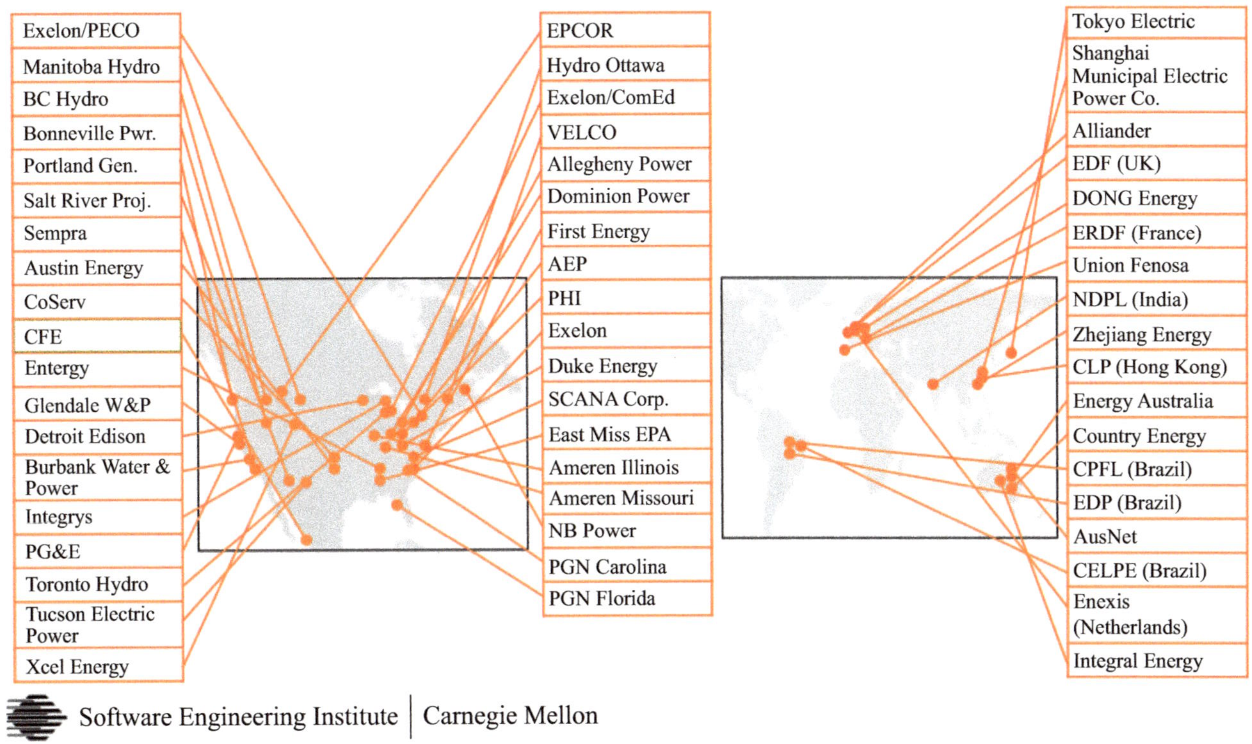

Figure 1.4 *Some organizations that use the SGMM (taken from the Carnegie Mellon Software Engineering Institute)*

is not the case, key points will be missed and efforts may be focused on irrelevant aspects. It is recommended to reevaluate after the first model attempt.

4. *Final results*: Results are obtained regarding the actual condition of the organization and the future condition. The results help identify gaps that must be filled in order to reach the desired condition. The cost and time required to do so should be carefully analyzed to assure the viability of the project. If the gap is large, a viable economic solution may be to extend the time or reduce the scope of the future vision.

1.4.5 Levels and domains of the SGMM

The characteristics and capabilities of an organization can be determined with the help of the Definition Matrix of SGMM as proposed in a document entitled "SGMM Model Definition. A framework for smart grid transformation" Version 1.2 of September 2011 of the Software Engineering Institute of the Carnegie Mellon University. The Definition Matrix of SGMM is composed of six levels in eight maturity domains.

1.4.5.1 Maturity levels of SGMM

The SGMM has six maturity levels that represent well-defined states. Each one describes the capabilities and the characteristics of the organization to achieve the Smart Grid vision regarding efficiency, automation, reliability, energy savings, interaction with the user, integration of distribution energy resources, and access to new opportunities of the business.

The lowest level (level 0) represents the default position of the organization when the study starts. An organization operating in a traditional way without modernization will be at this level. It is important for an organization to evaluate its condition to establish the future vision in a predefined time interval. Since level 0 represents the starting condition, the model does not have precise characteristics for this level.

- *Level 1*: It shows that the organization is in the starting process and exploring the Smart Grid technologies. At this level, the organization has a vision but does not have a clear strategy. At this level, the organization is capable of communicating its vision to the community and industry.
- *Level 2*: It shows the organization has a defined strategy and that it is already investing to attain the modernization of the electrical network. At this level, tests are performed with the business case already implemented, to assess the changes in the organization.
- *Level 3*: The organization integrates its Smart Grid program with the operating departments. The procedures should be made repeatable and the information should be shared within the whole organization.
- *Level 4*: The functionality and benefits of Smart Grid can be assessed. The organization performs analysis and makes corrections on real time.
- *Level 5*: The organization is on a permanent innovation state, develops standards, and improves procedures. The organization becomes a leader of the

industry. The vision and strategies of the organization fulfill national, regional, and local interests.

The organization will be in a higher maturity level when the implementation of changes starts. Each organization has to establish its own target maturity levels based on its own operating system, strategy, and timeline. It is obvious that higher levels of maturity in the model indicate a successful adoption from its grid modernization efforts. The target level is not the same for each organization. So achieving any level could be appropriate to an organization but not to other.

1.4.5.2 Domains of the SGMM

The SGMM has eight domains, the description of which is presented in the following paragraphs:

The Strategy, Management, and Regulatory (SMR) domain establishes the internal procedures of performance and government of the organization and encourages support relationships with the groups responsible to implement the vision and strategy.

The Organization and Structure (OS) domain represents the capabilities and characteristics that allows an organization to plan and operate in order to achieve an Smart Grid in place.

For the transformation efforts to be successful, the organizational structure must promote and reward the planning and operation in various functions. This domain focuses on changes in communications, culture, structure, training, and education within the organization. Maturity in this particular domain evidences an increased organizational proficiency to develop their decision-making initiatives focus on fact-based decision to meet its Smart Grid goals. Additionally, it exhibits the general vision of individuals committed to reach the Smart Grid goals.

The Grid Operations (GO) domain represents the functional practices that support reliable, safe, and efficient electric network operations. The organization employs new automation and communication solutions to improve key network elements view and decrease control action response time. The information gathered from implemented Smart Grid solutions provides organizations, which have successfully achieved high maturity levels, with valuable information for automation. Additionally, this enables managing of power flows to reduce losses and maximize generation at reduced costs. It also allows higher levels of automation and a larger view of the entire system. On the other hand, response times improve in communications and control avoiding cascading failures throughout the system. The benefits of such capabilities allow for larger grid improvements to reach the goal of serving customers with high-quality power, diversified generation, maximized asset utilization, and efficient operations.

The Work and Asset Management (WAM) domain represents the organization capabilities in managing assets and personnel. Maturity in the domain reflects improvements in predictive and reactive maintenance resulting in improving reliability, safety, and efficient operations. Advances in this domain represent an incremental ability to utilize information obtained from Smart Grid implementations to reduce causeless maintenance and out of service time, identify failure origin and indicate corrective actions, identify failures beforehand, minimize problem identification and solution time, and improve personnel resources and planning results.

The Technology (TECH) domain focuses in utilizing IT infrastructure that serves as a base to develop and supports services that open new markets. It reflects compliance with relevant standards and integrating a strategic technology to connect and support various data sources and users (applications, systems, and persons).

Advances in the technology domain represent compliance with industry and government standards and integration of applications in Smart Grid with overwhelming data. The utilization of the IT structures promotes new business solutions and promising new markets.

The transformation to Smart Grid provides benefits to the organization that requires efforts beyond the acquisition of new Smart Grid technologies that can provide a positive or negative support to the organization's efforts to materialize their Smart Grid plans. Smart Grid involves several technologies like wide-area monitoring, two-way digital communications, and advanced control. This comprehensive technological platform developed through advanced engineering and organization business initiatives requires processing large data sources and control systems that comprehend current Smart Grids and future applications.

The Customer (CUST) domain represents the organizational capabilities and characteristics that enable customer participation toward achieving the benefits of the Smart Grid transformation. The participation could be passive (the utility manages customer loads and select the power source) or active (provides tools to customer allowing them to manage the usage, sources, and energy cost based on cost and available options in the market).

Achieving important levels of maturity in the CUST domain represents big benefits for customer allowing them to decide and manage power usage selecting from different sources and energy cost, maintaining network security, and protecting customer privacy. Additionally, exemplifying high levels of maturity in the domain represents organizational ability to fulfill utility, regional and network wide goals, with the use of modern Smart Grids, in regards to grid stability, energy efficiency, peak load reduction, green energy integration employing distributed generation while minimizing the usage of foreign resources.

The Value Chain Integration (VCI) domain exhibits the utility potential to advance in reaching the goals set to successfully capitalize the Smart Grid initiatives by integrating the different utility departments with the production and delivery of energy demands. Smart Grid automations go beyond traditional boundaries (substation automation) offering innovation in load management, distributed generation, and market operations. Smart Grid serves as a platform for planning, implementing, and energy management from generation to final user consumption.

As a result of environmental concerns and the need for increased efficiencies, market forces and regulatory bodies will again force the industry to change, providing new opportunities for organizations with Smart Grid competence and causing new value chains to emerge. Automation will extend beyond traditional boundaries and across the entire value chain to provide opportunities for innovation and efficiencies in load management, distributed generation, and market structure. As utility matures, the cooperative planning, implementation, and management of electricity from the sources of production to end-use consumption will optimize profitability and improve performance of the utility's value chain. Networked IT and data sharing, aligned with value chain business units' requirements, are critical for success.

Finally, the Societal and Environmental (SE) domain represents the organizational capabilities and characteristics that allow contributing to meet social goals associated with reliability and safety of the electric network infrastructure, type of energy sources employed, and the environmental impact and quality of life.

Proper application of these social initiatives benefits the organization and solidifies the connection with users and regulators. Efficient operation enabled by enhanced Smart Grid solutions represents higher profits while reducing environmental impact. Continued prevention and risk mitigation of network security events is an important part of Smart Grid implementations.

1.4.6 Results and analysis obtained by SGMM

Once the results of the Maturity Model are obtained (the actual state and the desired future), a gap analysis is needed to determine what actions to take. Figure 1.5 helps one to illustrate an example of results obtained by applying the SGMM.

It is important to mention that establishing a high future condition is not precluded by a low present rating. On the other hand, not aiming for a level of 5 in a particular domain does not mean that the organization is not focused on the Smart Grid technology. It is possible for a business model not to involve all aspects in the domain or for the organization not to consider it profitable to improve their level in

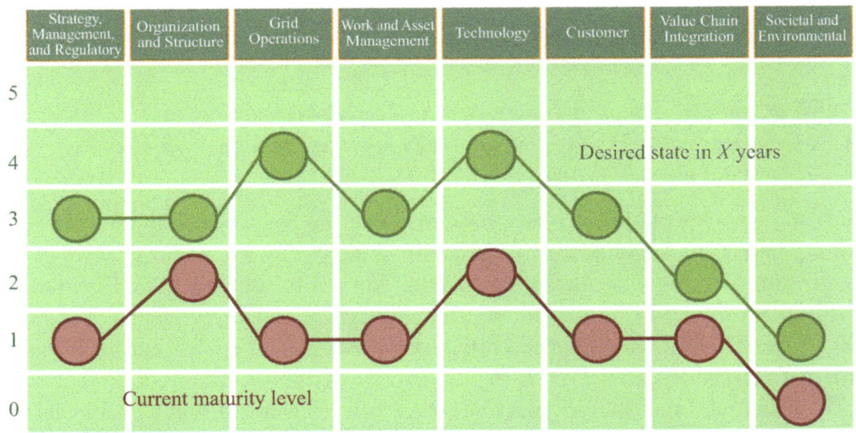

Figure 1.5 Example of results obtained from the SGMM (taken from the Carnegie Mellon Software Engineering Institute) [1]

that domain. Similarly to the software industry, where ideas are born from Maturity Models, some organizations consider an optimization condition acceptable in the processes in a domain.

1.4.7 Example case

The previous methodology was applied to the evaluation of a real utility, which has around 500,000 users. This utility has a vertical integration as it handles generation, transmission, distribution, and energy marketing. The results presented refer to some of the questions used in the SEI-SGMM model, which correspond to the GO domain and the answers given by the utility.

GO-1.3

Do you have proof-of-concept projects and/or component testing for grid monitoring and control underway?

(A) *No*
(B) *To some extent, not directly for Smart Grid*
(C) *To some extent, for Smart Grid*
(D) *To a great extent (i.e., numerous evaluations underway or completed)*

GO-2.3

Aside from SCADA, are you piloting remote asset monitoring of key grid assets to support manual decision-making?

(A) *No*
(B) *In documented plan including committed schedule and budget*
(C) *Piloting*
(D) *Pilots complete or technology being deployed*

GO-3.4

Have smart meters become important grid management sensors within your network?

(A) *No*
(B) *In documented plan including committed schedule and budget*
(C) *Moderately (<40% of a grid is using meters as management sensors)*
(D) *To a great extent (≥40%)*

It can be observed in the questions above that the degree of difficulty increases with the levels. At the end, a score is obtained indicating the maturity level of the organization. Figure 1.6 shows the results from this organization. The redline indicates the required limits to satisfy the requirements of each level. For the case of the example, the values are established by a certification institution by using particular standards.

It is assumed that the criterion of the certification institution to pass the first three levels establishes a score minimum of 0.6 for each one. After the third level, higher score is required to approve (level 4 specifically requires at least 0.7 and level 5 requires 0.8). Once the score gets below one of the defined limits at a level,

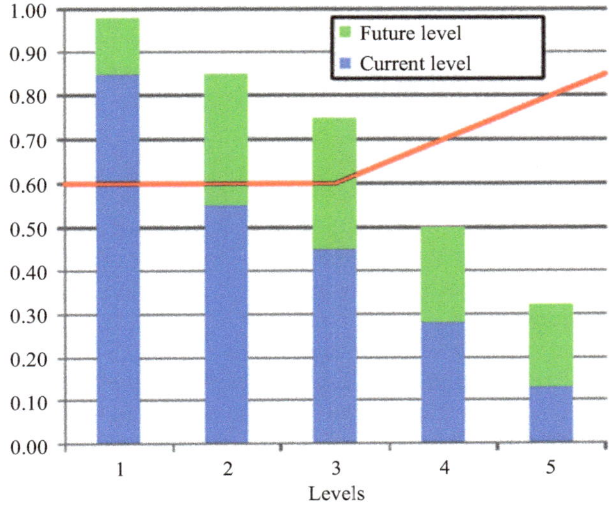

Figure 1.6 Example of results in Grid Operations (GO) domain

the total is then calculated with the summation of the scores obtained up to that level, including the score of the last one.

The GO domain for this organization indicates a maturity around 1.4, which indicates that the organization manages higher levels of operations automation and network optimization processes. The example shows that the current condition (blue bars) only reaches the redline in the first level with a 0.85 score (minimum is 0.6), and in the second level only 0.55 is achieved. Since the second level is the first level that does not approve the objective, the equivalent level will be the score summation up to this point (1.4). According to the results shown in Figure 1.6, this organization can reach at the most for the GO domain, the third level in 5 years, which could be acceptable for that company. In that case the total score would be 3.1, which is calculated from the results of the table (0.98 points for the first level, 0.85 for the second level, 0.75 for the third level, and 0.5 for the fourth level). The organization aims to improve its current condition by 1.7 points in a 5-year term for a total of 3.1.

This demonstrates that the organization currently has instituted Smart Grid initiatives where technologies and equipment have been elected for pilot testing, some of which may have concluded, and implementations are in process. For example, the Distribution Automation and the interoperable data system employ tools like CIM.

Once current and future state results have been obtained for the company under review, an analysis is conducted to evaluate what efforts must be made to reach such condition. It can also be defined based on the conditions the model expects for the particular level. For the particular example, efforts to reach the desired level must focus on the following:

- The organization must have functional-level approved business cases for improvements in the employee resources and Smart Grid assets in order to determine necessary changes in abilities of management personnel and assets, guaranteeing increased productivity, and better prediction of events. They are designed to reduce maintenance due to faults and optimization of life cycle costs, ensuring availability. For the expected level, the business cases do not need to be integrated with the rest of the organization.

- In the asset management area, the organization must have finalized the remote supervision usage evaluation. Remote monitoring offers more than the basic control using SCADA. This implies that a dual real-time communication channel must be established to obtain detailed asset information, in contrast with previously used performance reports that provided support to conduct corrective actions.

  At this level, the organization must have completed equipment and field personnel assessment along with focusing assets on the Smart Grid vision of the organization. This evaluation includes the geospatial equipment to connect assets with their geographical location. An example would be the field crew integration with the remote monitoring communication system.
- The organization should invest in technologies to support asset monitoring and field crew performance. A documented plan must be established to explore Smart Grid capabilities in order to create inventories and provide for event record keeping and asset follow-up. The equipment history should include tendencies and profile information based on actual data. Asset supervision must be developed by trial testing, and an integrated view of the GIS, state, and connectivity.
- Additionally, the organization should develop a strategy for field crew operations with the aim of generating a significant impact on the organization. It means that the organization has recognized the need to optimize mobile assets with specific performance objectives. Ideally, the field crew strategy must be interconnected with smart networks for higher performance.
- The organization should have started implementing Smart Grid solutions with asset management and field crews to employ the information available.
- The organization must have performance data, tendency data, and event control for at least 25% of individual components. Computer systems are SCADA and Remote Terminal Unit (RTU); physical elements include breakers, transformers, meters, etc. that support generation, transmission, and distribution.
- A maintenance plan must be implemented for a small percentage of equipment (smaller than 25%) to pilot the maintenance and replacement plans that will be developed based on real-time data obtained from an asset supervision program. This allows equipment capacity prediction and prevents damage to the system. It is important to integrate remote asset supervision with asset management while they are both in the planning stages. Integration at this level might be the ability to automate the order being processed and to relocate the crews based on the requirements in the network. GIS integration with the asset supervision system may be conducted based on location, state, and interconnectivity in order to improve the visual operability in at least one type of asset.
- The organization must make an effort to have a follow-up process with some level of automation for a small percentage (around 1%–25%) of assets from

when they leave the factory until they reach their destination. The follow-up must include information like asset location, usage state, scheduled maintenance date, and end-of-service date. Bar code and radio frequency identification are among the available technologies to accomplish this.

- The organization must have a documented plan to develop an asset investment model for key components based on Smart Grid data that allows integration and developments in asset control.

Table 1.1 summarizes the maturity level in each of the domains discussed in the model for this example, and Figure 1.7 shows a graphical representation of this score. Based on the results obtained, an analysis will identify the gaps to be improved. The distance from actual to future in the evaluated conditions represents

Table 1.1 Example of results after applying the SGMM

Domains	Current maturity level	Future maturity level (5 years)
SMR	*2.2*	*4.7*
OS	*2.6*	*3.6*
GO	*1.4*	*3.1*
WAM	*1.3*	*2.7*
TECH	*1.5*	*3.5*
CUST	*2.2*	*3.2*
VCI	*1.6*	*2.5*
SE	*1.5*	*3.5*
Mean maturity level	*1.80*	*3.36*

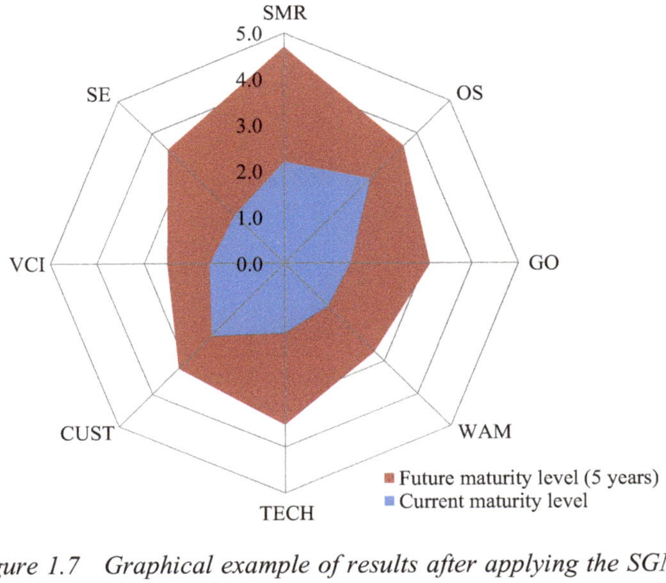

Figure 1.7 Graphical example of results after applying the SGMM

the efforts required to reach the desired future condition. The larger the gap between actual and future conditions (radar chart area differences), the more aspects need to be improved. Based on the estimated time and budget, necessary efforts must be defined and the probability of success evaluated.

1.5 Prioritization in Smart Grid projects

The decision of which projects have to go first is hard making inside any utility. Cost–benefit analyses are the most popular methods for making these decisions, which is discussed in the next section. Nonetheless, these methods are difficult to use when social and environmental impacts play an important role while evaluating projects. A list of different techniques is shown in the following pages.

Smart Grid evaluation system for the modernization of the electrical systems explained in this section considers the structure of the SGMM.

The structure of the Smart Grid decision problem considered in this paper consists of eight criteria and four subcriteria. The number of alternatives can vary, though it is recommended not to be greater than ten due to the number of pair comparisons needed.

The list of criteria for Smart Grid decision problems is based on the domains of the SGMM developed by Carnegie Mello University mentioned in the previous sections as shown in Table 1.2.

The list of subcriteria for Smart Grid decision problems is based on the type of benefits related to the Smart Grid, presented by the US Department of Energy (DOE) (Table 1.3).

Table 1.4 illustrates some of the alternatives needed to develop an integral Smart Grid into any utility. These alternatives are an example of some projects that must be evaluated in accordance with the SGMM results. Each alternative will be explained briefly afterward.

A1. *Vision and strategic planning*

Utilities must identify their needs and set their roadmap for the sequential implementation of new Smart Grid processes and technologies, based on a

Table 1.2 Criteria for prioritizing Smart Grid projects by using the AHP

Item	Criteria
C1	Strategy, Management, and Regulatory (SMR)
C2	Organization and Structure (OS)
C3	Grid Operations (GO)
C4	Work and Asset Management (WAM)
C5	Technology (TECH)
C6	Customer (CUST)
C7	Value Chain Integration (VCI)
C8	Societal and Environmental (SE)

Table 1.3 Subcriteria for prioritizing Smart Grid projects

Item	Subcriteria
C11	Economic benefits
C21	Power quality and reliability benefits
C31	Environmental benefits
C41	Security benefits

Table 1.4 Alternatives proposed for reaching the objectives of Smart Grid projects

Item	Alternatives
A1	Vision and strategic planning
A2	Business architecture
A3	Assets and work force management
A4	Physical and cyber security
A5	AMI
A6	Two-way communication
A7	Energy efficiency
A8	Advanced distribution automation
A9	Micrograds
A10	Distributed generation
A11	Energy storing
A12	Demand site management
A13	Transmission automation
A14	Smart mobility
A15	Smart street lighting

strategic and comprehensive vision that includes an understanding of the current state, resources, and future status to be achieved.

A2. *Business architecture*

A structured approach that will allow each company, from a strategic point of view, to understand and meet the needs of the Smart Grid project from an integral perspective, considering the business domain, data, planning, applications, and technological infrastructure.

A3. *Assets and work force management*

Represents the organizational capabilities and characteristics that support the optimal management of assets and work force resources.

A4. *Physical and cyber security*

Security must be inherent in the development of intelligent networks throughout their life cycle to ensure the integrity of the system. The data security has the privacy, integrity, and reliability of the same, and the access to the network refers to the authentication and identification of resources and users.

A5. *AMI*

Electricity meters that use two-way communication to collect use data of electricity and deliver information to customers. It allows utilities to balance supply, demand, and loadability, making a more intelligent and efficient network, promoting aspects of network supervision and control.

A6. *Two-way communication*

The Smart Grid development requires support for telecommunication platforms and protocols, enabling technological convergence, interoperability, data management, monitoring, surveillance, and mutual communication between company and customer. These structures are WiMax, MPLS, ZigBee, 3G, 4G LTE, WAN, LAN among others.

A7. *Energy efficiency*

This alternative aims is to reduce energy consumption, optimizing production processes, said otherwise produce the same or more with less electrical energy. Direct energy customers can reduce energy consumption to lower costs and promote economic, political, and environmental sustainability.

A8. *Advanced distribution automation*

Allows utilities to implement modern techniques in order to improve the reliability, efficiency, and quality of electric services.

A9. *Microgrids*

It refers to two-way electricity generation system that allows the distribution of electricity from suppliers to consumers, using digital technology and favouring the integration of sources of generation of renewable origin, with the aim of saving energy, reducing costs, and increasing reliability.

A10. *Distributed generation*

Small- and medium-scale energy generation systems that are connected to the electricity distribution network and characterized by being installed in points near consumption.

A11. *Energy storing*

Methods available to preserve a certain amount of electricity as other forms of energy to release it when required in the same way in which it was collected or in a different one.

A12. *Demand site management*

A set of initiatives that seek to influence the consumer to modify his/her pattern of demand in order to not only achieve net energy saving, but also a more efficient use of it, conducting into a more flexible and sustainable energy system.

A13. *Transmission automation*

A transmission system that makes use of monitoring and communication applications along with technologically compatible measurement, protection, and maneuvering equipment, which allow one to check, isolate, and correct the common faults of transmission assets in a faster and cheaper way.

A14. *Smart mobility*

In the context of Smart Grids, the electric vehicle is important as an energy storage system. V2G (vehicle to grid) technology is a system in which stored energy can be discharged to the grid. V2H (vehicle to home) technology is a variant of the V2G but specifically intended for housing. V2B (vehicle to building) is about the same concept as V2H only applied to buildings or industries.

A15. *Smart street lighting*

Public lighting in the context of Smart Grids comprises three levels. The first level corresponds to the luminaire and its control through wireless communications. The second level corresponds to the storage, combination, and processing of data. The third level means that each luminaire for exterior lighting can be converted into a node of the Internet of Things network, on a data-enabled lighting server.

1.6 Cost–benefit analysis

After the characterization of the Smart Grid projects through a selected methodology similar to those described, it is necessary to perform a quantitative analysis of the costs and benefits of the different projects under consideration, analysis which supports the decision to implement one project over another.

1.6.1 Definition of benefits

The term "benefit" is defined as an impact (of a Smart Grid project) that has value to a firm, a household, or society in general. If possible, benefits should be quantified to gauge their magnitude. In addition, to facilitate comparison, benefits may be expressed in monetary terms. A benefit is an outcome of a project that has value, which is not simply a project's performance or intermediate outcomes of the project.

Some examples of benefits may include the following:

- lower electricity costs to consumers
- the value of lower transmission and distribution losses
- lower operation and maintenance costs
- reduced cost of power interruptions
- better power quality
- reduction in greenhouse gas emissions

The Office of Electricity Delivery and Energy Reliability from the US Department of Energy (DOE) developed an Smart Grid computational tool (SGCT) in which the Smart Grid projects can be characterized by identifying the technology that will be deployed, also known as assets, their main functions, and mechanisms. Then, the economic, reliability, environmental, and security benefits are determined, as well as the corresponding monetary value. The process described in the SGCT is illustrated in Figure 1.8.

1.6.2 Cost–benefit analysis methodologies

Electric utilities invest large amounts of money aimed to comply with their regulatory or statutory obligations. For example, extending their services to new regions and planning for continuous growth and cost minimization. Smart Grid projects, on the other hand, may not fit into this proven paradigm in cost minimization time because they may be discretionary. For example, the decision to invest in Smart

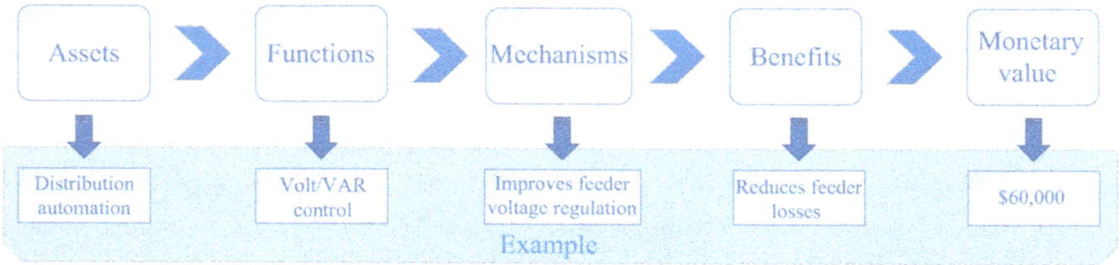

Figure 1.8 Definition of the Smart Grid benefits, by the SGCT

Grid projects to improve reliability beyond currently acceptable levels depends on how much money should be invested to obtain the improvement, and if the final results justify the amount of money to be invested.

Smart Grid applications offer new benefits beyond the basic services or lower costs. These applications can improve reliability and quality of service above currently accepted levels, in addition to providing customers with options and services never provided before. Eventually, Smart Grid solutions are realistic alternatives to address technical problems that may arise when technologies such as distributed generation or electric vehicles become more common in distribution networks.

Table 1.5 Summary of different approaches to estimate the Smart Grid cost–benefits

Reference	Authors
Sizing up the Smart Grid	Faruqui, A., Hledik, R., Davis, C.
Smart Grid Evaluation Metrics	GridWise Alliance
Metrics for Measuring Progress Toward Implementation of the Smart Grid	Office of Electricity Delivery and Energy Reliability
Smart Grid Metrics: Monitoring Our Progress	Iller, J.
Characterizing and Quantifying the Societal Benefits Attributable to Smart Metering Investments	EPRI (Hemphill, R., Neenan, B.)
The U.S. Smart Grid Revolution: KEMA's Perspectives for Job Creation	KEMA
Modern Grid Benefits	National Energy Technology Laboratory
Smart Grid: Enabling the 21st Century Economy	Pullins, S.
San Diego Smart Grid Study: Modernizing the Grid	Anders, Scott J.
Estimating the Benefits of the GridWise Initiative	Baer, W.S., Fulton, B., Mahnovski, S.
GridWise: The Benefits of a Transformed Energy System	Kannberg, L.D., Chassin, D.P., DeSteese, J.G., Hauser, S.G., Kintner-Meyer, M. C., Pratt, R.G., Schienbein, L.A., Warwick, W.M.
Methodological Approach for Estimating the Benefits and Costs of Smart Grid Demonstration Projects	EPRI (Wakefield, M.)

However, these technical issues are mostly in the future, so today it remains the priority for companies to conceive a positive business and economic scenario, showing sufficient benefits that offset costs.

There are different methodologies developed for the cost–benefit analysis of Smart Grid projects. Several studies have promoted benefits of Smart Grid projects, but few have not focused on developing a systematic way of defining and estimating those benefits. Some approaches found in the literature are summarized in Table 1.5.

Reference

[1] Software Engineering Institute. "SGMM Model Definition. A framework for smart grid transformation," Version 1.2. Technical Report CMU/SEI-2011-TR-025, ESC-TR-2011-025. CERT® Program Research, Technology, and System Solutions Program Software Engineering Process Management Program Carnegie Mellon University. September, 2011.

Further reading

[1] BChydro. Smart Grid at BChydro: Current Status, 2009.
[2] Chen, A. C. M. "Power/energy: Automated power distribution: Increasingly diverse and complex power operation and distribution systems will mean a larger role for microprocessor and communications technologies," IEEE Spectrum, vol. 19, no. 4, pp. 55–60, 1982.
[3] U.S. Congress. Energy independence and security act of 2007, Title XIII – Smart Grid. Public Law 110-140, 2007.
[4] Triantaphyllou, E. Multi-Criteria Decision Making Methods: A Comparative Study. Springer Science+Business Media, Dordrecht, 2000.
[5] Electric Power Research Institute (EPRI). Guidebook for Cost/Benefit Analysis of Smart Grid. Demonstration Projects Revision 1, Measuring Impacts and Monetizing Benefits. Electric Power Research Institute, Palo Alto, CA, USA, 2012.
[6] Electric Power Research Institute (EPRI). Methodological Approach for Estimating the Benefits and Costs of Smart Grid Demonstration Projects. Electric Power Research Institute, Palo Alto, CA, USA, 2010.
[7] Electric Power Research Institute (EPRI). Smart Grid Resource Center. 2014. Date accessed: 03-Sept.-2019. URL: https://smartgrid.epri.com/.
[8] XcelEnergy. "Xcel energy smart grid: A white paper," Denver Business Journal, pp. 1–13, 2007.
[9] European Commission. JRC Reference Reports/Guidelines for Conducting a Cost-Benefit Analysis of Smart Grid Projects. European Commission/Joint Research Centre Institute for Energy and Transport, Luxembourg, 2012.
[10] European Commission. JRC Scientific and Technical Research Guidelines for Cost Benefit Analysis of Smart Metering Deployment. European

Commission/Joint Research Centre Institute for Energy and Transport, Luxemburg, 2012.
[11] European Commission. International Initiatives Related to Smart Grid Standardisation: State of the Art. EU Commission Task Force for Smart, pp. 25–38, 2010.
[12] Gelling, C. The Smart Grid: Enabling Energy Efficiency and Demand Response. The Fairmont Press, Lilburn, GA, USA, 2009.
[13] Velasquez, M. and Hester, P. T. "An analysis of multi-criteria decision making methods," International Journal of Operational Research, vol. 10, no. 2, pp. 56–66, 2013.
[14] Mardani, A., Jusoh, A., Nor, K., Khalifah, Z., Zakwan, N., and Valipour, A. "Multiple criteria decision-making techniques and their applications – a review of the literature from 2000 to 2014," Economic Research-Ekonomska Istraživanja, vol. 28, no. 1, pp. 516–571, 2015.
[15] Momoh, J. Smart Grid: Fundamentals of Design and Analysis. Wiley-IEEE Press, Hoboken, NJ, USA, 2012.
[16] National Institute of Standards and Technology. NIST Smart Grid Standards Roadmap. National Coordinator for Smart Grid Interoperability, Gaithersburg, MD, USA, 2009.
[17] SMB Smart Grid Strategic Group. IEC Smart Grid Standardization Roadmap, pp. 14–32, 2010.
[18] U.S. Department of Energy. The Smart Grid: An Introduction. U.S. Department of Energy, Washington, DC, 2003.
[19] U.S. Department of Energy. User Guide for the U.S. Department of Energy Smart Grid Computational Tool (SGCT), Version 2.0, 2011.
[20] U.S. Department of Energy. Smart Grid System Report – Annex A and B, July 2009.
[21] U.S. Department of Energy. Smart Grid System Report, July 2009.
[22] U.S. Department of Energy. The Smart Grid Stakeholder Roundtable Group Perspectives, September 2009.
[23] Zardari, N. H., Ahmed, K., Shirazi, S. M., and Yusop, Z. B. Weighting Methods and their Effects on Multi-Criteria Decision Making Model Outcomes in Water Resources Management, 2015.

Chapter 2
Distribution automation functions

Distribution automation started in the 1970s. It allows utilities to implement modern techniques in order to improve the reliability, efficiency, and quality of electric service. Distribution automation is also referred to as feeder automation. It has been defined by the IEEE as follows: "Distribution Automation is a system that enables an electric utility to remotely monitor, coordinate and operate distribution components in a real-time mode from remote locations."

Distribution automation's main applications could be categorized into four groups: the first group is fault location and automatic sectionalizing/service restoration, which primarily depends on a switchgear as illustrated in Figure 2.1.

Figure 2.1 Typical installation of a switch on an overhead distribution feeder

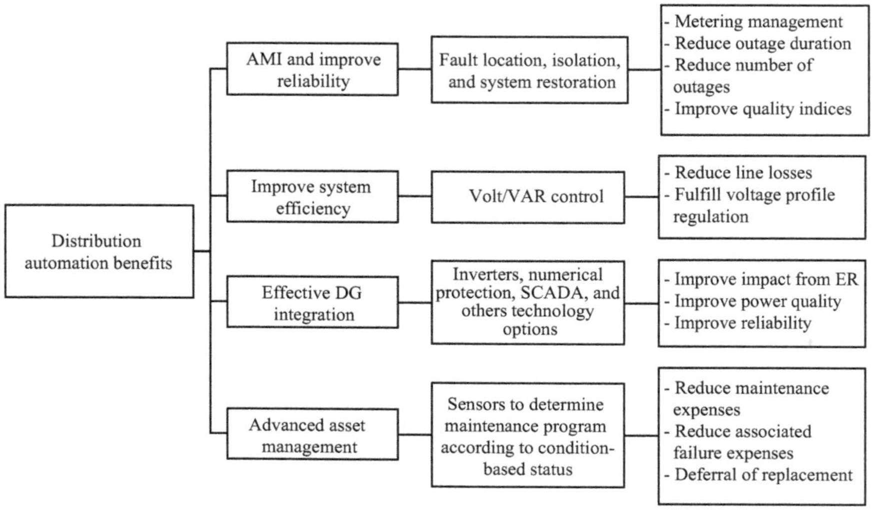

Figure 2.2 Main benefits of distribution automation

The second is Volt/VAR control and optimization, which mainly employs voltage regulators and capacitors. The third is the integration of distributed generation, which requires appropriate protection equipment and a robust supervisory control and data acquisition (SCADA) system among other components. Finally, the last group is associated with the advanced asset management, where sensors and metering devices are necessary. Figure 2.2 shows the applications and the benefits rendered by the implementation of the four groups.

Distribution automation, commonly known as DA, has evolved into advanced distribution automation, known as ADA, which incorporates advanced communication schemes, new computer technology, state-of-the-art equipment technologies, and high-speed power electronic devices.

2.1 Electrical system automation

Automation of power systems can be handled from control centers connected to generation and transmission systems on one hand and to distribution systems on the other. It could be said that they overlap at the HV/LV substations that also feature their own automation. This is shown in Figure 2.3.

Control centers for generation/transmission systems and for distribution systems are handled by using software applications customized for each application. They are referred to as energy management systems (EMS) and distribution management systems (DMS), respectively. These are illustrated in Figure 2.4.

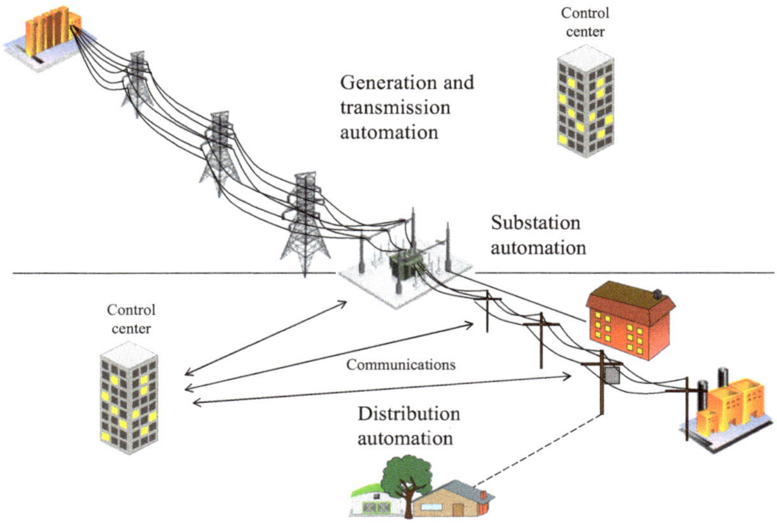

Figure 2.3 Power system automation components

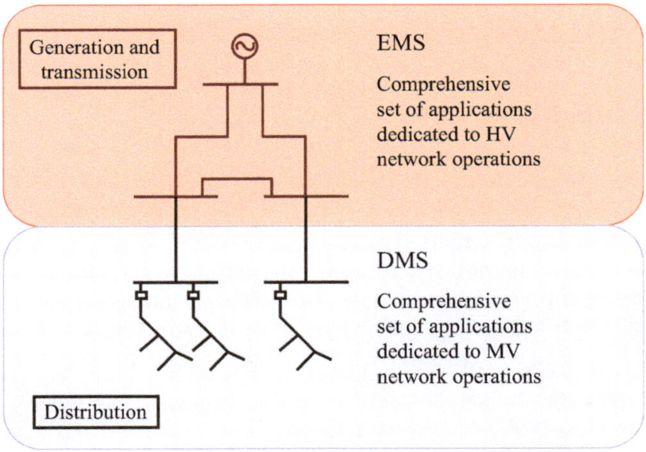

Figure 2.4 Network management EMS/DMS

2.2 EMS functional scope

The EMS is based on two main subsystems, SCADA and network analysis. Each one incorporates several applications as follows:

The SCADA subsystem incorporates applications like multiple remote terminal unit (RTU) protocols in a nonproprietary environment, load shed, sequence-switching

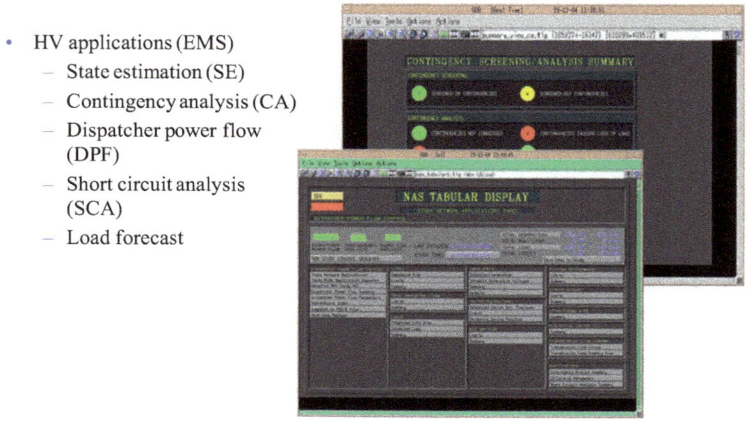

Figure 2.5 Screenshots of typical EMS

management, and disturbance storage and analysis. The network analysis applications offer the possibility of obtaining power flow, state estimation, contingency analysis, short circuit, security enhancement, optimal power flow, and Volt/VAR dispatch information. Figure 2.5 presents a typical screenshot of an EMS.

2.3 DMS functional scope

A DMS is a collection of applications designed to monitor and control the entire distribution network efficiently and reliably. It is a decision support system that assists the control room and field operating personnel with the monitoring and control of the electric distribution system. Improving the reliability and quality of service in terms of reducing outages, minimizing outage time, maintaining acceptable frequency and voltage levels are the key deliverables of a DMS.

The SCADA subsystem offers functions like feeder topology, coloring and circuit tracing, and mesh detection. The network analysis offers functions like real-time and power flow (balanced or unbalanced), short circuit, Volt/VAR control and optimal VVC, automatic online feeder reconfiguration, fault location and automatic service restoration, protection system, substation automation, customer load management, advanced metering infrastructure (AMI)/automated meter reading, demand side management (DSM), and load management. Figure 2.6 presents a typical screenshot of a DMS.

2.4 Functionality of DMS

The functionality of DMS can be divided into steady-state performance improvement and dynamic performance improvement.

- MV applications (DMS)
  - Connectivity analysis (CA)
  - Load flow (LF)
  - Demand estimation (DME)
  - Fault level analysis (FLA)
  - Fault isolation and system Restoration (FISR)
  - Voltage and VAR control (VVC)
  - Trouble call and outage Management (TCOM)

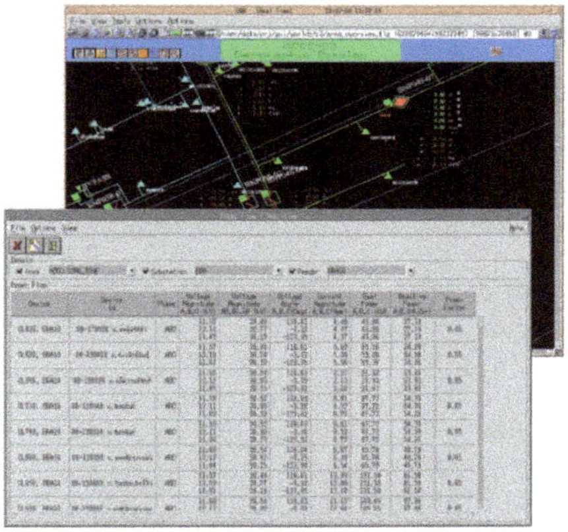

Figure 2.6 Screenshots of a typical DMS

2.4.1 Steady-state performance improvement

In this section the main functions associated with the analysis of the steady-state performance of the system are considered. These functions are characterized by fulfilling targets previously defined in planning studies.

2.4.1.1 Voltage/VAR control

Volt/VAR control deals mainly with the detection and prediction of voltage violations which require close control of the relationship between the voltage and the reactive power equipment. The equipment involved mainly includes capacitor, static VAR control devices, load tap changers, and voltage regulator controls.

2.4.1.2 Feeder reconfiguration

Feeder reconfiguration refers to the operation of switches, breakers, or reclosers to reconfigure the topology of feeders and improve the operating condition of the system. It is done under normal conditions in order to reduce losses and increase reliability. It also helps one to minimize voltage drop on the feeders. Feeder reconfiguration represents great benefits. However, it is not always possible to achieve all the objectives simultaneously.

2.4.1.3 Demand side management

DSM refers to a method for controlling the load at the user's premises, based on an agreement previously established with the utility.

2.4.1.4 Advanced metering infrastructure/automatic meter reading

AMI/automatic meter reading plays an important role as it aids utilities in establishing competitive strategies that conduct load profile gathering in order to characterize the value of individual customers to the utility. It identifies the major customers (aggregated load) and offers load information to customers as a special service.

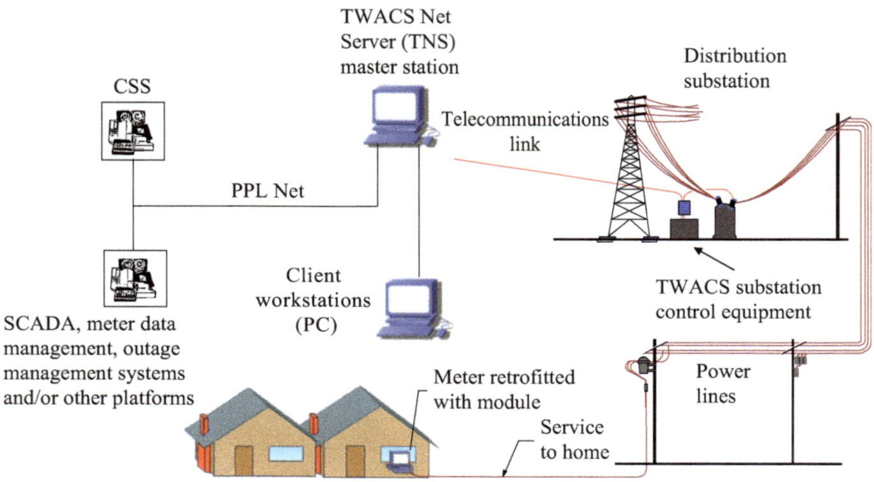

Figure 2.7 Illustration of power line communication (PLC)

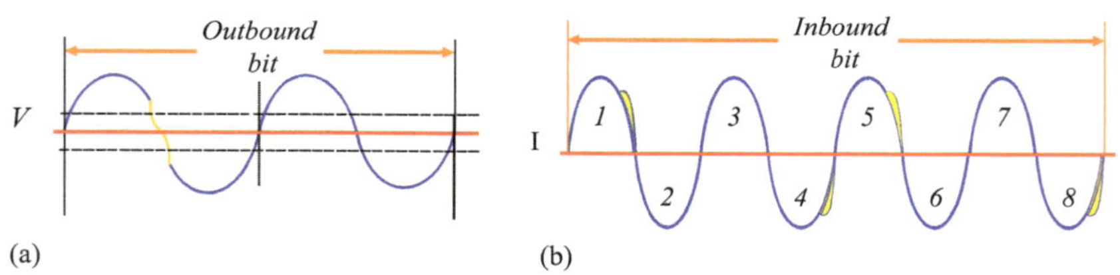

Figure 2.8 (a) Modulation for the outbound signal and (b) modulation for the inbound signal

One of the technologies used by AMI is power line communication—PLC—that uses the wires of the distribution system, including the customer's incoming lines. The signal rides on the fundamental power line frequency and modulation occurs at the zero-cross. It does not require repeaters or line conditioning equipment, resulting in lower installation costs. This technology does not affect power quality. Figure 2.7 illustrates a typical PLC system that uses the technology known as TWACS (two-way automatic communication system). This system uses modulation for the outbound signals and modulation for inbound signals as shown in Figure 2.8.

2.4.2 Dynamic performance improvement

In this section the main functions associated with the analysis of the dynamic performance of the system are considered. These functions are characterized by actions to be taken during faults, unpredicted events, and emergency conditions.

2.4.2.1 Fault location, isolation, and service restoration

The use of automated feeder switching aids in detecting feeder faults, determining the fault location (between two or more switches), isolating the faulted section of the feeder, and providing the possibility of restoring service to "healthy" portions of the feeder.

Service restoration allows one to find alternate options to quickly restore power to healthy parts of the system making sure that voltage levels are within accepted ranges and that any overload is avoided. Cold load pickup has to be considered for long outages in order to avoid tripping when reenergizing the feeders. Customer prioritization is carefully considered.

Automating this process renders immense benefits as the outage time is greatly reduced, as shown in Figure 2.9.

2.4.2.2 Trouble call system

When a fault occurs, protective relays and fault locators should raise the corresponding alarms that outages have occurred. However, the report of users affected is very important not only to reaffirm the fault occurrence but also to help locating it more quickly.

Trouble call systems (TCSs) are part of the outage management system (OMS) that enables electric utility employees to discover, locate, and resolve power outages in a more informed, orderly, efficient, and timely manner. The following sections present the main ideas of OMS.

Utilities now have modern facilities where the calls from users are readily received, classified, and matched with the database of the billing system to identify other users who may be associated with the problem. Thus, they can achieve solution schemes that identify and involve all the customers affected by the event. These facilities are TCSs that are manned but assisted with state-of-the-art

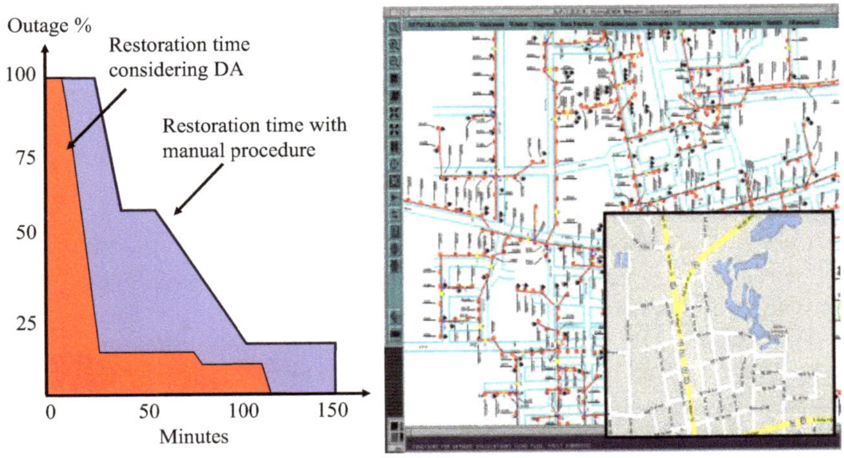

Figure 2.9 Comparison of restoration time with and without DA

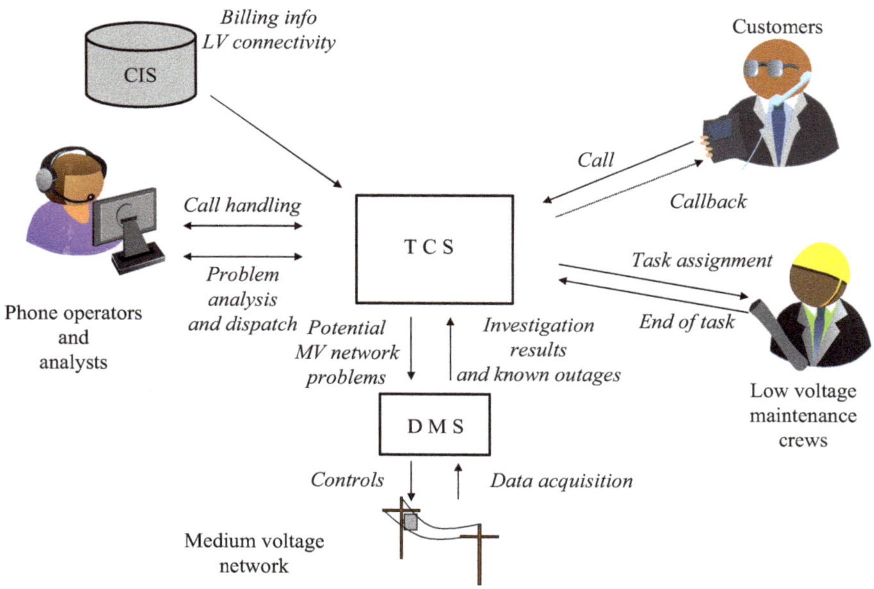

Figure 2.10 Trouble call system

communication and software capabilities. A typical process of a TCS is illustrated in Figure 2.10, which requires a close relationship with the Customer Information System (CIS).

2.4.2.3 Alarm triggering

Alarms are triggered by the operation of the protection relays. They send coded information to distribution control centers so that operators are aware of what is happening on the network. Other alarms indicate the state of the power system, for example, voltages at various locations and load flows on the more important circuits. These alarms provide one of the main sources of information flowing in real time into a distribution control center and are normally channeled to one printer in the control room where a hard copy can be produced. The alarm streams are also channeled to the operator's or control engineer's console where they can be displayed on computer screens.

A third avenue for alarm streams is storage in a data logger where the principal function is to retain a history of the alarm streams. This can be used for post-system fault analysis if it is needed later.

It is worth nothing that, with this arrangement, the alarms are not processed. Several events may occur simultaneously, or at a close time proximity with respect to each other, and each incident may trigger many alarms resulting in a large number of alarms flowing into the control center in close succession. The operator would then have to use his judgment to determine what has happened to the system. Subsequent telephone calls from customers may also help one to determine the

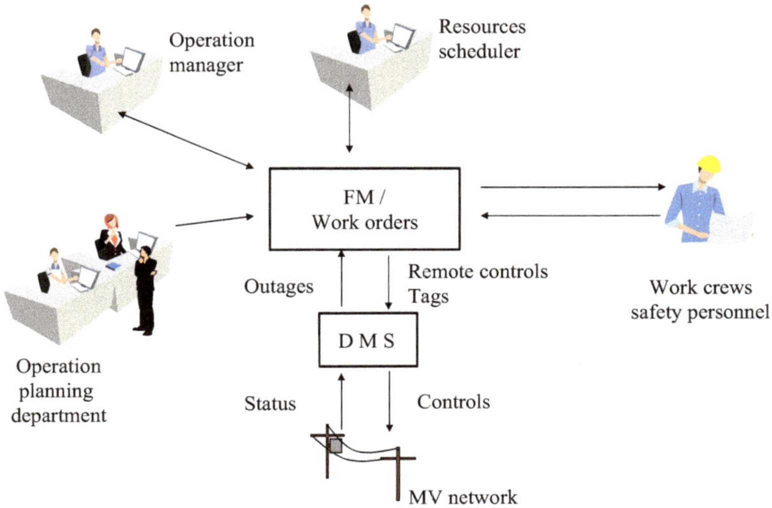

Figure 2.11 Work order illustration

exact location of these incidents such as a blackout in a certain area of the system. The aim of an alarm processor is to help the operator to arrive at a sensible conclusion speedily and to discard redundant information in the alarm streams.

2.4.2.4 Work orders

In order to expedite the coordination of crews responsible for the maintenance or feeder fixing, a proper work order system is essential. After a fault is reported, the system should be capable of automatically locating the most convenient crew and traveling means and of determining the essential elements required to reestablish service as quickly as possible. This requires a proper coordination with the DMS, the resource scheduler, and the operation manager of the utility, as shown in Figure 2.11.

Table 2.1 illustrates the functionalities for traditional DMS-based DA applications for the Smart Grid. These functionalities can be either available or not, limited or a custom feature according to the system being implemented by the utility.

2.5 Outage management systems

An OMS implements connectivity models and graphical user interfaces (GUIs) with functions such as trouble-call handling, outage analysis and prediction, crew management, and reliability reporting. Connectivity maps of the distribution system assist operators with outage management, including partial restorations and detection of nested outages. OMS also helps one to identify all affected customers and to maintain a complete, accurate, time-synched record of all data collected during an outage event.

Table 2.1 Traditional and DMS-based DA applications

Application	Traditional DA	DMS-based DA
Basic Monitoring and Control	✓	✓
Monitor Equipment on Feeder	✓	✓
Network Switching Management/ Analysis/Optimization	Limited	✓
Relay Protection Coordination	Custom	✓
Integrated Volt/VAR Control	Custom	✓
Direct Customer Load Control	✓	✓
Interface with AMI/OMS/GIS Systems	Custom	✓
Distribution System Real-Time Analysis Tools	Limited	✓
Multi-level Feeder Reconfiguration	Limited	✓
Emergency System Restoration Support	✓	✓
Dispatcher Training and Simulation	✓	✓
Power Quality Assessments	Limited	✓
Demand Response Analysis	✗	✓
Load Forecasting	Custom	✓
Predictive Equipment Maintenance/ Asset Management	Limited	✓

OMS have become more integrated with other operational systems such as Geographic Information Systems (GISs), CISs, Work Management Systems (WMSs), Mobile Workforce Management, SCADA, and AMI among others. Integration of OMS with these systems results in improved workflow efficiency and enhanced customer service. At some utilities, it can be utilized simultaneously by hundreds of users. It integrates information about customers, system status, and resources such as crews, providing a platform for operational decision support. This is shown in Figure 2.12.

Integration of OMS with DMS aims to assists operators in performing their responsibilities better, compared to separated DMS and OMS, bringing some extra benefits. The operator training is minimized, since operators only need to learn the features of one graphic user interface or GUI, being more comprehensive system visualization.

The integration of DMS applications in the OMS improves outage performance. For instance, a fault location algorithm uses the operated electric network model, including the location of open switches, along with an electrical model of the distribution system with lengths and impedances of the lines in order to estimate fault location. The DMS fault location functionality therefore uses the electrical DMS model but ultimately improves the OMS process.

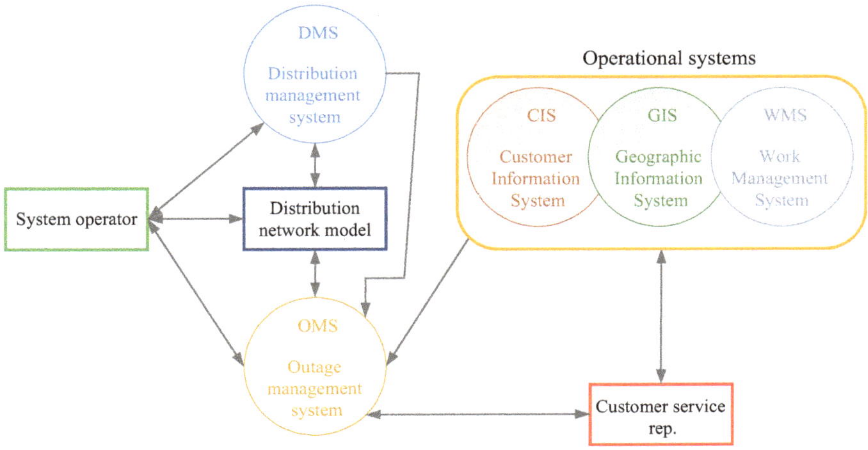

Figure 2.12 DMS/OMS integration

In a restoration switching analysis, the DMS/OMS system evaluates the possible isolation and restoration switching actions that can be done upon occurrence of a permanent fault. The application executes an unbalanced load flow to determine overloaded lines for each switching action, and the operator is provided with a listing of recommended switching actions. The functionality utilizes the DMS model of the system but improves the OMS process and functionalities.

With DMS/OMS working with the same operational model of the distribution system, circuit analysis can be considered with time changes. This includes circuit tracing, trouble call and outage analysis, loop and parallel source detection, fault location, and load flow. The result is a more comprehensive and accurate understanding of system conditions at any moment.

Data maintenance processes are simplified if the DMS and OMS are operating from the same model in their geographic information system, since the distribution operators make planned updates on their models. Hence, there are no data links or model synchronization concerns.

Figures 2.13 and 2.14 illustrate a typical GUI for a work request of an OMS. The OMS report indicates the location and type of operation, start time and the time of completion, the respective crew in charge and some final remarks, as well as the geographic location. The system provides a trouble ticket originated from a user call, which includes the trouble ID, status of the system, number of interruptions, type of priority and coordinates provided to the crew.

2.6 Geographic information systems

A geographical information system (GIS) can integrate, store, edit, and display geographically referenced information. A GIS may be linked to the online environment of a DMS. An example is shown in Figure 2.15.

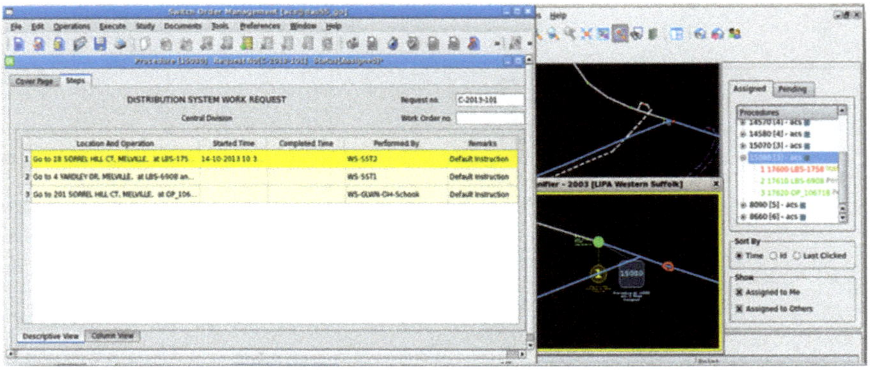

Figure 2.13 Outage management system report

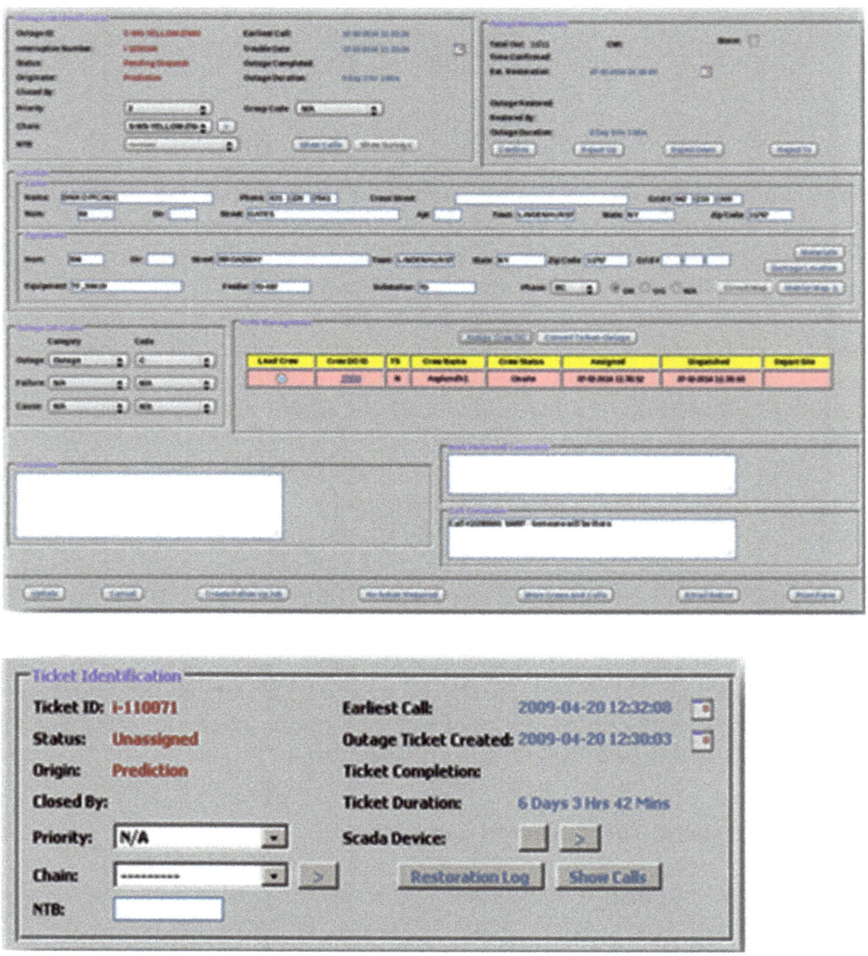

Figure 2.14 OMS trouble call ticket

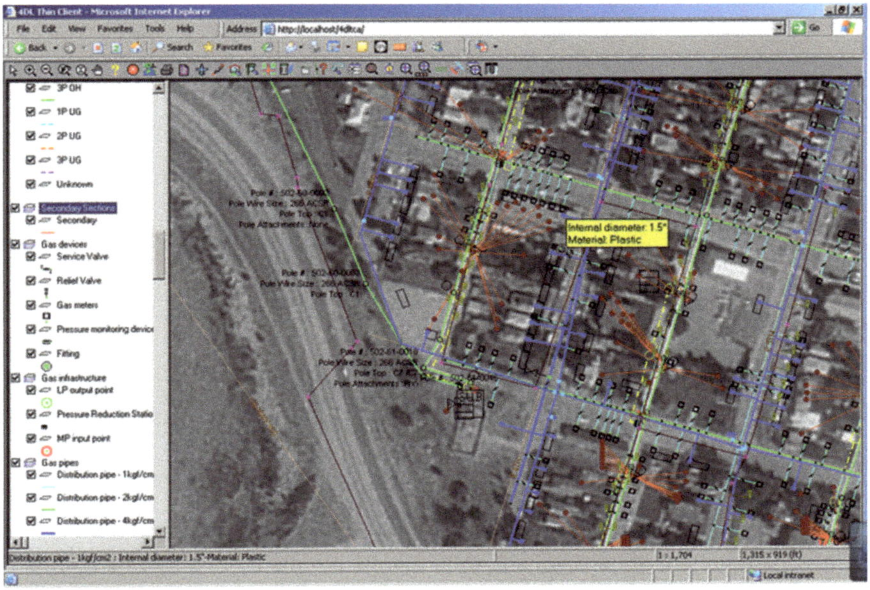

Figure 2.15 Example of a GIS

2.6.1 AM/FM functions

Automated mapping/facility management (AM/FM) is a specific application of the GIS aimed at providing database query and GUI for efficient management of distribution system devices. It involves activities related to geographical and electrical maps.

2.6.2 Database management

GIS systems require a computerized mapping system with the capability of storing and displaying the geographic locations of customers, feeder sections, substations, and/or pieces of equipment.

2.7 Communication options

DMS may employ a variety of communication mediums such as cooper wires, fiber optics, VHF and UHF radio, satellites, microwaves, cellular communications, Wi-Fi, and WiMAX.

Rapid advances in wireless, cellular, and satellite systems will soon provide cost-effective solutions for enterprise connectivity. Figure 2.16 illustrates different types of communication.

2.8 Supervisory control and data acquisition

SCADA is designed to achieve control and monitor processes remotely. This system is based on remote data acquisition.

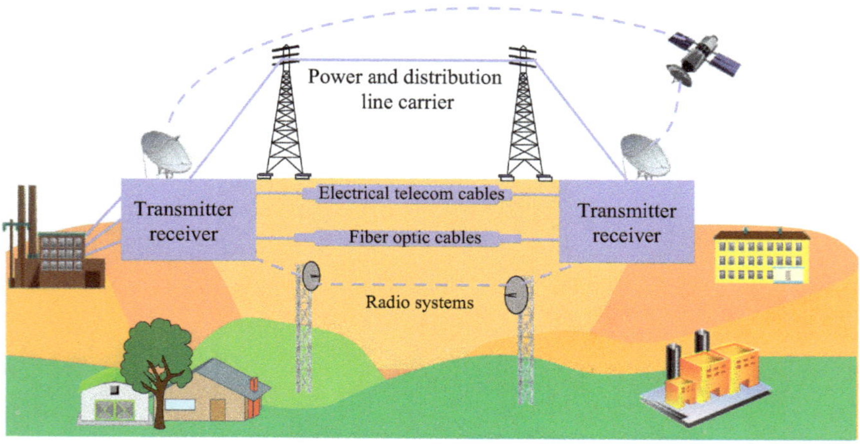

Figure 2.16 Typical communication methods

SCADA applications are globally accepted as a real-time power system monitoring and control tool, especially in generation and transmission systems. The RTU collects telemetry analog data and the status of devices in the substations, and communicates control commands. It is installed in a central location such as the control center of the utility with GUI, engineering applications managing history software, and other components.

SCADA describes a technique whereby computers are used to collect real-time data from plant machinery and/or to control plant machinery in real time. Typically this is done through the use of programmable logic controllers (PLCs). Figure 2.17 presents a common SCADA arrangement.

Recent trends in SCADA systems provide greater situational awareness through improved GUI, data presentation and information, intelligent alarm processing, improved integration with other engineering processes and business systems, and enhanced security functions.

2.8.1 SCADA functions

DMS/EMS integration with SCADA is a growing trend. The main functions of SCADA systems are as follows:

- supervisory control,
- data acquisition and processing,
- sequence of events (SOE) registry,
- misoperation revision,
- tagging,
- alarm processing,
- historical information system (HIS).

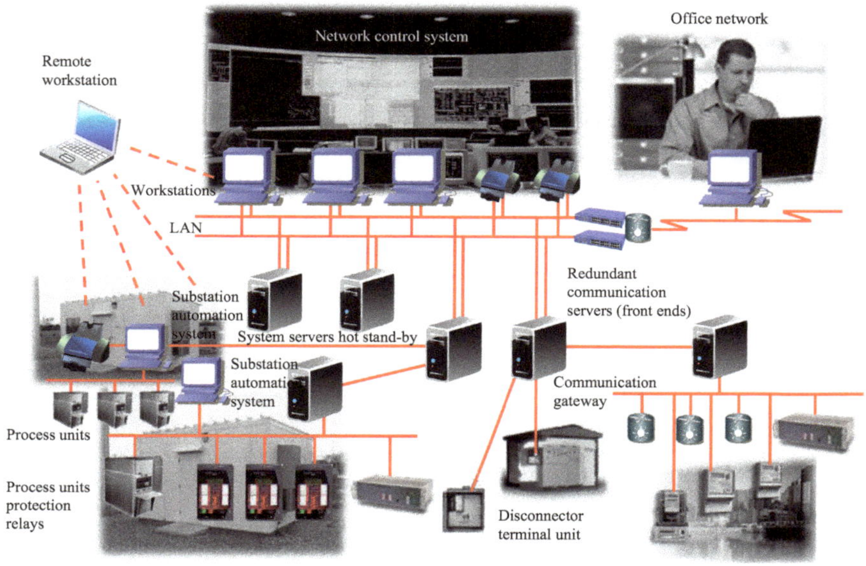

Figure 2.17 SCADA illustration

2.8.1.1 Supervisory control

SCADA controls the operation equipment (switches, breakers) in locations remote from the control center. This includes operation, commands, open/close commands, and set point control.

The control should include a selection sequence and verification before operation (select–check–back–before–operate—SCBO). In this case the user selects the equipment to be controlled using SCADA and the operation to be conducted. The operator observes the result of the operation in SCADA. The option to cancel the command should be available. The user is able to manually update the state of all control points that are not telemetry related.

Commands executed by the user are registered at the time they occur and they are stored in a historic file for future use, and in the HIS. Each event is identified by date, time, device name, device ID, and operation time. All failed remote command actions are identified as alarms.

The selected intelligent electronic devices (IEDs) use a process that meets the supervisory functional control requirements. The following factors should be taken into consideration:

- The time delay after the command has been sent or received from a communication point.
- The time delay after a command has been executed before another command can be initiated.
- Backup for multiple control points per command.

- SCBO control command backup.
- IEDs block or single or multiple output identification.

Figure 2.18 is a good illustration of a SCADA operating on different pieces of equipment in a distribution system, following orders received from application packages.

2.8.1.2 Data acquisition and processing

Figure 2.19 depicts a display of a SCADA system showing data acquisition in a distribution utility.

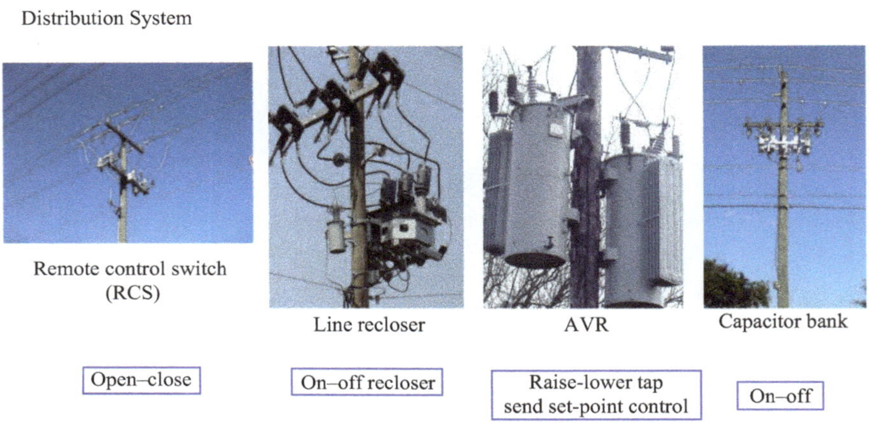

Figure 2.18 SCADA functions: supervisory control

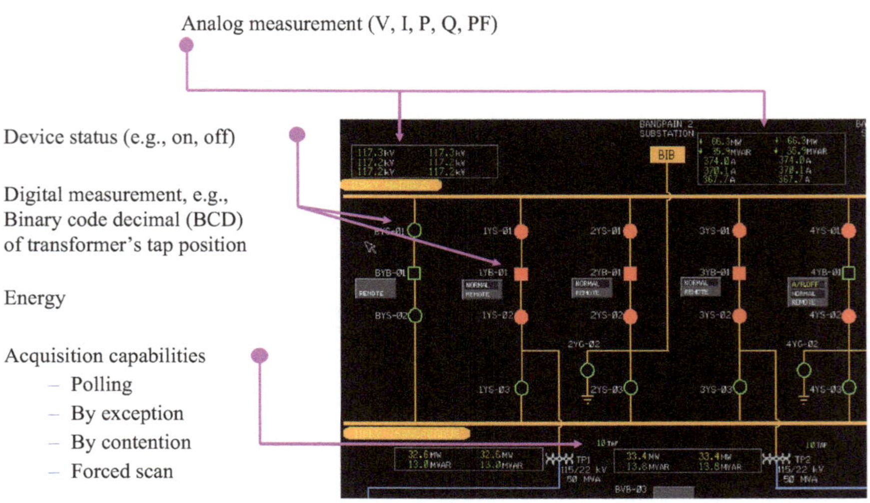

Figure 2.19 SCADA functions: data acquisition

SCADA systems require a reliable connection with databases that have information about the system and the users. Connection with packages having the "intelligence" or the software is essential to performing the appropriate operations. Figure 2.20 shows the interaction of a SCADA with other information packages of the system.

The data acquisition function is employed to supervise the state of the equipment and to measure different equipment parameters and external SCADA systems. These states and measurement data are viewable to the operators and also available for other functions.

The SCADA/DMS/EMS/GMS system master station obtains data from the data concentrators. Data acquisition is conducted through a private WAN-dedicated network for IEDs and SCADA/DMS/EMS communications and a wireless data network with IP technology dedicated to IEDs.

These two networks should support TCP/IP or UDP/IP protocols. IEC 60870-5-104 or IEC 61850 protocols are normally used for all substations. The substation data concentrator in addition to obtaining data from auxiliary services should be designed to question the distribution network IEDs. The communication protocol is typically DNP 3.0 or IEC 61850.

The data obtained from each IED includes the following:

- States (operation state, breaker position, switch position, recloser, alarms, etc.)
- SOE data (high precision event data timestamp)
- Metering (currents, voltages, MW, MVAR, frequency, transformer MW and MVAR, transformer tap position, etc.)
- Alarms (overvoltage, undervoltage, high frequency, low frequency, etc.)
- Input change processing time.

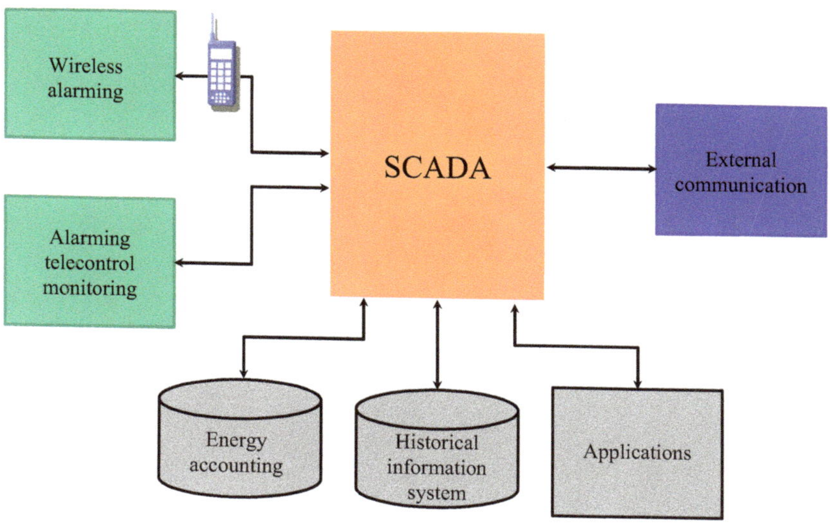

Figure 2.20 Relationship of SCADA with databases

2.8.1.3 Sequence of events (SOE) registry

The distribution system SCADA has the ability to store SOE data for specific events. The SOE information includes relay operation analysis and alarms as part of the equipment operation analysis. The SOE events messaged are similar to the alarm messages, but they will be accurate to the millisecond, for example, 17:31:19.509.

The SOE data obtained are stored in chronological order in the HIS database and reported individually.

2.8.1.4 Misoperation revision

The misoperation revision consists of the creation and continuous update of the event file. Every 10 s, a picture of the operator-selected points or the real-time database is stored in the event file that consists of three files: the pre-event, event, and post-event period.

The file size is determined by the time period covered by the recording, the amount of data to be stored, and the available space on the disk. The recommended time frame is 10 min for the pre-event, 15 for the event period, and 15 min for the post-event period.

2.8.1.5 Tagging

The control center SCADA labels each specific event, monitored or not, on a graph. Each label represents the attention of a device using a graphical symbol indicating what the remote control is for such device. Different labels are normally used to classify devices in a priority order that refers to the open and close controls.

2.8.1.6 Alarm processing

The alarms detected by the data acquisition functions are managed in such a way that the predetermined alarm conditions are reported clearly and promptly only for the work station that needs the information.

The alarm management is supported by priority levels. When an alarm is presented, its responsibility area and priority will be displayed.

The alarms and events are saved daily and stored in the HIS for later reference. The alarms are presented in a single line diagram with screen visualization using symbols or color changes.

2.8.1.7 Historical information system

Traditional methods to collect historical data included chart recorders, operator logs, and alarm printers. Trends and historical databases can be constructed with the help of SCADA data. Reports, for example, can include compliance reports, operation reports, asset management, and work order generation.

2.8.2 System architecture

The system architecture has a layout and several components that are presented in the following paragraphs.

2.8.2.1 Master station (control center)

The master station has a distributed software and hardware architecture. Modern master station SCADA systems employ open architecture characteristics allowing interconnection with other systems. Open system standards allow interaction with other products from suppliers.

To ensure openness, the system must meet international standards or industry standards like Microsoft Windows and computer applications related products, IEC 60870-6 (TASE.2) for communications with other control centers, and IEEE 1379 or IEC 61850 for the communications with RTU. IEEE has launched the IEEE Standard 2030 providing alternative approaches and best practices for achieving Smart Grid interoperability.

The main SCADA system elements are shown in Figure 2.21 and should include the following components.

2.8.2.2 Human–machine interface

This interface consists of the mimic board and the multi-video display unit. The board map shows a simplified scheme of the power system. This board presents the most detailed representation of the electric system. It is organized to geographically represent the power system, which is useful for observing large disturbances that cover an extended geographical area.

Human–machine interface or HMI is a software application which provides situational awareness of the automation processes to a plant operator such as process values, alarms, and data trends. An HMI can operate on various platforms, including desktop computers, tablets, smart phones, or dedicated flat panel screens.

The multi-video display interface is employed to observe the condition of the power system devices in more detail. In modern SCADA systems, multiple screen

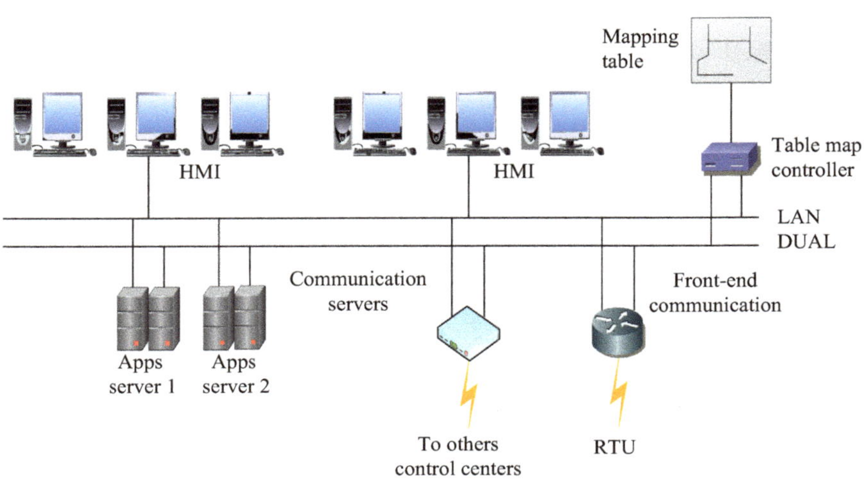

Figure 2.21 Control center general scheme

workstations offer operators easy access to a wide variety of application functions and controls. These workstations combine graphical capabilities with multiple window functions like zoom, detachable menus, and drag-and-drop operations.

2.8.2.3 Application servers

- Main SCADA subsystem: This server is mainly used for data-processing functions and real-time process control.
- Database subsystem: This server backs up the historic database and other databases.
- Configuration and operation: This server is used for control, operation, and maintenance of the SCADA system.
- Front-end communication: This system is used for data acquisition from the RTU and field equipment. Besides RTU data acquisition functions, it also includes protocol conversion, security check, digital and analog data temporary storage, and analog values and digital state changes.
- External communications server: This server provides interchange with other control centers. A standard protocol such as IEC 60870 (TASE.2) must be used for real-time interchange and data storage.

2.8.2.4 Remote stations

Remote stations are measuring devices or IEDs connected to the plant that are monitored and controlled using an interface.

The remote station could be an RTU or PLC. The RTU is mainly used when communications are difficult. The RTU has digital and analog inputs, and outputs to an LED indicator (selected by channel), with optical insulation for protection against overvoltage or short circuits.

The remote station has RTU that collects the corresponding data for transmission to the control center and sends commands from the control center to the devices. Additionally, the data concentrators collect data from all IEDs, and, if necessary, provide data interchange with other systems (possibly with other SCADA systems).

2.8.2.5 Architecture selection

The architecture of the overall system control can have different levels of complexity according to the size of the system and the number of users and applications. Not every system needs to have all the elements or the same number of devices. The following figures show different levels starting from simple systems based on RTUs to the more complex ones. Figure 2.22 shows a single master station with multiple RTUs. Figure 2.23 shows a single master station with data concentrator or gateway, and Figure 2.24 shows a multiple master station, LAN/WAN substation connection using routers.

2.8.2.6 SCADA for electrical distribution networks

Some of the main advantages of implementing SCADA systems for electrical distribution networks include the continuous monitoring and control of the distribution network performed from remote locations; costs savings by reducing manual operation,

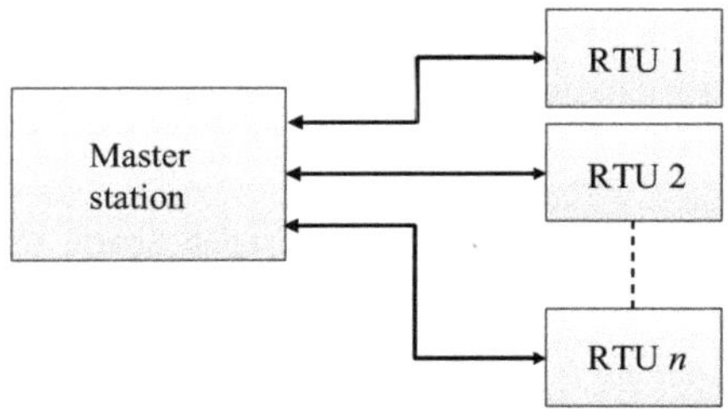

Figure 2.22 Single master station, multiple RTU, radial circuit

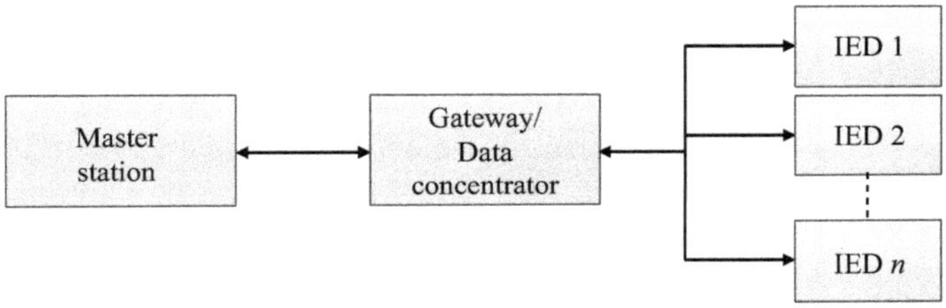

Figure 2.23 Single master station, data concentrator, or gateway

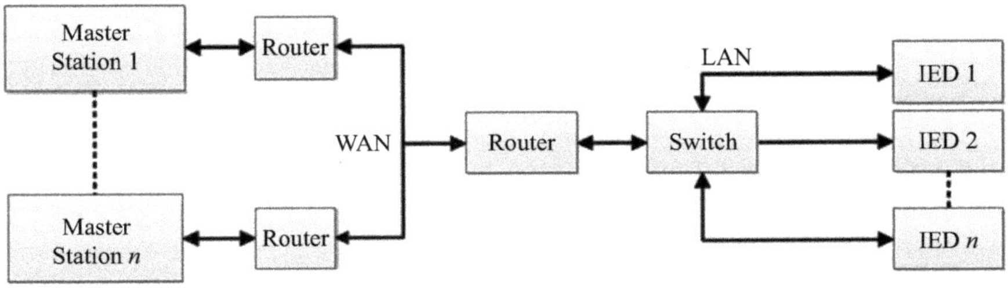

Figure 2.24 Multiple master stations, LAN/WAN substation connection using routers

outage time reduction, improvement on the continuity of service, voltage profile improvements by power factor correction and Volt/VAR control, etc.

Nowadays, distribution SCADA or D-SCADA provides a modular system option for interaction with automated and intelligent distribution field devices. Since the D-SCADA model is not required to manage the distribution system connectivity, it can be generated quickly and efficiently to support rapid deployment initiatives and ongoing maintenance. The D-SCADA may be implemented to communicate directly with the field through a variety of methods to meet specific customer requirements. Combined with DMS applications, the D-SCADA enables the replacement of multiple SCADA stations that may be in place for capacitor control, DA switching, direct load control, and other functionalities.

2.9 Synchrophasors and its application in power systems

Due to the needs of the energy market and in order to better use company assets, key tools are employed. These include communication technology, signal processing, global synchronization diffusion among monitoring, protection, and control applications in the electric power system aiming for efficient operations within the existing electrical, economic, and regulatory margins.

In the past, it was not possible to monitor in real time the relative phase angle in all voltages and intensities in the network due to processing constraints and difficulties in information collection, coordination, and network data synchronization. Changes have been made in new communication solutions, metering, digital data processing, and data synchronization through satellites that accompanied by synchrophasors provide a new scenario to implement a great number of applications that allow one to increase network efficiency, tools to detect the power system network conditions, and reliable operation near electrical and physical limits.

In this sense, for example, the idea of real-time system monitoring is recommended to detect stability risk situations to make necessary and accurate decisions limiting the impact on the affected areas avoiding total collapse of voltages in the system, increasing the efficiency of using transmission assets, and reducing investments in new circuits.

New developments in synchronized phasor metering (synchrophasors) are presented as an important evaluation tool for system operators for real-time metering presenting the operator with sufficient information to take appropriate actions in the interconnection system when violations are presented and oscillations could endanger the system stability.

The first phasor measurement unit (PMU) was created in 1988 allowing data synchronization using the clock pulses from a GPS; 7 years later the IEEE 1344 Standard was developed where the GPS synchronized PMU is designated as the synchrophasors. The IEEE 1344 denominated as "the synchrophasor standard" was superseded by the IEEE C37-118 norm in 2005. This norm has been widely accepted as the preferred method for synchrophasor's metering. The norm defines the synchronized phasor samples employed in power system applications and provides methods to quantify metering samples, test to guarantee metering samples, test error limits, and real-time data communication protocols.

2.9.1 Definition

A phasor is the electric network sinusoidal wave representations (voltages or currents) through the projection in the real axis of a vector rotating at angular speed ω at a moment $t = 0$ producing an angle φ with respect to the real axis. The time instant of the rotational vector "picture" define the maximum values of a reference cosine wave in way the phasor phase X represents the displacement between the reference wave and the observed cosine wave. In addition to obtaining a mathematical simplification, the system reduces complex electronic system needs and processing capabilities enabling network supervision employing PMUs. Figure 2.25 shows the elaboration of the sinusoidal waveform using its phasorial representation.

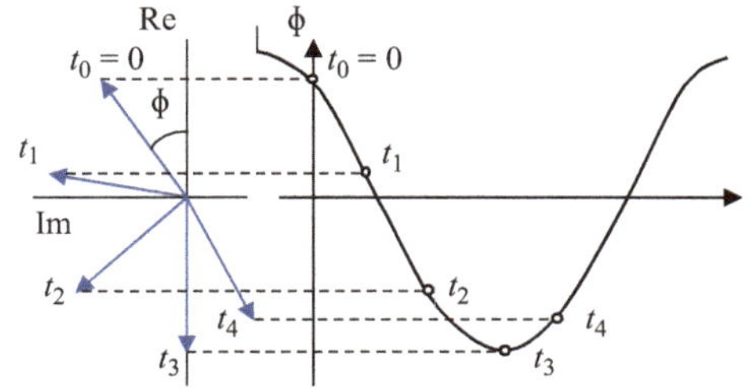

Figure 2.25 Sinusoidal waveform from its phasorial representation

2.9.2 Application of PMUs

In general terms, synchrophasors are phasor samples that are synchronized with UTC (universal time coordinated) time that creates a nominal frequency (50 or 60 Hz) cosine wave as reference. IEEE C37.118 norm from 2005 defines the synchrophasors as a phasor calculated from sampled data using a standard time signal as sampling reference such that phasors sampled in remote locations have a defined common phase relation.

PMUs measure voltage and currents from the power system, using the instrumentation transformers installed in substations, filter and attenuators are employed in the connection of this equipment. Once the phasor is metered, the PMU generates a message with timestamp and phasor data in a format defined in 1995 by the IEEE 1344 norm (replaced in 2005 by IEEE C37.118) in such a way that could be transmitted to a distant location across any available communications link. The format established by the IEEE C37.118 is not required for historic records or monitoring activities, but it is required in real-time control systems.

A simple phasorial network consists of two nodes: one is connected to a PMU linked to a phasorial data concentrator (PDC) in the second node. The communication, as illustrated in Figure 2.26, can be conducted employing any of the technologies currently available as direct wiring, radio network, microwave, telephone, digital radio, or the combination among those mentioned. The communication protocol is described in the IEEE C37.118 Standard.

Applications employing synchrophasors have been developed to detect synch loss, multi-area state estimation, oscillation mode identification, voltage stability protection, and dynamic system monitoring. The latest application is real-time wide-area control. PMU offers attractive solutions for protection and control action improvements in modern power systems solving the distributed data time incoherence, including a timestamp to the data aligning samples with a common time for processing.

Wide area networks extent across large geographical areas like a country or continent. The main function is focused in network interconnection or terminals distantly located. They operate on the basis of commutation nodes that conduct elements interconnection that allow a continuous large data flow.

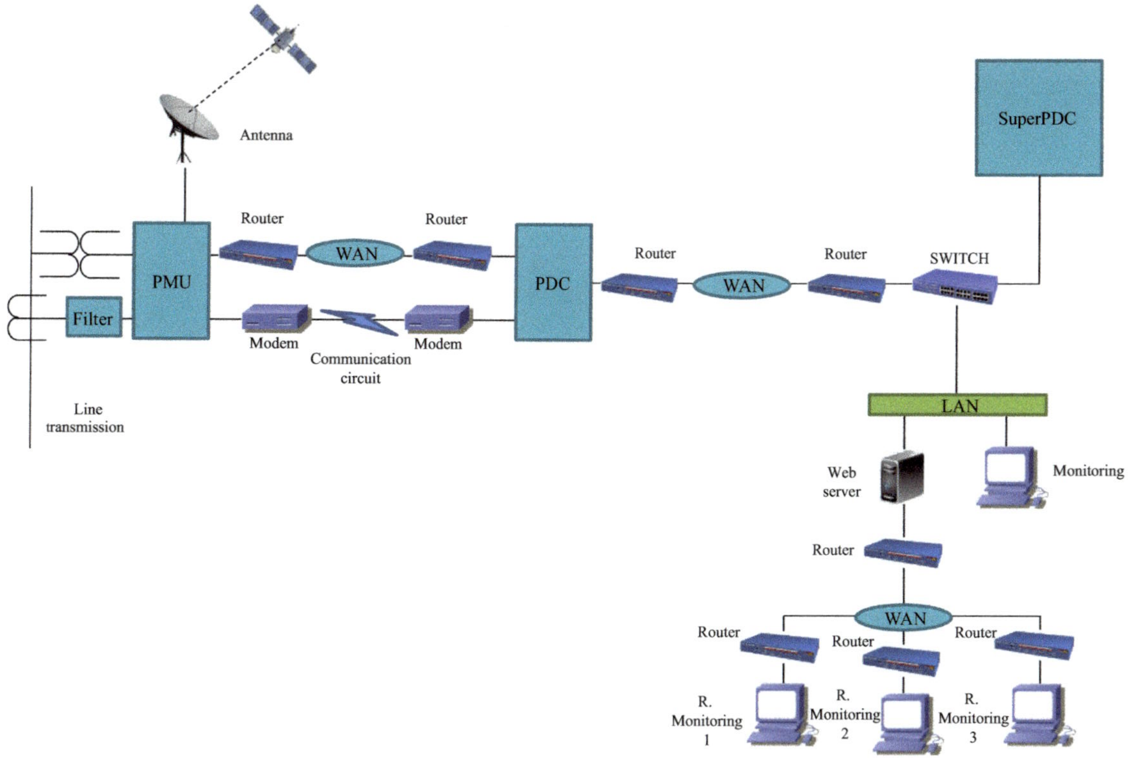

Figure 2.26 PMU's integration with the current communication system

The wide area monitoring system's main purpose is to provide system operators with a large information system with analysis tools that increase detection speeds and response time in the event of risky situations allowing the operators to make appropriate decisions efficiently preventing uncontrollable events or cascaded outages. This allows one to convert such applications as components of protection and control applications being the first step in Wide Area Monitoring Protection and Control (WAMPC) projects.

The wide area (WAMPC) protection and control schemes have not been widely applied yet and research groups continue to develop new applications generally divided into three groups: power systems monitoring, advanced network protections (wide area), and advanced control schemes.

WAMPC applications comprehend two categories: applications in local substations and applications in system or regional control centers. This classification takes advantage of local intelligence and the substation PMU decision power to conduct corrective actions locally in case power system events occur while maintaining general system control in control centers.

2.9.2.1 Line parameters calculation

Figure 2.27 shows the transmission line model and its parameters. Transmission line impedance parameters can be calculated using PMU on both ends of the line eliminating error sources. PMU measures voltage and current synchrophasors in the terminals: V_S, V_R, I_S, and I_R. Line parameters R, X_L, Y_1, and Y_2 can be calculated. Different load and ambient profiles can be obtained from these results calculating distributed π sections allowing determining line voltage profiles in any transmission conditions.

2.9.2.2 State estimation

This application allows the operator to know relative angle values in different network zones making decisions with the objective of maintaining power system integrity during events that cause unexpected generation or transmission outages as the increase in angle difference represents an increase in the static voltage between two points resulting in instability and partial or total blackouts.

2.9.2.3 Transmission lines thermal monitoring

Line resistance calculations based on PMU samples offer an economic approximation to thermal monitoring allowing efficient usage of the full transmission

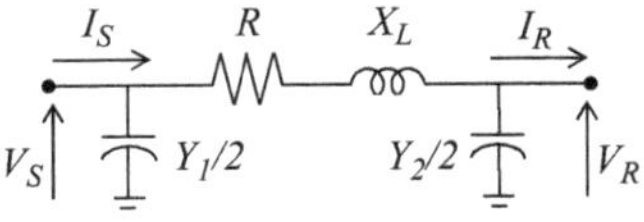

Figure 2.27 Transmission line model

capability. By simply obtaining voltage magnitude and angle in the line connection nodes and the current magnitude and angle flowing at the beginning and end points, the line impedance can be obtained. Based on the conductor resistance at 25°C or 50°C (manufacturer data), the average conductor temperature can be calculated extrapolating the resistance and temperature relation. With this data and in an economical manner, warnings can be received for overloads, transmission line capacity dynamic control, and/or direct estimation of line depression.

$$Z_1 = \frac{V_S^2 - V_R^2}{I_S \cdot V_R - I_R \cdot V_S}, \quad Z_1 = R + jX \tag{2.1}$$

2.9.2.4 Voltage instability

Voltage instability refers to power systems' capacity to maintain stationary voltages in all busbars after a disturbance; this ability directly depends on a balance between power generated and demanded by loads. The instability is reflected as a continued deterioration of voltage magnitudes in certain busbars or nodes in the system and it could be visualized as the relationship between voltages and demanded power in the busbar; consequently, it can be determined that voltage stability phenomena are directly related to reactive power flow in the network, load behavior in the event of voltage variations, and automatic voltage regulation device actions.

Voltage stability deficiencies represent a high risk for safe and reliable power system operations; consequently, indicators have been established based on certain calculations to determine system state in the event of a possible stability perturbance from the voltage point of view. However, complex mathematical processes needed to evaluate the power system state have been a limiting factor in the implementation of monitoring schemes and voltage instability prevention. Synchrophasors are a solution to the time delay issues with data acquisition since the phasorial metered samples from the PMUs, correctly located, present the system state without involving complex data processing; additionally, the time stamped by the PMU in each sample allows one to establish a temporary pattern for each indicator which guarantees that all samples obtained belong to the same system state.

Some indicators applicable to this case are as follows:

- SDI and SDC: Calculations are based on samples metered in a transmission element ends, in two consecutive time instants and the power increment in one of the terminals.
- ISI: Based on consecutive voltage and current samples measured at a substation considering the system Thévenin equivalent.
- VSLBI: Applies the maximum power transfer theorem in a calculation based on substation samples and Thévenin equivalent impedance.

2.9.2.5 Power-transfer stability

Generator power output increase is only possible up to the maximum electrical power value to be transferred. This power value is referred to as stable state stability limit, and it occurs at 90° maximum angular displacement between the reception busbar and supply busbar. Any attempt to increase the power to be

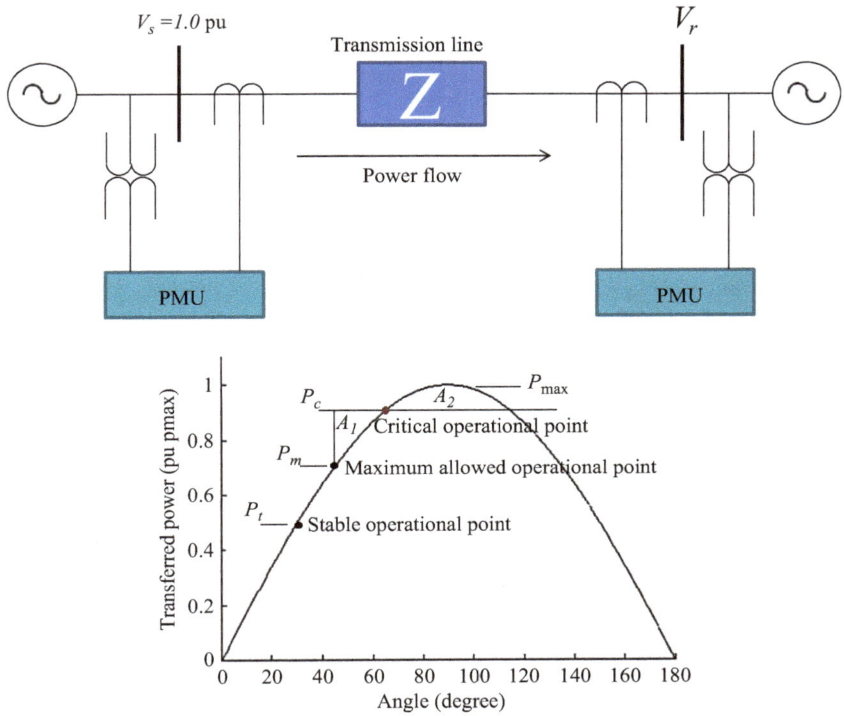

Figure 2.28 Transferred power through a line

transferred from this maximum transfer point causes a decrease in the electric power to be transferred and consequently a collapse due to the lack of sufficient power to supply the demand, as is shown in Figure 2.28.

The voltage synchrophasor and the transmission line impedance model could be used to illustrate the power transfer curve and the system operational point. When the operational point P_T reaches an unstable condition, preventive actions can be taken.

2.9.2.6 Power oscillations

Significant changes in power transfer cause large voltage variations in the system resulting in angle oscillations that can produce hard oscillations and system instability. To prevent the collapse, synchrophasor-based schemes have been developed to detect imminent unstable operation conditions and inform the system operator. These algorithms monitor oscillation speed to predict instability; hence, their implementation in control centers requires data reception at rates higher than 60 Hz with delays close to 0—in any case it can be locally detected by PMUs at substations.

A hard oscillation metering method is the modal analysis which is a signal processing technique that requires samples uniformly spaced; consequently, traditional data acquisition systems are not recommended due to their asynchronous nature. Mode measuring provides valuable information regarding frequency and oscillation damping in the system; it provides dynamic system nature inherent information of the power system.

2.9.2.7 Mode control governor

When a generator operates in island mode, the governor control system must be in asynchronous mode so that the generation system can control and maintain the frequency in the event of demand changes. Traditional systems that detect synchronism employ frequency samples not aligned in time, so they are susceptible to false synchronism claims when the systems operate at the same frequency but not connected. A more reliable method is to employ angle data in addition to the frequency to detect if an island condition has occurred.

Synchrophasor provides vector data with chronologic stamps to determine absolute angles across the power system. In such case, a substation relay with PMU capabilities can communicate angle metering and frequency to the cogeneration control center. With this information, the control system is able to automatically commutate the governor control mode if a separation is detected and it transforms into isochronous mode to regulate frequency.

2.9.2.8 Distributed generation control

Alternate sources generation has converted as an attractive window for new development projects. However, for implementations like photovoltaic to be widely accepted, their connection to the transmission system must be reliable. When a source is disconnected from the main transmission system, the source must be isolated from the portion of the system where it was separated; a fault in the source trip generates a high risk in personnel safety, in power quality, and in an out-of-phase reclosing.

Traditional methods employing island sensing local metering may not detect an island mode early enough, 2 s maximum according to IEEE 1547, for all load/generation conditions. These methods employ local voltage and frequency information verifying if their magnitude is outside predetermined ranges. The inconvenience with current local detection schemes is that they cannot detect a network separation promptly if the difference between power generated and consumed is small.

The synchrophasors provide the means to detect island for almost all load/generation conditions as they obtain precise metering from all areas. This vision provides a platform for solutions that could maintain the generation online during transient conditions. Solutions involving two relays, as is shown in Figure 2.29, are available, which take phasorial voltage samples and communicate between them (60 times per second). As an example, the second relay can receive remote data from relay one and calculate the angle difference between local and remote, the difference variation with respect to time determines the frequency displacement, and the displacement difference with respect to time indicates the acceleration between terminals.

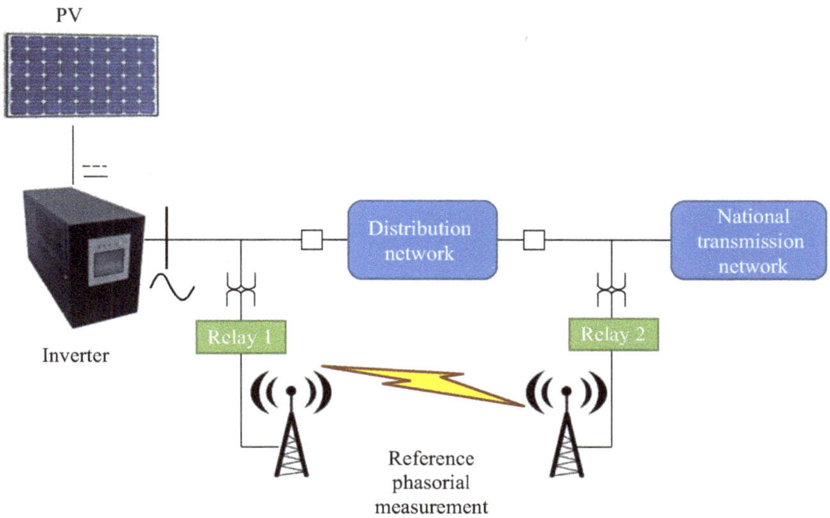

Figure 2.29 Integration of a distributed generation source using PMUs

Further reading

[1] EUCI. Distribution Automation Course Notes, 2009.

[2] IEC 61850 Ed 2. Communication Networks and Systems for Power Utility Automation Scheduled for 2010, 2010.

[3] IEEE. IEEE Std C37.1-2008 – IEEE Standard for SCADA and Automation Systems, 2008.

[4] Lin, C., Chen, C., Ku, T., Tsai, C., and Ho, C. “A Multiagent-based distribution automation system for service restoration of fault contingencies,” European Transactions on Electrical Power, vol. 21, pp. 239–253, 2010.

[5] Momoh, J. Electric Power Distribution, Automation, Protection, and Control, 1st ed. CRC Press, Boca Ratón, FL, USA, September 2008.

[6] Northcote-Green, J. and Wilson, R. Control and Automation of Electrical Power Distribution Systems. Taylor & Francis, Boca Raton, FL, USA, 2007.

[7] Uluski, R. W. “The Role of Advanced Distribution Automation in the Smart Grid,” in Power and Energy Society General Meeting, 2010 IEEE, 25–29 July 2010, pp. 1, 5.

[8] Uluski, R. W. “Using Distribution Automation for a Self-Healing Grid,” in Transmission and Distribution Conference and Exposition (T&D), 2012 IEEE PES, 7–10 May 2012, pp. 1, 5.

[9] Venkata, S. S. Distribution Automation – Course Notes. Minneapolis, September 2007.

Chapter 3

Fundamentals of distribution system analysis

Distribution automation requires a deep knowledge of the system to where it is applied. For this, a proper handling of analysis techniques is very important. Distribution system analysis is a part of a broader concept referred to power system analysis. In distribution system analysis, only some fields of the overall picture are studied and of course those referring mainly to feeders and radial systems.

The main topics to be analyzed correspond to the modeling of the elements, the analysis of load flows for ring and radial systems, and short circuit conditions. Load flows are applied for different applications involving not only the analysis of power flows and voltage regulation but also feeder reconfiguration and losses reduction. Short circuit as always is essential in sizing equipment and protective relay setting. Studies like transient and small signal stability are not usually conducted at distribution levels and therefore will not be considered in this book.

3.1 Electrical circuit laws

The most important concepts of electrical circuits are treated in this chapter. The three basic laws of electrical circuits are described next.

3.1.1 Ohm's law

Ohm's law establishes the relationship between voltage, current, and impedance in electrical circuits, as follows:

$$V = I \cdot Z \tag{3.1}$$

3.1.2 Kirchhoff's voltage law

For any closed path in a network, Kirchhoff voltage law states that the algebraic sum of the voltages is zero. Some of the voltages will be sources, while the other will result from current in passive elements creating a voltage, which is sometimes referred to as a voltage drop. The law applies equally well to circuits driven by constant sources, DC, time variable sources, $v(t)$ and $i(t)$, and to circuits driven by sources.

3.1.3 Kirchhoff's current law

The connection of two or more circuit elements creates a junction called a node. The junction between two elements is called a simple node and no division of current results. The junction of three of more elements is called a principal node, and here current division does take place. Kirchhoff's current law states that the algebraic sum of the currents at a node is zero. It may be stated alternatively that the sum of the currents entering a node is equal to the sum of the currents leaving that node. The basis for the law is the conservation of electric charge.

3.2 Circuit theorems

Circuit theorems are derived from the circuit laws. The three most commonly used for system analysis are as follows:

3.2.1 Thévenin's theorem

This is useful for replacing part of a network that is not of particular interest. Any active network viewed from any two terminals can be replaced by a single driving voltage in series with single impedance where the Thévenin equivalent voltage is the driving voltage that equals the open circuit voltage between terminals. Likewise, the Thévenin equivalent impedance is equal to the impedance of the network as viewed from the two terminals with all source voltages short-circuited. These concepts are illustrated in the drawings of Figure 3.1:

where

$$E' = \frac{z_3}{z_1 + z_3} \cdot E_1 \tag{3.2}$$

and

$$z' = \frac{z_1 \cdot z_3}{z_1 + z_3} \tag{3.3}$$

3.2.2 Star/Delta transform

Figure 3.2 shows the circuit equivalent in a Star/Delta configuration. The expressions to transform from one to the other configuration are the following:

$$Z_{12} = Z_{10} + Z_{20} + \frac{Z_{10} \cdot Z_{20}}{Z_{30}} \tag{3.4}$$

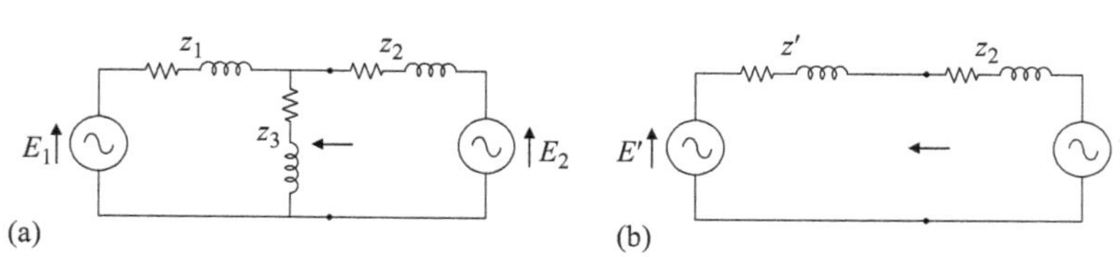

Figure 3.1 (a) Circuit before using Thévenin's theorem and (b) circuit after using the Thévenin's theorem

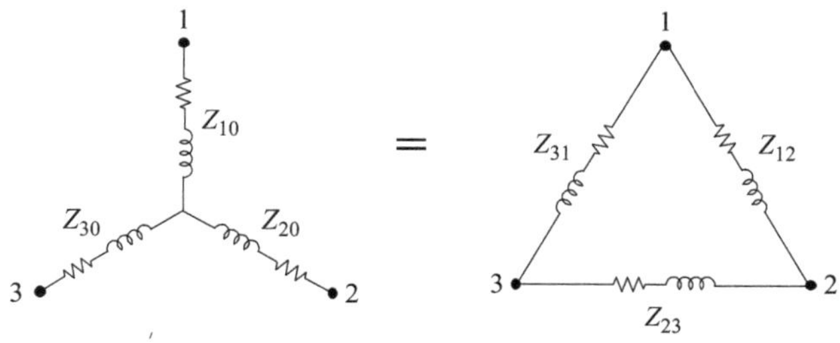

Figure 3.2 Star/Delta equivalent

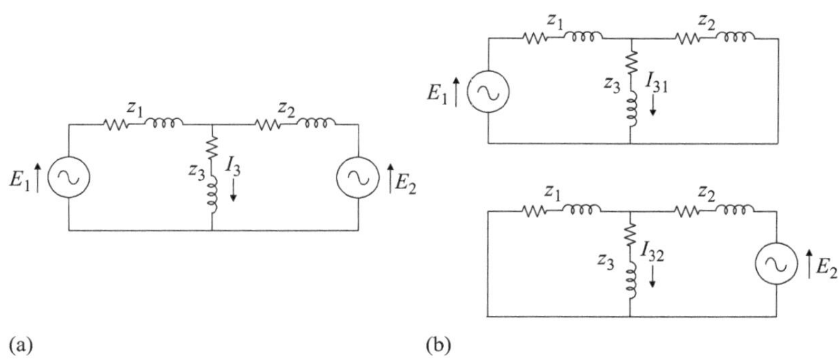

Figure 3.3 (a) Circuit complete and (b) solution using superposition theorem

$$Z_{10} = \frac{Z_{12} \cdot Z_{31}}{Z_{12} + Z_{23} + Z_{31}} \tag{3.5}$$

3.2.3 Superposition theorem

In any linear network the current in any branch is equal to the vector sum of the currents caused by each source voltage acting alone with the others short-circuited. An example case is illustrated in Figure 3.3:

$$I_3 = I_{31} + I_{32} \tag{3.6}$$

The Superposition theorem is highly used for short circuit calculations, in order to include the pre-fault current values that are normally taken from load flow programs.

3.3 Power AC circuits

Let $v(t) = V_m \sin \omega t$ and $i(t) = I_m \sin(\omega t - \theta)$ be the steady-state voltage and current in an AC circuit, where θ is the angle of the impedance of the circuit. The instantaneous power p is then

$$p(t) = v(t) \cdot i(t) = V_m I_m \sin(\omega t)\sin(\omega t - \theta) \tag{3.7}$$

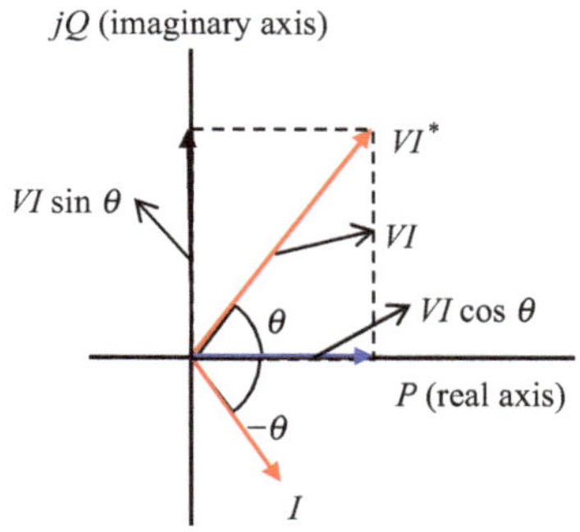

Figure 3.4 Relationship of current, voltage, and power in electrical circuits

If T is the period of the voltage, or current, waveform, then the average power P is given by

$$P = \frac{1}{T}\int_0^T p(t)\cdot dt = \frac{1}{2}V_m I_m \cos(\omega t) = VI\cos(\omega t) \tag{3.8}$$

where V and I are the rms voltage and rms current, respectively. The definition now becomes meaningful: the apparent power VI must be multiplied by the factor $\cos\theta$ to obtain the actual average power.

Sometimes, P is referred to as the true power, real power, in-phase power, or active power. Figure 3.4 helps one to understand easier this concept (drawn for the case $\theta > 0$), from which the following definitions of (average) power emerge:

Note that whereas active power is never negative, reactive power can have either sign. The three power units, W, V A, and VAR, are physically equivalent.

All three kinds of power may be derived from a single complex power, defined as VI^*(V A). In fact, since $V\angle 0°$ and $I\angle - \theta°$.

$$VI^* = (V\angle 0°)\cdot(I\angle + \theta°) = VI\cos\theta + jVI\sin\theta = P + jQ$$

And so

$$VI = \|VI^*\| P = \mathrm{Re}(VI^*) Q = \mathrm{Im}(VI^*)$$

Example 3.1 Two voltage sources $V_1 = 110\angle - 3°$ and $V_2 = 100\angle 50°$ are connected by a short line of impedance $Z = 2 + j5\ \Omega$ as shown in Figure 3.5, where δ is the phase angle of source 1. Write a MATLAB® program for the system such that the phase angle source 1 is changed from its initial value of $-3°$ within an angle range of $-35°$ to $35°$, applying steps of 2.5°. Voltage magnitudes of the two sources and the voltage phase angle of source 2 are to be kept constant. Compute the complex power for each source and the line loss. Tabulate the real power and plot P_1, P_2, and P_L versus voltage phase angle δ.

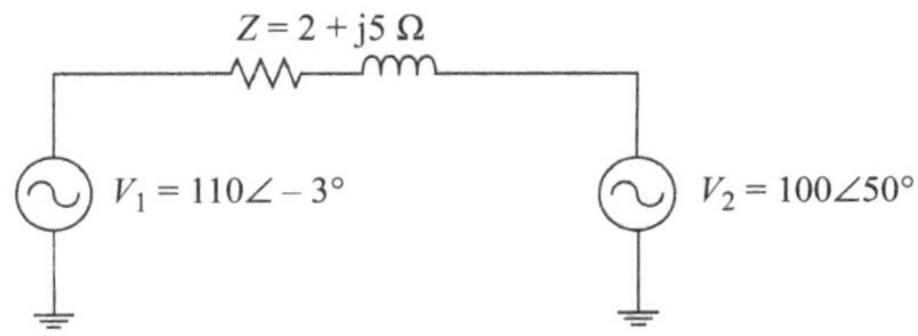

Figure 3.5 System for Example 3.1

In general terms,

$$I_{12} = \frac{V_1 - V_2}{Z}$$
$$I_{21} = \frac{V_2 - V_1}{Z}$$
$$S_{12} = P_{12} + jQ_{12} = V_1 I_{12}^*$$
$$S_{21} = P_{21} + jQ_{21} = V_2 I_{21}^*$$
$$S_L = P_L + jQ_L = S_{12} + S_{21}$$

The code of MATLAB for the general case:

```
V_1_mag=input('Write the voltage magnitude of V1 [V] = ');
V_1_ang=input('Write the voltage phase angle of V1 (degree) = ');
V_2_mag=input('Write the voltage magnitude of V2 [V] = ');
V_2_ang=input('Write the voltage phase angle of V2 (degree) = ');
var=input('Write the highest value of the angle variation range for V1
(degree) = ');
step=input('Write the angle step variation (degree) = ');
R= input ('Write the resistance in the line between both sources [ohm] = ');
X= input ('Write the reactance in the line between both sources [ohm] = ');
Z=R+j*X;
angle_1=[V_1_ang-var:step:V_1_ang+var]*pi()/180;
n=length(angle_1);
angle_2=ones(1,n)*V_2_ang*180/pi();
V1=V_1_mag.*(cos(angle_1)+j*sin(angle_1));
V2=V_2_mag.*(cos(angle_2)+j*sin(angle_2));

I12=(V1-V2)./Z;
I21=(V2-V1)./Z;

S1=V1.*conj(I12);
P1=real(S1);
Q1=imag(S1);

S2=V2.*conj(I21);
P2=real(S2);
Q2=imag(S2);

SL=S1+S2;
PL=real(SL);
QL=imag(SL);

result=[angle_1'*180/pi(), P1', P2', PL'];
disp('  Delta-1        P1          P2          PL')
disp(result)
plot(angle_1*180/pi(), P1, angle_1*180/pi(), P2, angle_1*180/pi(),
PL)
legend('P1','P2','PL')
ylabel('P [W]')
xlabel(' \delta [°]')
```

The results of calculations for Example 3.1 are given in the following list. The graphs for P_1, P_2, and P_L versus voltage phase angle δ are shown in Figure 3.6.

δ	P_1	P_2	P_L
−38, 0	−481, 751	565, 527	83, 776
−35, 5	−412, 364	476, 711	64, 347
−33, 0	−340, 602	388, 300	47, 697
−30, 5	−266, 605	300, 463	33, 858
−28, 0	−190, 511	213, 366	22, 855
−25, 5	−112, 466	127, 176	14, 711
−23, 0	−32, 618	42, 057	9, 439
−20, 5	48, 880	−41, 829	7, 051
−18, 0	131, 874	−124, 323	7, 551
−15, 5	216, 205	−205, 267	10, 937
−13, 0	301, 713	−284, 508	17, 204
−10, 5	388, 235	−361, 895	26, 340
−8, 0	475, 606	−437, 280	38, 327
−5, 5	563, 661	−510, 519	53, 142
−3, 0	652, 231	−581, 475	70, 757
−0, 5	741, 149	−650, 010	91, 139
2, 0	830, 244	−715, 995	114, 248
4, 5	919, 346	−779, 305	140, 042
7, 0	1,008, 288	−839, 818	168, 470
9, 5	1,096, 898	−897, 420	199, 478
12, 0	1,185, 009	−952, 001	233, 009
14, 5	1,272, 453	−1,003, 457	268, 997
17, 0	1,359, 063	−1,051, 689	307, 374
19, 5	1,444, 675	−1,096, 607	348, 067
22, 0	1,529, 125	−1,138, 125	390, 999
24, 5	1,612, 252	−1,176, 164	436, 088
27, 0	1,693, 899	−1,210, 651	483, 249
29, 5	1,773, 911	−1,241, 520	532, 390
32, 0	1,852, 134	−1,268, 714	583, 420

Example 3.2 A balanced three-phase source with the following instantaneous phase voltages:

$$v_{an} = 1,250 \cos(\omega t)$$

$$v_{bn} = 1,250 \cos(\omega t - 120^\circ)$$

$$v_{cn} = 1,250 \cos(\omega t - 240^\circ)$$

This source supplies a balanced load of impedance $Z = 125\angle 30^\circ \Omega$ per phase, as shown in Figure 3.7. Using MATLAB, plot instantaneous powers p_a, p_b, p_c and their sum versus ωt over a range of 0 to 2π with steps of 0.1. Comment in the nature of the instantaneous power in each phase and the total three-phase real power.

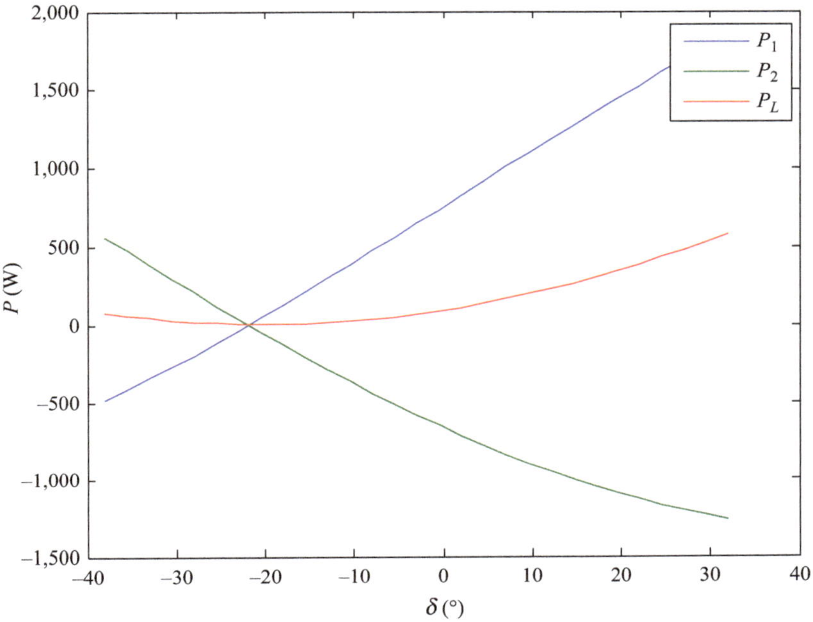

Figure 3.6 Power losses for different voltage angles

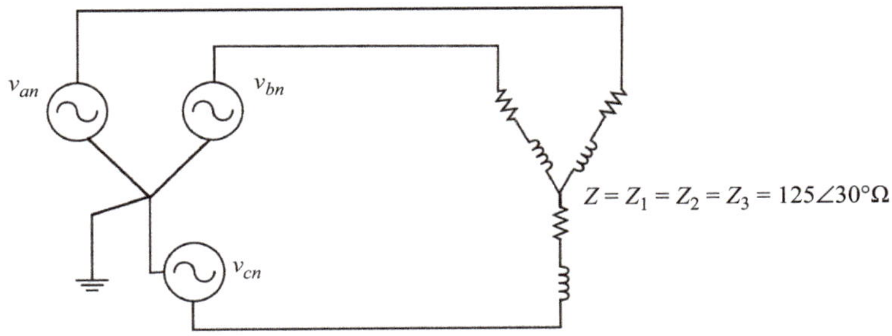

Figure 3.7 System for Example 3.2

In general terms,

$$v_n(t) = V_m \cos(\omega t + \beta)$$

$$i_n(t) = I_m \cos(\omega t + \beta - \theta) = \frac{V_m}{\|Z\|} \cos(\omega t + \beta - \theta)$$

$$p_n(t) = V_m I_m \cos(\omega t + \beta)\cos(\omega t + \beta - \theta)$$

A useful relation is

$$\cos(A)\cos(B) = \frac{1}{2}\cos(A + B) + \frac{1}{2}\cos(A - B)$$

The new expression of power is

$$p_n(t) = \frac{V_m I_m}{2} [\cos(2\omega t + 2\beta - \theta) + \cos(\theta)]$$

The expression can be easily calculated and plotted using MATLAB. The following code can apply for this purpose:

```
V_1_m=input('Write the voltage magnitude of V1 [V] = ');
V_1_a=input('Write the voltage phase angle of V1 (degree) = ');
V_2_m=input('Write the voltage magnitude of V2 [V] = ');
V_2_a=input('Write the voltage phase angle of V2 (degree) = ');
V_3_m=input('Write the voltage magnitude of V3 [V] = ');
V_3_a=input('Write the voltage phase angle of V3 (degree) = ');
Za=input('Write the impedance for phase a [complex number, ohms] = ');
Zb=input('Write the impedance for phase b [complex number, ohms] = ');
Zc=input('Write the impedance for phase c [complex number, ohms] = ');

wt=[0:0.1:2*pi()];
I1_m=V_1_m/abs(Za);
I2_m=V_2_m/abs(Zb);
I3_m=V_3_m/abs(Zc);
P1=(V_1_m*I1_m/2).*(cos((2*wt)+(2*(V_1_a*pi()/180))-
(cos(angle(Za))))+cos(angle(Za)));
P2=(V_2_m*I2_m/2).*(cos((2*wt)+(2*(V_2_a*pi()/180))-
(cos(angle(Zb))))+cos(angle(Zb)));
P3=(V_3_m*I3_m/2).*(cos((2*wt)+(2*(V_3_a*pi()/180))-
(cos(angle(Zc))))+cos(angle(Zc)));

Pt=P1+P2+P3;

plot(wt, P1, wt, P2, wt, P3, wt, Pt)
legend('P1','P2','P3', 'Pt')
ylabel('P [W]')
xlabel(' \omega * t')
```

Figure 3.8 is obtained, and it shows the constant three-phase power.

3.4 PU normalization

In power system calculations a normalization of variables called per unit normalization is generally used. It is especially convenient when several transformers and voltage levels are involved.

The idea is to pick base values for quantities such as voltages, currents, impedances, power, and so on and to define the quantity in per unit as follows:

$$Quantity\ in\ per\ unit = \frac{actual\ quantity}{base\ value\ of\ quantity}$$

Base variables are picked to satisfy the same kind of relationship as the variables. Ohm's law establishes that $V = Z \cdot I$.

If the equation is expressed by using the base values, it becomes

$$V_B = Z_B \cdot I_B \tag{3.9}$$

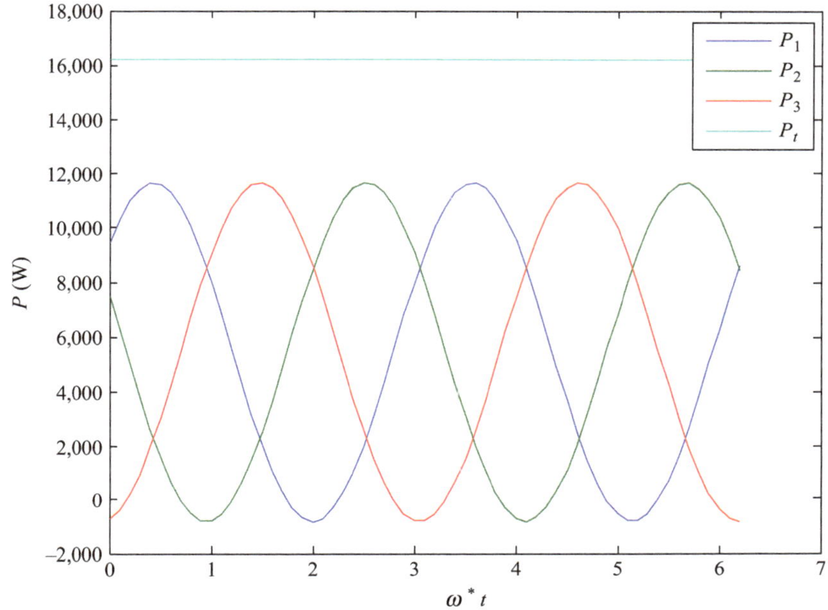

Figure 3.8 Power losses for different voltage angles

Dividing both equations, the following is obtained:

$$\frac{V}{V_B} = \frac{Z \cdot I}{Z_B \cdot I_B} \tag{3.10}$$

or

$$V_{\text{pu}} = Z_{\text{pu}} \cdot I_{\text{pu}} \tag{3.11}$$

Equation (3.10) has the same form as (3.9), which implies that circuit analysis can be done exactly as with (3.9). The pu subscript indicates per unit values.

As it is done for Ohm's law can also be done in the case of power calculations. For example, corresponding to

$$S = V \cdot I^* \tag{3.12}$$

$$S_B = V_B \cdot I_B \tag{3.13}$$

And

$$S_{\text{pu}} = V_{\text{pu}} \cdot I_{\text{pu}}^* \tag{3.14}$$

Since (3.9) and (3.13) involve four base quantities V_B, I_B, Z_B, and S_B, the selection of any two base quantities determines the remaining two base quantities. If V_B and

S_B are selected, then $I_B = S_B/V_B$ can be replaced in Z_B as follows:

$$Z_B = \frac{V_B}{I_B} = \frac{V_B^2}{S_B} \tag{3.15}$$

The extension to other related variables is routine. For example,

$$S_{\text{pu}} = P_{\text{pu}} + jQ_{\text{pu}} \tag{3.16}$$

where

$$P_{\text{pu}} = \frac{P}{S_B} \quad Q_{\text{pu}} = \frac{Q}{S_B}$$

For the impedance,

$$Z_{\text{pu}} = R_{\text{pu}} + jX_{\text{pu}} \tag{3.17}$$

where

$$R_{\text{pu}} = \frac{R}{Z_B} \quad X_{\text{pu}} = \frac{X}{Z_B}$$

Note that

$$Y_B \triangleq \frac{I_B}{V_B} = \frac{1}{Z_B} \tag{3.18}$$

With several items of equipment, with different ratings, it is not usually possible to pick base values so that they are always the same as the nameplate ratings. It is the necessary to recalculate the per unit values on the new basis. The key idea is that Z_{pu} depends on Z_B but, of course, Z_{actual} does not. The relationship between old and new values is

$$Z_{actual} = Z_{\text{pu}}^{old} \cdot Z_B^{old} = Z_{\text{pu}}^{new} \cdot Z_B^{new} \tag{3.19}$$

Then

$$\begin{aligned} Z_{\text{pu}}^{new} &= Z_{\text{pu}}^{old} \cdot \frac{Z_B^{old}}{Z_B^{new}} \\ Z_{\text{pu}}^{new} &= Z_{\text{pu}}^{old} \cdot \left[\frac{V_B^{old}}{V_B^{new}}\right]^2 \cdot \frac{S_B^{new}}{S_B^{old}} \end{aligned} \tag{3.20}$$

Note: In applying (3.20), it can be substituted three-phase and/or line–line values.

Example 3.3 The system shown in Figure 3.9 has the impedances in the bases of the corresponding elements. Find the pu values considering that at the generator, the base values are 11 kV and 50 MV A.

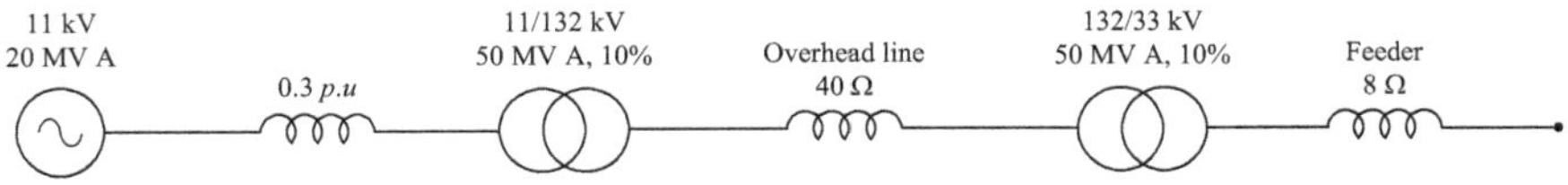

Figure 3.9 Electrical system for Example 3.3

At the generator:

$$Base\ \mathrm{kV}_B = 11$$

$$Base\ \mathrm{MV\ A}_B = 50$$

$$Z_B = \frac{(11\ \mathrm{kV})^2}{50\ \mathrm{MV\ A}} = 2.42\ \Omega$$

$$I_B = \frac{50\ \mathrm{MV\ A}}{\sqrt{3}\cdot 11\ \mathrm{kV}} = 2,624\ \mathrm{A}$$

$Z_{GEN\mathrm{pu}}$ on common base

$$Z_{GEN\mathrm{pu}} = 0.3\cdot\frac{50\ \mathrm{MV\ A}}{20\ \mathrm{MV\ A}} = 0.75\ \mathrm{pu}$$

At the transformers: 0.1 pu
At the line,

$$\mathrm{Base\ kV}_B = 132$$

$$\mathrm{Base\ MV\ A}_B = 50$$

$$Z_B = \frac{(132\ \mathrm{kV})^2}{50\ \mathrm{MV\ A}} = 348.5\ \Omega$$

$$I_B = \frac{50\ \mathrm{MV\ A}}{\sqrt{3}\cdot 132\ \mathrm{kV}} = 218.7\ \mathrm{A}$$

$$Z_{OHL\mathrm{pu}} = \frac{40\ \Omega}{348.5\ \Omega} = 0.115\ \mathrm{pu}$$

At the feeder,

$$\mathrm{Base\ kV}_B = 33$$

$$\mathrm{Base\ MV\ A}_B = 50$$

$$Z_B = \frac{(33\ \mathrm{kV})^2}{50\ \mathrm{MV\ A}} = 21.78\ \Omega$$

$$I_B = \frac{50\ \mathrm{MV\ A}}{\sqrt{3}\cdot 33\ \mathrm{kV}} = 874.77\ \mathrm{A}$$

$$Z_{FEEDER\mathrm{pu}} = \frac{8\ \Omega}{21.78\ \Omega} = 0.367\ \mathrm{pu}$$

3.5 Load flow

Load flows, also called power flows, are mathematical procedures performed on power system networks, which allow the calculation of the voltage magnitudes and angles of all buses in the system. Voltage magnitudes are only known for those buses having generators that allow setting the voltage magnitude if appropriate regulators are available. These buses are referred to as voltage buses. For the strongest bus of the system, the voltage magnitude can be determined and the angle is normally defined equal to zero as this is the reference for the rest of the system buses. This bus is referred to as the slack bus. The other buses where the active and reactive power are defined as *PQ* buses. The bus classification will be mentioned later.

Once the voltage and angles are calculated at all buses, other parameters can easily be determined by means of the expressions already reviewed earlier in this chapter. Those parameters include among others, the following: power (total, active and reactive) transferred, current, including angles, losses, power factors, and loading of the different elements.

Load flows are analyzed by using very efficient algorithms that conduct almost invariably to software packages. This should not though prevent hand calculations, which are appropriate especially for small systems.

Up to the mid-1960s before use of computer, big systems were analyzed with the aid of board simulators. Nowadays there are flexible and affordable software packages that allow comprehensive analysis, and Figure 3.10 illustrates the big changes from the old simulators to the new representations achieved from software packages.

The most known algorithms to implement load flow programs are the Newton–Raphson method and Gauss–Seidel methods. These can be used for the solution of HV network systems that invariably correspond to transmission systems and big complexes with internal generation. In HV systems normally there is a high *X*/*R*

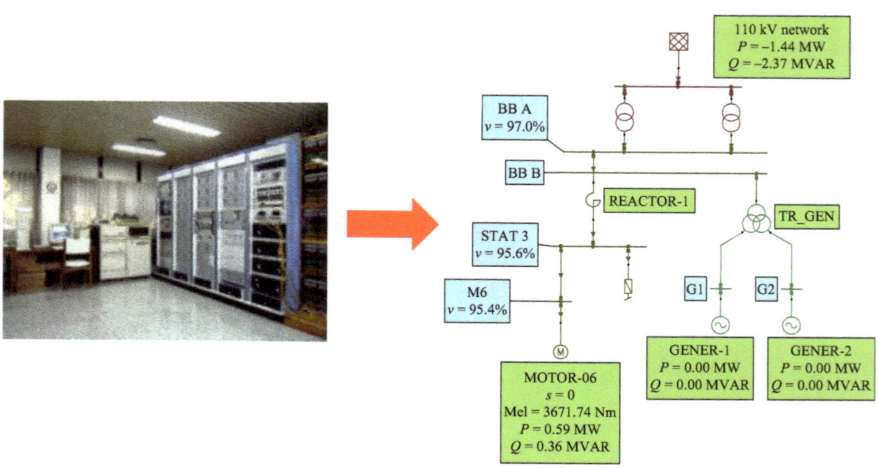

Figure 3.10 Illustration of a power system board simulator

ratio, and the loads are symmetric. Therefore, the analysis assumes three-phase balanced system operation.

These algorithms can also be used in distribution systems, although in these cases convergence could be a problem since the *X*/*R* ratio is lower. For these cases then other algorithms are more appropriate. Mathematical algorithms for distribution systems are normally simpler than those developed for high voltage systems because of the radial condition. They will be treated in Section 3.6.

In this section the basis and structure of the algorithms for load flows are reviewed, as well as some examples to illustrate their application in a comprehensive case study.

3.5.1 Formulation of the load flow problem

For load flow analysis, all machines at any one bus are treated as a single machine and represented by a single emf and a series reactance. The basic equations that will be the basis for the nodal analysis are the following:

The current is $I = Y \cdot V$ which becomes for the entire system $I = Y_{bus} \cdot V$.
The complex power is $S = V \cdot I^*$ and therefore $S^* = V^* \cdot I$.
In an electrical system Y_{bus} is symmetrical around the principal diagonal.

A simple electrical system with three nodes is shown in Figure 3.11 to illustrate the concepts. Here each generator is replaced by an equivalent current source and shunt admittance.

At each node it is known that $I_{ik} = Y_{ik} \cdot (V_i - V_k)$. Therefore, the following expressions can be obtained for the figure shown before:

$$I_0 = Y_0 V_0 + Y_{10}(V_0 - V_1)$$
$$I_1 = Y_1 V_1 + Y_{10}(V_1 - V_0) + Y_{12}(V_1 - V_2)$$
$$I_2 = Y_2 V_2 + Y_{12}(V_2 - V_1)$$

Regrouping these terms, the following equations are obtained:

$$I_0 = (Y_0 + Y_{01})V_0 - Y_{01}V_1$$
$$I_1 = -Y_{01}V_0 + (Y_1 + Y_{01} + Y_{12})V_1 - Y_{12}V_2$$
$$I_2 = -Y_{12}V_1 + (Y_2 + Y_{12})V_2$$

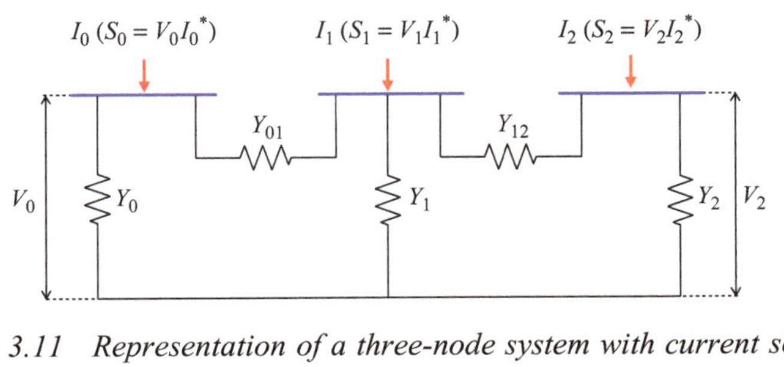

Figure 3.11 Representation of a three-node system with current sources

In matrix form those equations can be expressed as

$$\begin{bmatrix} I_0 \\ I_1 \\ I_2 \end{bmatrix} = \begin{bmatrix} Y_0 + Y_{01} & -Y_{01} & 0 \\ -Y_{01} & Y_1 + Y_{01} + Y_{12} & -Y_{12} \\ 0 & -Y_{12} & Y_2 + Y_{12} \end{bmatrix} \begin{bmatrix} V_0 \\ V_1 \\ V_2 \end{bmatrix} \tag{3.21}$$

The equations can be expressed as

$$\begin{aligned} &\sum Y_{ik} V_k = I_i \\ &\sum Y_{ik} V_k = \frac{S_i^*}{V_i^*} = I_i \\ &V_i^* \sum Y_{ik} V_k = S_i^* \end{aligned} \tag{3.22}$$

With $i = 1,2,\ldots,n$, being n the total number of nodes.

3.5.2 Newton–Raphson method

The former equations are nonlinear and require for their solution of numerical iterative methods. There are a number of them but the most well-known ones are those of Newton–Raphson method and Gauss–Seidel. Here the method of NR will be developed using Taylor's series. The Taylor series for a function *f(x)* that has derivatives of all orders at $x = a$ is expressed as

$$y = f(x) = f(a) + f'(a)(x-a) + \frac{f''(a)}{2!}(x-a)^2 + \cdots + \frac{f^{(n)}(a)}{n!}(x-a)^n \tag{3.23}$$

Neglecting the second-order terms and higher, the expression becomes

$$y = f(x) = f(a) + f'(a)(x-a)$$

The series converges rapidly for values of x near to a. If x_0 is the initial estimate, the expression becomes

$$y = f(x_0) + f'(x_0)(x_1 - x_0) \tag{3.24}$$

where x_1 is the new value of x which is a closer estimate. This curve crosses the axis at the new value x_1. Thus,

$$0 = f(x_0) + f'(x_0)(x_1 - x_0) \tag{3.25}$$

Normally $x_0 - x_1$ is referred to as Δx. Therefore,

$$\Delta x = -\frac{f(x_0)}{f'(x_0)} \tag{3.26}$$

$$x_1 = x_0 - \Delta x = x_0 - \frac{f(x_0)}{f'(x_0)} \tag{3.27}$$

Figure 3.12 explains this concept.

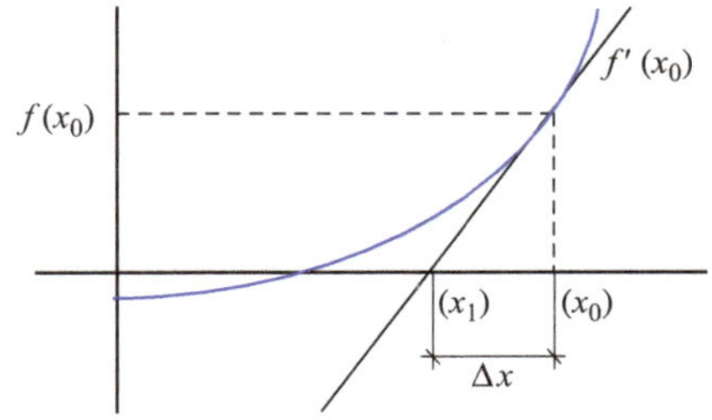

Figure 3.12 Illustration of the Newton–Raphson concept

Furthermore, the Newton–Raphson method is used for solving systems of simultaneous equations. The Taylor series is applied to n nonlinear equations in n unknowns:

$$x_1, x_2, \ldots, x_n$$

The differentiable functions are expressed as

$$\begin{aligned} f_1(x_1, x_2, \ldots, x_n) &= 0 \\ f_2(x_1, x_2, \ldots, x_n) &= 0 \\ &\vdots \\ f_n(x_1, x_2, \ldots, x_n) &= 0 \end{aligned}$$

The matrix of partial derivatives is called a Jacobian matrix J. By using the same concept of (3.27), this result can be written as

$$[X_{k+1}] = [X_k] - [J^{-1}][F(X_k)] \tag{3.28}$$

The square matrix J is the Jacobian matrix of $F(x)$, whose (i,k)th element is defined as $\partial f_i/\partial x_k$. This algorithm can be used to solve load flows. In that case, it is required to write the equations as a set $F(x) = 0$. In general it can be written that

$$x_{k+1} = x_k - \frac{f(x_k)}{f'(x_k)} \tag{3.29}$$

Example 3.4 Find the solution for the following equation using the expression for quadratic equations and compare the results with Newton–Raphson method:

$$f(x) = 0 = x^2 - 20x + 100$$

This equation is of the form $f(x) = y = a \cdot x^2 + b \cdot x + c$. The solution applying the mathematical formula is obtained as

$$\begin{aligned} x &= \frac{-b \pm \sqrt{b^2 - 4ac}}{2a} \\ &= \frac{20 \pm \sqrt{400 - 400}}{2} = 10 \end{aligned}$$

Now the Newton–Raphson method is used and therefore,

$$f'(x) = 2x - 20$$

The first approximation is taken as $x = 12$:

$$f(x) = (12)^2 - 20(12) + 100 = 4$$

$$f'(x) = 2(12) - 20 = 4$$

$$\Delta x_1 = -\frac{f(12)}{f'(12)} = -\frac{4}{4} = -1$$

$$x_1 = x_0 + \Delta x_1 = 12 + (-1) = 11$$

The second iteration:

$$f(x) = (11)^2 - 20(11) + 100 = 1$$

$$f'(x) = 2(11) - 20 = 2$$

$$\Delta x_2 = -\frac{f(11)}{f'(11)} = -\frac{1}{2} = -0.5$$

$$x_2 = x_1 + \Delta x_2 = 11 + (-0.5) = 10.5$$

Third iteration:

$$f(x) = (10.5)^2 - 20(10.5) + 100 = 0.25$$

$$f'(x) = 2(10.5) - 20 = 1$$

$$\Delta x_3 = -\frac{f(10.5)}{f'(10.5)} = -0.25$$

$$x_3 = x_2 + \Delta x_3 = 10.5 + (-0.25) = 10.25$$

Fourth iteration:

$$f(x) = (10.25)^2 - 20(10.25) + 100 = 0.0625$$

$$f'(x) = 2(10.25) - 20 = 0.5$$

$$\Delta x_4 = -\frac{f(11)}{f'(11)} = -\frac{0.0625}{0.5} = -0.125$$

$$x_4 = x_3 + \Delta x_4 = 10.25 + (-0.125) = 10.125$$

The last result is very close to the exact solution.

Example 3.5 Use the Newton–Raphson method to resolve the following equation:

$$x = 2 - \sin x$$

Table 3.1 Values for Example 3.5

Iteration	X_k	X_{k+1}
1	0.0000000	1.0000000
2	1.0000000	1.1029207
3	1.1029207	1.1060571
4	1.1060571	1.1060602

With the Newton–Raphson method (Table 3.1):

$$f(x) = 0 = x - 2 + \sin x$$
$$f'(x) = 1 + \cos x$$
$$\Delta x = -\frac{f(x)}{f'(x)} = \frac{2 - (x + \sin x)}{1 + \cos x}$$
$$x_{k+1} = x_k + \Delta x$$
$$x_{k+1} = x_k + \frac{2 - (x + \sin x)}{1 + \cos x}$$

3.5.3 Type of buses

The buses are classified in three different categories for load flow analysis as follows:

- Slack buses: In this type of node the values voltage magnitude and angle are known. The slack generator covers system losses. The angle is used as reference for all voltage angles. Usually this node is set as the supply for the largest generator.
- *PV* buses: In this type of node, the values of P and V are known. Q is a free variable within limits ($Q_{min} < Q < Q_{max}$). This type of node is used for voltage-controlled generators.
- *PQ* buses: In this type of node, the values of P and Q are known. V is a free variable within limits ($V_{min} < V < V_{max}$).

Figure 3.13 illustrates the different types of buses.

3.5.4 Application of the Newton–Raphson method to solve load flows

The theory seen in the previous sections will be applied to the full solution of load flows with the method of NR.

Equation (3.22) can be expressed as follows, indicating the k nodes connected to node I denoted by $k\varepsilon i$:

$$S_i^* = V_i^* \sum_{k\varepsilon i} Y_{ik} V_k \tag{3.30}$$

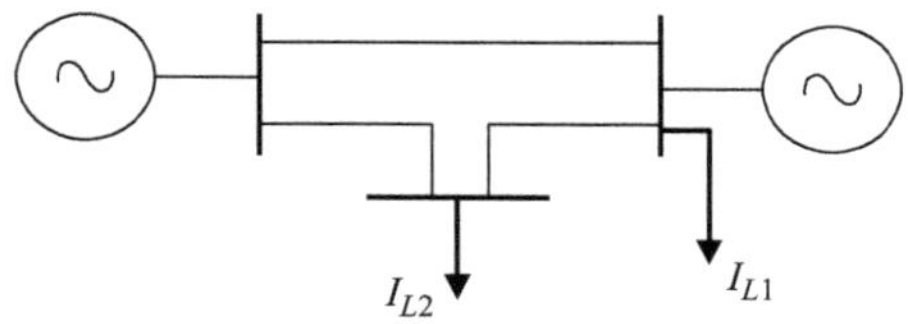

Figure 3.13 Illustration of types of buses for load flow analysis

From (3.30), the following derivation is obtained:

$$S_i^* = V_i \angle -\theta_i \sum Y_{ik} V_k \angle\theta_k$$
$$S_i^* = V_i \sum Y_{ik} V_k \angle(\theta_k - \theta_i)$$

Since $Y_{ik} = G_{ik} + jB_{ik}$ and $S^* = P_i - jQ_i$, then,

$$S_i^* = V_i \sum(G_{ik} + jB_{ik})(V_k \cos\theta_{ik} - jV_k \sin\theta_{ik})$$
$$S_i^* = V_i \sum(G_{ik} \cos\theta_{ik} + B_{ik} \sin\theta_{ik})V_k + jV_i \sum(B_{ik} \cos\theta_{ik} - G_{ik} \sin\theta_{ik})V_k \tag{3.31}$$

Now the values of P and Q calculated are the following:

$$P_i\ (calculated) = V_i \sum(G_{ik} \cos\theta_{ik} + B_{ik} \sin\theta_{ik})V_k$$
$$P_i = V_i^2 G_{ii} + V_i \sum_{kei,\ k\neq i}(G_{ik} \cos\theta_{ik} + B_{ik} \sin\theta_{ik})V_k$$

$$Q_i\ (calculated) = V_i \sum(G_{ik} \sin\theta_{ik} - B_{ik} \cos\theta_{ik})V_k$$
$$Q_i = -V_i^2 B_{ii} + V_i \sum_{kei,\ k\neq i}(G_{ik} \sin\theta_{ik} - B_{ik} \cos\theta_{ik})V_k$$

$$\Delta P_i^j = P_i\ (specified) - P_i^j\ (calculated)$$
$$\Delta Q_i^j = Q_i\ (specified) - Q_i^j\ (calculated)$$

With $i = 1,2,\ldots,n-1$, where P_1, jQ_1 specified are net power at bus i in pu ΔP and ΔQ are calculated from the equations of P and Q as follows:

$$X^{r+1} = X^r - \Delta X$$
$$\Delta X = J^{-1} F(X^r)$$

$$\begin{bmatrix} \Delta\theta \\ \Delta V \end{bmatrix} = [J^{-1}] \begin{bmatrix} \Delta P \\ \Delta Q \end{bmatrix}$$

$$[J]_{Jacobian} = \begin{bmatrix} H & N \\ M & L \end{bmatrix}$$

$$\begin{bmatrix} \Delta P \\ \Delta Q \end{bmatrix} = [J] \begin{bmatrix} \Delta\theta \\ \dfrac{\Delta V}{V} \end{bmatrix} = \begin{bmatrix} H & N \\ M & L \end{bmatrix} \begin{bmatrix} \Delta\theta \\ \dfrac{\Delta V}{V} \end{bmatrix}$$

$$H_{ii} = \frac{\partial \Delta P_i}{\partial \theta_i} \qquad H_{ik} = \frac{\partial \Delta P_i}{\partial \theta_k}$$

$$N_{ii} = V_i \, \frac{\partial \Delta P_i}{\partial V_i} \qquad N_{ik} = V_k \, \frac{\partial \Delta P_i}{\partial \theta_k}$$

$$M_{ii} = \frac{\partial \Delta Q_i}{\partial \theta_i} \qquad M_{ik} = \frac{\partial \Delta Q_i}{\partial \theta_k}$$

$$L_{ii} = V_i \, \frac{\Delta Q_i}{\partial V_i} \qquad L_{ik} = V_k \, \frac{\Delta Q_i}{\partial V_k}$$

The following trigonometric functions have to be considered:

$$\frac{\partial \cos\theta_{ik}}{\partial \theta_i} = -\sin\theta_{ik} \qquad \frac{\partial \cos\theta_{ik}}{\partial \theta_k} = \sin\theta_{ik}$$

$$\frac{\partial \sin\theta_{ik}}{\partial \theta_i} = \cos\theta_{ik} \qquad \frac{\partial \sin\theta_{ik}}{\partial \theta_k} = -\cos\theta_{ik}$$

By applying the trigonometric functions, the values of the Jacobian become

$$H_{ii} = \frac{\partial \Delta P_i}{\partial \theta_i} = -V_i \sum_{kei,k\neq i} (G_{ik}\sin\theta_{ik} - B_{ik}\cos\theta_{ik}) V_k = -Q_i^{calculated} - |V_i|^2 \, B_{ii}$$

$$H_{ik} = \frac{\partial \Delta P_i}{\partial \theta_k} = V_i \, (G_{ik}\sin\theta_{ik} - B_{ik}\cos\theta_{ik}) V_k$$

$$M_{ii} = \frac{\partial \Delta Q_i}{\partial \theta_i} = V_i \sum_{kei,k\neq i} (G_{ik}\cos\theta_{ik} + B_{ik}\sin\theta_{ik}) V_k = P_i^{calculated} - |V_i|^2 \, G_{ii}$$

$$M_{ik} = \frac{\partial \Delta Q_i}{\partial \theta_k} = -V_i \, (G_{ik}\cos\theta_{ik} + B_{ik}\sin\theta_{ik}) V_k$$

$$N_{ii} = V_i \, \frac{\partial \Delta P_i}{|\partial V_i|} = P_i^{calculated} + |V_i|^2 \, G_{ii} = M_{ii} + 2|V_i|^2 \, G_{ii}$$

$$N_{ik} = V_k \, \frac{\partial \Delta P_i}{|\partial V_k|} = -\, M_{ik}$$

$$L_{ii} = V_i \, \frac{\Delta Q_i}{|\partial V_i|} = Q_i^{calculated} - |V_i|^2 \, B_{ii} = -H_{ii} - 2|V_i|^2 \, B_{ii}$$

$$L_{ik} = V_k \, \frac{\Delta\theta}{|\partial V|_k} = H_{ik}$$

The algorithm of Newton–Raphson method can be implemented following the chart shown in Figure 3.14. It is schematic but illustrates the procedure. The equations used are those shown in the previous paragraph.

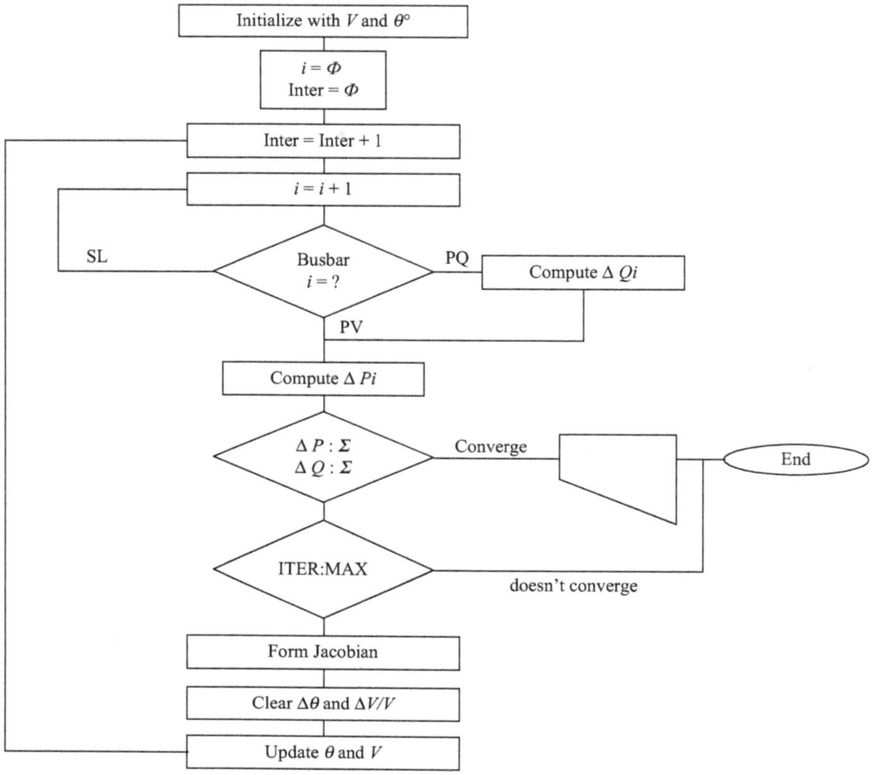

Figure 3.14 Flow chart for load flow analysis using NR algorithm

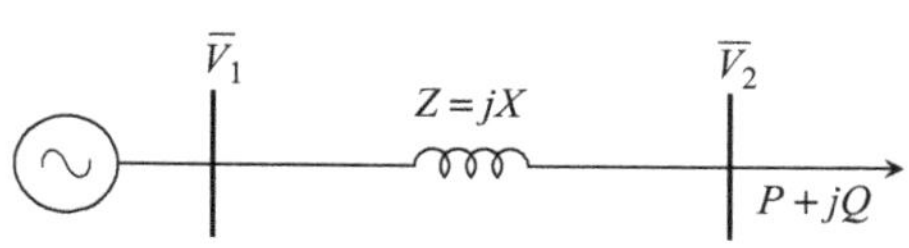

Figure 3.15 Two-bus system with load at the end bus

3.5.5 Decoupling method

Brian Stott proposed a very clever procedure to solve load flows, known as the Decupled Method taking advantage of an inherent characteristic of power systems that is strong interdependence between active power and bus voltage angles, and between reactive power and voltage magnitudes. Decoupling makes reference to the separate solution of the P–Θ and Q–V problems. These facts can be appreciated in Figure 3.15 showing a two-node system.

This method does not need to invert the Jacobian J, which is its most important advantage. However, the method is slow to converge.

Assuming $\bar{V}_1$ as reference (1 pu and $\theta_1 = 0$) and $\bar{Z} = jX$, since $R \ll z$, then,

$$\bar{V}_2 = \bar{V}_1 - \bar{I}\bar{Z} = \bar{V}_1 - \bar{I}(jX)$$

It has to be taken into account that

$$\bar{I} = \frac{(P - jQ)}{\bar{V}_1^{\;*}}$$

Since $\bar{V}_1 = 1\angle 0^\circ = \bar{V}_1^{\;*}$, therefore,

$$\bar{V}_2 = \bar{V}_1 - \frac{(P - jQ)}{\bar{V}_1^{\;*}}(jX) = \frac{|\bar{V}_1|^2 - (P - jQ)(jX)}{\bar{V}_1^{\;*}} = 1 - XQ - jXP$$

It can be seen from the phasor diagram that a change in $P(\Delta P)$ changes the angle by $\Delta\theta$ but has a small effect in the voltage magnitude, as shown in Figure 3.16. On the other hand, a change in $Q(\Delta Q)$ changes the voltage by ΔV but has a small effect in the angle, as shown in Figure 3.17.

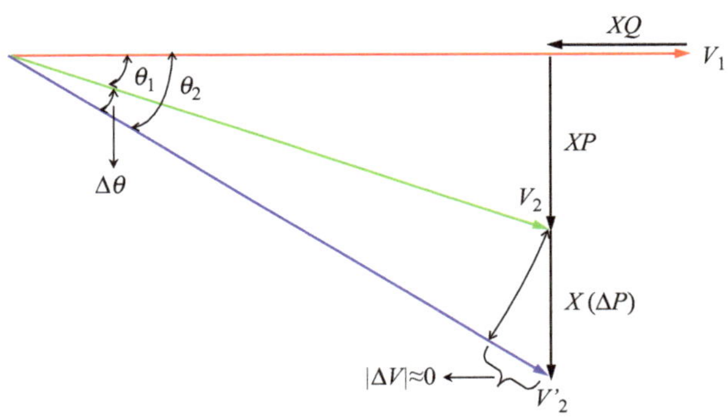

Figure 3.16 Illustration of the relationship between P and θ

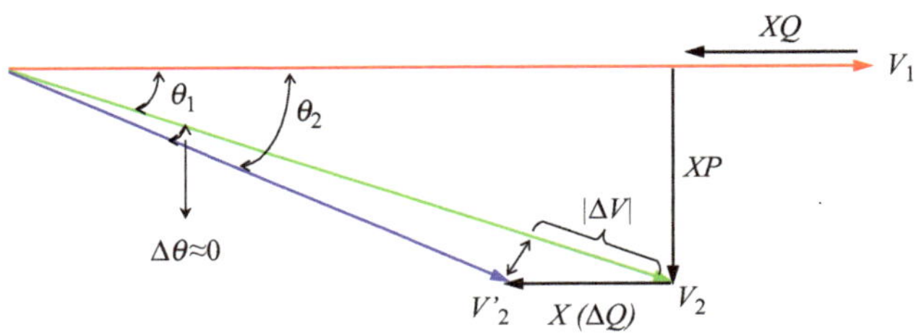

Figure 3.17 Illustration of the relationship between Q and V

Algebraically the above approximate relation may be stated as

$$\Delta P = H\Delta\theta$$
$$\Delta Q = L'\Delta V/V$$

Therefore, applying the decoupling phenomena to an electrical system with several nodes, the sub-matrices N and J in equation can be neglected resulting in

$$\Delta P_i = \left[\sum_{k=1,k\neq i} V_i V_k (G_{ik}\sin\theta_{ik} - B_{ik}\cos\theta_{ik})\Delta\theta_k\right] - (Q_i^{calculated} + V_i^2 B_{ii})\Delta\theta_i$$

$$\Delta Q_i = \left[\sum_{k=1,k\neq i} V_i V_k (G_{ik}\sin\theta_{ik} - B_{ik}\cos\theta_{ik})\frac{|\Delta V_k|}{V_k}\right] - (Q_i^{calculated} + V_i^2 B_{ii})\frac{|\Delta V_k|}{V_i}$$

θ_{ik} is small and so $\cos\theta_{ik} = 1.0$ and $\sin\theta_{ik} = 0.0$, $G_{ik}\sin\theta_{ik} \ll B_{ik}\cos\theta_{ik}$. As a product of engineering studies, it is also found that $Q_i \ll V_i^2 B_{ii}$. Therefore, Q_i can be neglected:

$$\Delta P_i = \left[\sum_{k=1,k\neq i} -V_i V_k B_{ik}\Delta\theta_k\right] - (V_i^2 B_{ii})\Delta\theta_i$$

$$\frac{\Delta P_i}{V_i} = \left[\sum_{k=1,k\neq i} -V_k B_{ik}\Delta\theta_k\right] - (V_i B_{ii})\Delta\theta_i$$

$$\Delta Q_i = \left[\sum_{k=1,k\neq i} -V_i V_k B_{ik}\frac{|\Delta V_k|}{V_k}\right] - (V_i^2 B_{ii})\frac{|\Delta V_k|}{V_i}$$

$$\frac{\Delta Q_i}{V_i} = \left[\sum_{k=1,k\neq i} -V_i V_k B_{ik}\frac{|\Delta V_k|}{V_i V_k}\right] - (V_i^2 B_{ii})\frac{|\Delta V_k|}{V_i^2} = \left[\sum_{k=1,k\neq i} -B_{ik}|\Delta V_k|\right] - (B_{ii})|\Delta V_k|$$

Further approximation can be made by setting voltage on the right-hand side of equation to unit results in

$$\frac{\Delta P_i}{V_i} = -\left[\sum_{k=1,k\neq i} B_{ik}\Delta\theta_k + (\Delta\theta_i B_{ii})\right] = -B'\Delta\theta$$

$$\frac{\Delta Q_i}{V_i} = -\left[\sum_{k=1,k\neq i} B_{ik}|\Delta V_k| + (B_{ii}|\Delta V_k|)\right] = -B''|\Delta V|$$

The application of this simplification, which solves separately the P–θ and Q–V, renders a 4:1 saving on storage of the Jacobian matrix. This results in more iterations, which is compensated by saving in inversion time and shorter individual iteration time.

Example 3.6 For the system shown in Figure 3.18, calculate the voltages and powers at nodes 1 and 2 for the first two iterations using the Newton–Raphson method.

Bear in mind the following information:

Base power: 100 MV A
Convergence value: 0.1 MV A
Reactive power limits for bus 1: −5 MVAR and 10 MVAR

The voltages and active/reactive power at each node are indicated in Table 3.2. The line impedances are indicated in Table 3.3.

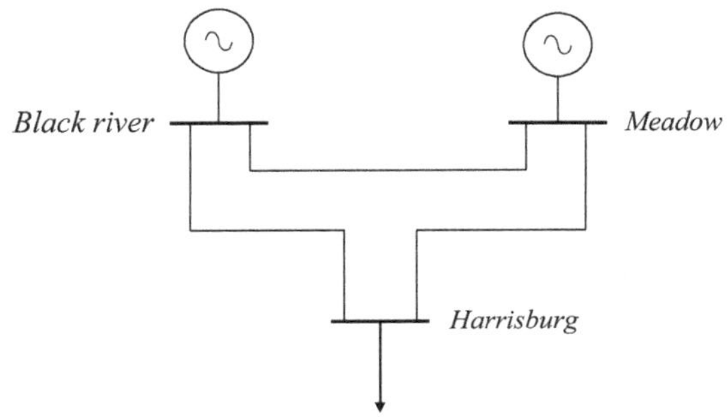

Figure 3.18 Power system for Example 3.6

Table 3.2 Values of voltages for Example 3.6

Nodes	Name	Voltages	Power
0	Black River	1.01	–
1	Meadow	1.00	$P_G = 55$ MW
2	Harrisburg	–	$P_c = 105$ MW, $Q_c = 65$ MVAR

Table 3.3 Values of impedances for Example 3.6

Lines	Impedances in pu
0–1	0.015 + j0.035j
0–2	0.015 + 0.055j
1–2	0.0075 + 0.0175j

$$y_{01} = \frac{1}{0.015 + j0.035} = 10.34 - j24.14$$

$$y_{02} = \frac{1}{0.015 + j0.055} = 4.62 - j16.92$$

$$y_{12} = \frac{1}{0.0075 + j0.0175} = 20.69 - j48.28$$

The bus admittance matrix is

$$Y = \begin{bmatrix} y_{01} + y_{02} & -y_{01} & -y_{02} \\ -y_{01} & y_{01} + y_{12} & -y_{12} \\ -y_{02} & -y_{12} & y_{02} + y_{12} \end{bmatrix}$$

$$Y = \begin{bmatrix} 14.96 - j41.06 & -10.34 + j24.14 & -4.62 + j16.92 \\ -10.34 + j24.14 & 31.03 - \text{j}72.42 & -20.69 + \text{j}48.28 \\ -4.62 + j16.92 & -20.69 + \text{j}48.28 & 25.31 - \text{j}65.20 \end{bmatrix}$$

With the data given Table 3.4 can be formed:

The power equations injected at each node are

$$P_i^{cal} = V_i \sum_j V_j \left(G_{ij} \cos\theta_{ij} + B_{ij} \sin\theta_{ij} \right)$$

$$Q_i^{cal} = V_i \sum_j V_j \left(G_{ij} \sin\theta_{ij} - B_{ij} \cos\theta_{ij} \right)$$

$$\Delta P = P_i^{esp} - P_i^{cal} = 0$$

$$\Delta Q = Q_i^{esp} - Q_i^{cal} = 0$$

$$P_i^{esp} = P_{Gi} - P_{Ci} \text{ and } Q_i^{esp} = Q_{Gi} - Q_{Ci}$$

For the first iteration:

$$V^0 = \begin{bmatrix} 1.01 \\ 1.00 \\ 1.00 \end{bmatrix} \qquad \theta^0 = \begin{bmatrix} 0 \\ 0 \\ 0 \end{bmatrix}$$

Table 3.4 Initial input data for Example 3.6

Node	**Type**	**Data**	**Unknown**
0	Slack	$V_0 = 1.01$ $\theta_0 = 0.0$	P_0 Q_0
1	*PV*	$V_1 = 1.00$ $P_1 = 0.55$	θ_1 Q_1
2	*PQ*	$P_2 = -1.05$ $Q_2 = -0.65$	θ_2 V_2

$$P_1{}^{(0)} = \left(V_1{}^2 \cdot G_{11}\right) + \left(V_0 \cdot (G_{10}\cos\theta_{10} + B_{10}\sin\theta_{10}) + V_2 \cdot (G_{12}\cos\theta_{12} + B_{12}\sin\theta_{12})\right) \cdot V_1$$

$$P_1{}^{(0)} = \left((1.0)^2 \cdot (31.03)\right) + \begin{pmatrix} (1.01)\cdot(-10.34\cos(0) + 24.14\sin(0)) + \\ (1.0)\cdot(-20.69\cos(0) + 48.28\sin(0)) \end{pmatrix} \cdot (1.0)$$

$$P_1{}^{(0)} = -0.103$$

$$P_2{}^{(0)} = \left(V_2{}^2 \cdot G_{22}\right) + \left(V_0 \cdot (G_{20}\cos\theta_{20} + B_{20}\sin\theta_{20}) + V_1 \cdot (G_{21}\cos\theta_{21} + B_{21}\sin\theta_{21})\right) \cdot V_2$$

$$P_2{}^{(0)} = \left((1.0)^2 \cdot (25.31)\right) + \begin{pmatrix} (1.01)\cdot(-4.62\cos(0) + 16.92\sin(0)) + \\ (1.0)\cdot(-20.69\cos(0) + 48.28\sin(0)) \end{pmatrix} \cdot (1.0)$$

$$P_2{}^{(0)} = -0.046$$

$$Q_2{}^{(0)} = \left(V_2{}^2 \cdot B_{22}\right) + \left(V_0 \cdot (G_{20}\sin\theta_{20} + B_{20}\cos\theta_{20}) + V_1 \cdot (G_{21}\sin\theta_{21} + B_{21}\cos\theta_{21})\right) \cdot V_2$$

$$Q_2{}^{(0)} = -\left((1.0)^2 \cdot (-65.20)\right) + \begin{pmatrix} (1.01)\cdot(4.62\sin(0) - 16.92\cos(0)) + \\ (1.00)\cdot(20.69\sin(0) - 48.28\cos(0)) \end{pmatrix} \cdot (1.0)$$

$$Q_2{}^{(0)} = -0.169$$

Therefore,

$$\begin{aligned} &0.55 - P_1^{cal} = 0 \\ &-1.05 - P_2^{cal} = 0 \\ &-0.65 - Q_2^{cal} = 0 \end{aligned}$$

The Jacobian for the first iteration is

$$J = \begin{bmatrix} \frac{\partial P_1}{\partial \theta_1} & \frac{\partial P_1}{\partial \theta_2} & \frac{\partial P_1}{\partial |V_2|} \\ \frac{\partial P_2}{\partial \theta_1} & \frac{\partial P_2}{\partial \theta_2} & \frac{\partial P_2}{\partial |V_2|} \\ \frac{\partial Q_2}{\partial \theta_1} & \frac{\partial Q_2}{\partial \theta_2} & \frac{\partial Q_2}{\partial |V_2|} \end{bmatrix}$$

Then, it can be obtained the following:

$$\Delta P = \begin{bmatrix} - \\ 0.653 \\ -1.004 \end{bmatrix} \qquad \Delta Q = \begin{bmatrix} - \\ - \\ -0.481 \end{bmatrix}$$

The Jacobian components are as follows:

$$\frac{\partial P_1}{\partial \theta_1} = -(1.00) \cdot \begin{pmatrix} (-10.34\sin(0) - 24.14\cos(0)) \cdot (1.01) + \\ (-20.69\sin(0) - 48.28\cos(0)) \cdot (1.00) \end{pmatrix} = 72.66$$

$$\frac{\partial P_1}{\partial \theta_2} = \frac{\partial P_2}{\partial \theta_1} = (1.00) \cdot (-20.69\sin(0) - 48.28\cos(0)) \cdot (1.00) = -48.28$$

$$\frac{\partial P_2}{\partial \theta_2} = -(1.00)\cdot\begin{pmatrix}(-4.62\sin(0) - 16.92\cos(0))\cdot(1.01)+\\(-20.69\sin(0) - 48.28\cos(0))\cdot(1.01)\end{pmatrix} = 65.37$$

$$\frac{\partial Q_2}{\partial \theta_1} = -(1.00)\cdot(-20.69\cos(0) + 48.28\sin(0))\cdot(1.00) = 20.69$$

$$\frac{\partial Q_2}{\partial \theta_2} = (1.00)\cdot\begin{pmatrix}(-4.62\cos(0) + 16.92\sin(0))\cdot(1.01)+\\(-20.69\cos(0) + 48.28\sin(0))\cdot(1.00)\end{pmatrix} = -25.36$$

$$\frac{\partial P_2}{\partial |V_2|} = (1.00)\cdot(-20.69\cos(0) + 48.28\sin(0))\cdot(1.00) = -20.69$$

$$\frac{\partial P_2}{\partial |V_2|} = \frac{\partial Q_2}{\partial \theta_2} + 2|V_2|^2 G_{22} = -25.36 + 2|1|^2 25.31 = 25.26$$

$$\frac{\partial Q_2}{\partial |V_2|} = -\frac{\partial P_2}{\partial \theta_2} - 2|V_2|^2 B_{22} = -65.37 + 2|1|^2(65.20) = 65.03$$

$$J = \begin{bmatrix} 72.66 & -48.28 & -20.69 \\ -48.28 & 65.37 & 25.26 \\ 20.69 & -25.36 & 65.03 \end{bmatrix}$$

$$b = \begin{bmatrix} \Delta P_1 \\ \Delta P_2 \\ \Delta Q_2 \end{bmatrix} = \begin{bmatrix} 0.654 \\ -1.004 \\ -0.481 \end{bmatrix}$$

Solving for $J\Delta x = b$, with $x = [\Delta\theta_1\ \Delta\theta_2\ \Delta V_2/V_2]T$:

$$\Delta x = \begin{bmatrix} -0.0030 \\ -0.0131 \\ -0.0116 \end{bmatrix}$$

$$\theta_1{}^{(1)} = \theta_1{}^{(0)} + \Delta\theta_1{}^{(0)} = 0 - 0.0030 = -0.0030 = -0.172^\circ$$
$$\theta_2{}^{(1)} = \theta_2{}^{(0)} + \Delta\theta_2{}^{(0)} = 0 - 0.0131 = -0.0116 = -0.751^\circ$$
$$V_2{}^{(1)} = V_2{}^{(0)}\left(1 + \frac{\Delta V_2{}^{(0)}}{V_2{}^{(0)}}\right) = 1.00(1 - 0.0116) = 0.988$$

$$V^{(1)} = \begin{bmatrix} 1.01 \\ 1.00 \\ 0.988 \end{bmatrix}_V \qquad \theta^{(1)} = \begin{bmatrix} 0 \\ -0.0030 \\ -0.0131 \end{bmatrix}_{rads}$$

Then,

$$P_0{}^{(1)} = \left((1.01)^2\cdot(14.97)\right) + \begin{pmatrix}(1.00)\cdot(-10.35\cos(0.0030) + 24.14\sin(0.0030))+\\(0.988)\cdot(-4.62\cos(0.0131) + 16.93\sin(0.0131))\end{pmatrix}\cdot(1.01)$$

$$P_0{}^{(1)} = 15.271 - 10.38 - 4.389 = 0.502$$

$$P_1{}^{(1)} = \left((1.00)^2 \cdot (31.04)\right)$$

$$+\begin{pmatrix}(1.01)\cdot(-10.35\cos(-0.0030)+24.14\sin(-0.0030))+\\(0.988)\cdot(-20.69\cos(0.0101)+48.28\sin(0.0101))\end{pmatrix}\cdot(1.00)$$

$$P_1{}^{(1)} = 31.04 - 10.527 - 19.959 = 0.554$$

$$P_2{}^{(1)} = \left((0.988)^2 \cdot (25.31)\right)$$

$$+\begin{pmatrix}(1.01)\cdot(-4.62\cos(-0.0131)+16.93\sin(-0.0131))+\\(1.00)\cdot(-20.69\cos(-0.0101)+48.28\sin(-0.0101))\end{pmatrix}\cdot(0988)$$

$$P_2{}^{(1)} = 24.706 - 4.831 - 20.922 = -1.047$$

$$Q_0{}^{(1)} = -\left((1.01)^2 \cdot (-41.07)\right)$$

$$+\begin{pmatrix}(1.00)\cdot(-10.35\sin(0.003)-24.14\cos(0.003))+\\(0988)\cdot(-4.62\sin(0+0.0131)-16.93\cos(0.0131))\end{pmatrix}\cdot(1.01)$$

$$Q_0{}^{(1)} = 41.896 - 24.413 - 16.953 = 0.53$$

$$Q_1{}^{(1)} = -\left((1.00)^2 \cdot (-72.42)\right)$$

$$+\begin{pmatrix}(1.01)\cdot(-10.35\sin(-0.003)-24.14\cos(-0.003))+\\(0.988)\cdot(-20.69\sin(0.0101)-48.28\cos(0.0101))\end{pmatrix}\cdot(1.00)$$

$$Q_1{}^{(1)} = 72.42 - 24.35 - 47.905 = 0.165$$

$$Q_2{}^{(1)} = -\left((0.988)^2 \cdot (-65.21)\right)$$

$$+\begin{pmatrix}(1.01)\cdot(-4.62\sin(-0.0131)-16.93\cos(-0.0131))+\\(1.00)\cdot(-20.69\sin(-0.0101)-48.28\cos(-0.0101))\end{pmatrix}\cdot(0.988)$$

$$Q_2{}^{(1)} = 63.654 - 16.832 - 47.492 = -0.67$$

$$P_i^{cal} = \begin{bmatrix} 0.502 \\ 0.554 \\ -1.047 \end{bmatrix} \quad Q_i^{cal} = \begin{bmatrix} 0.530 \\ 0.165 \\ -0.670 \end{bmatrix}$$

$$\Delta P = \begin{bmatrix} - \\ -0.004 \\ -0.003 \end{bmatrix} \quad \Delta Q = \begin{bmatrix} - \\ - \\ 0.02 \end{bmatrix}$$

These are the results for the first iteration. The residues exceed the limit of 10^{-1}. Furthermore, the maximum reactive power that the generator 1 could produce is 10 MVAR and the Q_{ical} obtained is 16.5 MVAR. It makes that the generator 1 is out of the limits. For that reason, the node 1 changes to a *PQ* node on the next iteration, maintaining the reactive power in $Q_{ical} = 0.165$ pu. For this change, the Jacobian that needs to be found on the second iteration would be the corresponding to two *PQ* nodes.

The same process is followed for the next iterations until the algorithm converges. Final results are obtained using a software package and are presented in Figure 3.19. They correspond to the same pu values that are given in the problem formulation. However, they are shown in kV, MW, and MVAR since the package used required inputs in real magnitudes.

In order to present the results with realistic values, the buses are assumed with voltage levels of 115 kV, and the base magnitudes as 100 MV A and 115 kV. The line impedances are also expressed in the same values. By making these adjustments, the new load flow obtained is presented in Figure 3.20 that of course has equivalent values of those of the previous figure.

Example 3.7 An example taken from a power system at 230/115/34.5/13.2 is used here to illustrate the application of load flow. Figure 3.21 shows the single line of the system. Rated bus voltages are given in Table 3.5. Tables 3.6 and 3.7 give the HV and LV transformer data. Table 3.8 gives the generator data. Bus and line data are given in Tables 3.9 and 3.10.

Load flow analysis is required to assess the voltage levels and determine remedial actions where required.

Load flows will be run to analyze the behavior of the system under normal conditions. If violations occur, measures to overcome them have to be proposed.

The load flow for the base condition was run using a software package. The results are given in graphical form in Figure 3.22 for a portion of the system from which it is clear that the voltages at buses Willow 34.5 kV and Willow 13.2 kV are 0.91 and 0.902, respectively, which are below the range allowed of $\pm 5\%$. The following actions will be considered to overcome the problem:

- Installation of capacitors at Willow 13.2 kV bus.
- Operation of the tap changer of the transformer connecting buses 34.5 kV Willow and 13.2 kV Willow.
- Adding another line between nodes 34.5 kV Emerald and 34.5 kV Willow.

One of the remedial actions is the installation of capacitors. A capacitor of 3.36 MVAR has to be installed at 13.2 kV Willow bus, in order to push the voltage to 1.0 pu. The results are shown in Figure 3.23, and they indicate that the new voltages at buses Willow 34.5 kV and Willow 13.2 kV are 0.963 and 1.0, respectively, which demonstrates the validity of the approach.

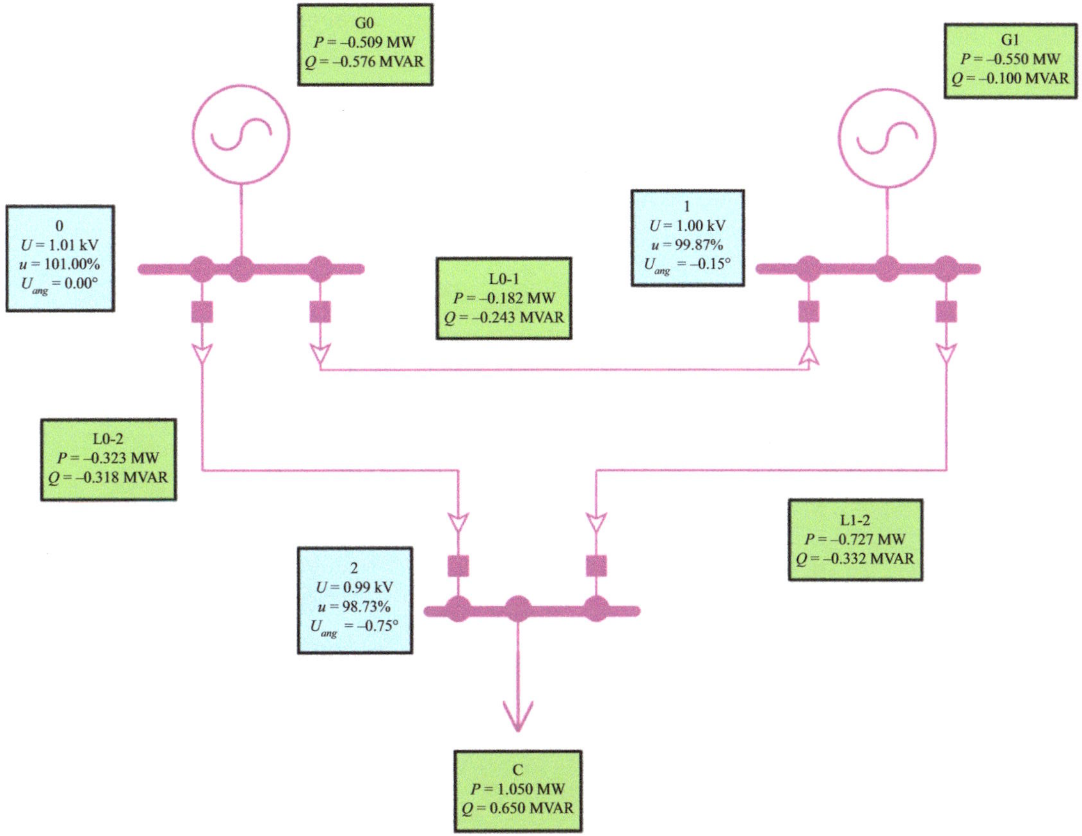

Figure 3.19 Load flow results in pu for the system of Example 3.6

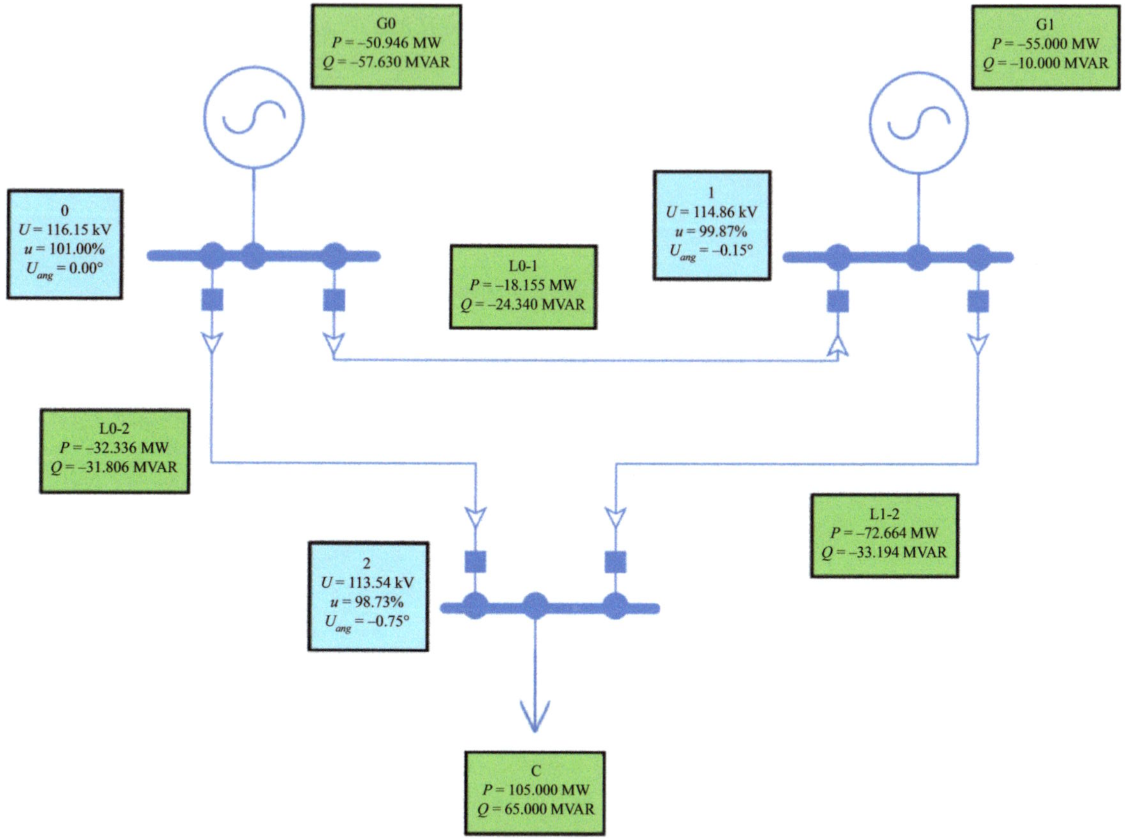

Figure 3.20 Load flow results in real magnitudes for the system of Example 3.6

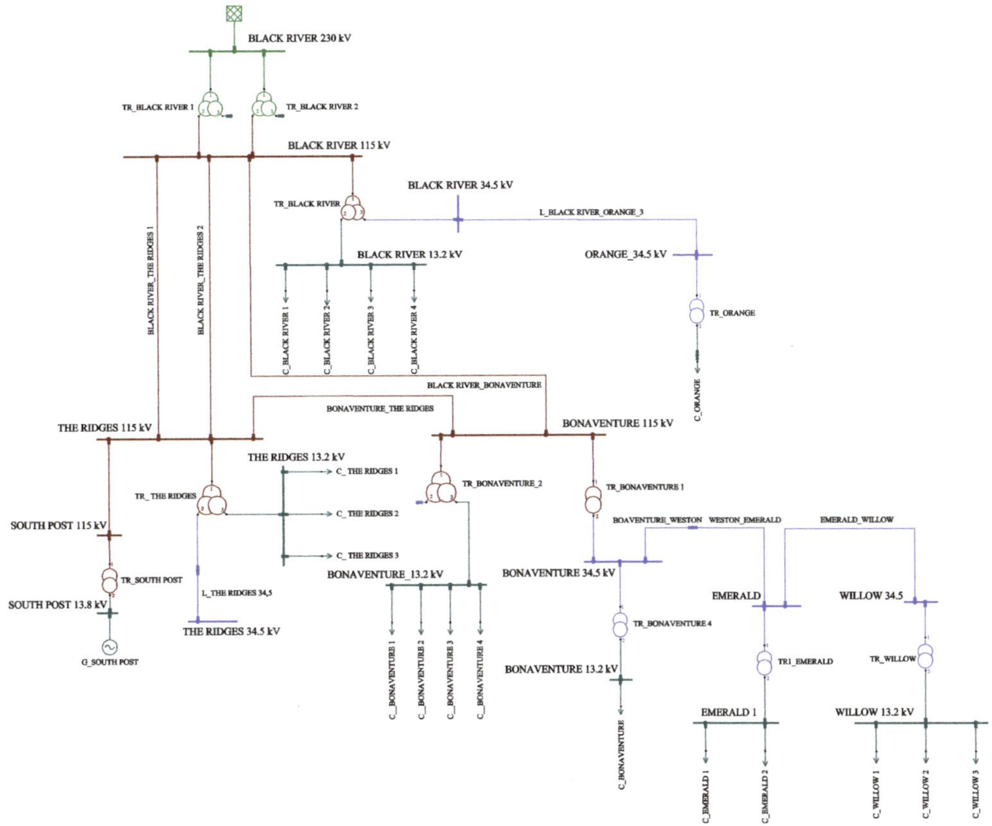

Figure 3.21 Single-line diagram for Example 3.7

Table 3.5 Rated bus voltages for Example 3.7

Bus	V_n (kV)	Bus	V_n (kV)
THE RIDGES 115 kV	115	BONAVENTURE_13.2 kV	13.2
THE RIDGES 13.2 kV	13.2	BONAVENTURE2_1	34.5
THE RIDGES 34.5 kV	34.5	EMERALD	34.5
BLACK RIVER	13.2	EMERALD 1	13.1
BLACK RIVER 115 kV	115	ORANGE_34.5 kV	34.5
BLACK RIVER 13.2 kV	13.2	ORANGE 13.2	13.2
BLACK RIVER 230 kV	230	SOUTH POST 115 kV	115
BLACK RIVER 34.5 kV	34.5	SOUTH POST 13.8 kV	13.8
BLACK RIVER 13.8 kV	13.8	THE RIDGES1 34.5 kV	34.5
BONAVENTURE 115 kV	115	WESTON	34.5
BONAVENTURE 13.2 kV	13.2	WILLOW 13.2 kV	13.2
BONAVENTURE 34.5 kV	34.5	WILLOW 34.5	34.5

The other action that can be considered is the operation of the tap changer of the transformer connecting buses 34.5 kV Willow and 13.2 kV Willow. As shown in Figure 3.24 a tap setting at 7 is required to take the voltage at the low voltage side to 1 pu. In this case although the voltage at the bus 34.5 kV Willow is not 1.00, the improvement is clear.

Finally the option of adding another line between nodes Emerald 34.5 kV and Willow 34.5 kV was considered. The results are shown in Figure 3.25. Here the voltage of Willow 34.5 kV improves but still does not fit within the accepted range. The voltage of 13.2 kV Willow is also below the accepted range. Actually this option is mainly considered when the initial line is overloaded.

Other actions can be analyzed in addition to those mentioned before, like increasing the wire size of the transmission line connecting buses 34.5 kV Emerald and 34.5 kV Willow if this is practical, or adding series shunt capacitors. Normally to find the optimal case, a number of cases have to be run and the solution eventually could be a combination of several actions where economic reasons should be considered too.

3.6 Radial load flow concepts

Power systems are analyzed by using very efficient algorithms that conduct almost invariably to software packages. This should not though prevent hand calculations, which are appropriate especially for small systems. Up to the mid-1960s before use of computers, big systems were analyzed with the aid of board simulations.

Nowadays there are flexible and affordable software packages that allow comprehensive analysis. In this section the basis and structure of the two main tools are covered focused to distribution systems. These tools are load flow and short circuit calculations.

Table 3.6 HV transformer data for power system of Example 3.7

Transformer	S_r12 (MV A)	S_r23 (MV A)	S_r31 (MV A)	V_r1 (kV)	V_r2 (kV)	V_r3 (kV)	$Z_{cc}12$ (%)	$Z_{cc}23$ (%)	$Z_{cc}31$ (%)	Vector group	Tap oper.	Tap min.	Tap nom.	Tap max.	ΔV (%)
TR_THE RIDGES	30	30	15	115.0	34.5	13.2	8.8	8.8	8.8	YNd11yn0	4	−7	0	7	−1.25
TR_BLACK RIVER 1	90	90	90	230.0	115.0	13.8	10.0	12.0	25.0	YNyn0d5	1	−10	0	14	−1.20
TR_BLACK RIVER 2	90	90	90	230.0	115.0	13.8	10.0	12.0	25.0	YNyn0d5	1	−10	0	14	−1.20
TR_BLACK RIVER	12	12	12	115.0	34.5	13.8	7.6	3.3	11.3	YNd1yn2	3	−4	0	4	−1.25
TR_BONA-VENTURE_2	22	22	22	115.0	34.5	13.8	5.0	7.0	9.6	Dyn5yn5	0	−7	0	7	−1.25

Table 3.7 MV transformer data for power system of Example 3.7

Transformer	**S_r (MV A)**	**V_r1 (kV)**	**V_r2 (kV)**	**Z_{cc}(1) (%)**	**Vector group**	**Tap oper.**	**Tap min.**	**Tap nom.**	**Tap max.**	**ΔV (%)**
TR_ORANGE	3	34.5	13.2	4.9	Dyn5	0	−3	0	3	−1.25
TR_BONAVENTURE 1	30	115.0	34.5	9.0	YNyn0	10	1	7	15	−1.25
TR1_EMERALD	10	34.5	13.8	6.1	Dyn5	2	−5	0	5	−1.25
TR_WILLOW	10	34.5	13.8	10.3	Dyn11	0	−10	0	10	−1.25
TR_BONAVENTURE 4	10	34.5	13.8	12.5	Dyn5	0	−3	0	3	−1.25
TR_SOUTH POST	50	115.0	13.8	8.0	YNd1	2	−1	0	3	2.5

Table 3.8 Generator data for power system of Example 3.7

Generator	S_r **(MV A)**	V_r **(kV)**	**cos**ϕ	P_{Gen} **(MW)**	Q_{Gen} **(MVAR)**
G_SOUTH POST	68.8	13.8	0.8	15	13

Table 3.9 Bus data for power system of Example 3.7

Load	P **(MW)**	Q **(MVAR)**	S **(MV A)**	I **(kA)**	**cos(**ϕ**)**
C_ORANGE	1.0	0.6	1.2	0.1	0.9
C_EMERALD 1	3.7	2.7	4.6	0.2	0.8
C_EMERALD 2	3.1	2.3	3.8	0.2	0.8
C_BONAVENTURE	2.0	1.0	2.2	0.1	0.9
C_BLACK RIVER 2	1.3	1.0	1.6	0.1	0.8
C_BLACK RIVER 1	3.5	2.6	4.3	0.2	0.8
C_BLACK RIVER 3	0.1	0.1	0.1	0.0	0.9
C_BLACK RIVER 4	0.0	0.0	0.0	0.0	0.0
C_WILLOW 2	1.1	0.8	1.3	0.1	0.8
C_WILLOW 1	1.1	0.8	1.3	0.1	0.8
C_WILLOW 3	1.1	0.8	1.3	0.1	0.8
C_BONAVENTURE 4	4.7	3.5	5.9	0.3	0.8
C_BONAVENTURE 3	4.3	3.2	5.4	0.2	0.8
C_BONAVENTURE 2	1.5	1.1	1.9	0.1	0.8
C__BONAVENTURE 1	4.0	2.9	5.0	0.2	0.8
C_THE RIDGES 1	1.1	0.8	1.4	0.1	0.8
C_THE RIDGES 3	5.2	3.9	6.5	0.3	0.8
C_THE RIDGES 2	4.9	3.6	6.1	0.3	0.8

Standard algorithms like NR or Gauss–Seidel can be used for the solution of distribution systems. However, it is better for these cases to use programs designed for radial systems for two main reasons:

- Programs designed for power systems normally assume a high X/R ratio.
- A program developed specifically for distribution systems will be more efficient and simpler than those developed for high-voltage systems.

This method has low memory requirement and good accuracy. On the other hand, it is simple to be implemented and has good convergence speed.

3.6.1 Theoretical background

The simplest way to start the analysis of radial feeders is by examining Figure 3.26:

$$\bar{V}_s - V_r = \bar{I} \cdot (R + jX) \tag{3.32}$$

where s is the source node, r is the receiving node, V_s is the voltage magnitude in source node, V_r is the voltage magnitude in receiving node, P, Q is the active and reactive load, R, X is the branch resistance and reactance.

Table 3.10 Line data for power system of Example 3.7

LINE	Long (km)	$R(1)$ (Ω/...)	$X(1)$ (Ω/...)	$C(1)$ (μF/...)	$B(1)$ (μS/...)	$R(0)$ (Ω/...)	$X(0)$ (Ω/...)	$B(0)$ (μS/...)	I_r max (low) (A)	Bus to Bus	
L_BLACK RIVER_ ORANGE_3	9.6	0.4	0.5	0.0	4.6	0.5	1.9	2.1	340	BLACK RIVER 34.5 kV	ORANGE_34.5 kV
BOAVENTURE_ WESTON	2.5	0.2	0.4	0.0	4.4	0.5	1.7	2.4	434	WESTON	BONAVENTURE 34.5 kV
WESTON_EMERALD	1.6	0.2	0.4	0.0	4.4	0.5	1.7	2.4	434	WESTON	EMERALD
EMERALD_WILLOW	20.2	0.2	0.4	0.0	4.4	0.5	1.7	2.4	434	EMERALD	WILLOW 34.5
BLACK RIVER_THE RIDGES1	12.0	0.1	0.5	0.0	3.4	0.2	0.9	2.4	0	THE RIDGES 115 kV	BLACK RIVER 115 kV
L_THE RIDGES_ SOUTH POST 115 kV	0.1	0.1	0.5	0.0	3.4	0.2	0.9	2.4	0	SOUTH POST 115 kV	THE RIDGES 115 kV
EMERALD_WILLOW 2	20.2	0.2	0.4	0.0	4.4	0.5	1.7	2.4	434	EMERALD	WILLOW 34.5
L_THE RIDGES 34.5	0.3	0.2	0.4	0.0	4.4	0.5	1.7	2.4	434	THE RIDGES1 34.5 kV	THE RIDGES 34.5 kV
BLACK RIVER_ BONAVENTURE	10.0	0.1	0.5	0.0	3.4	0.2	0.9	2.4	0	BLACK RIVER 115 kV	BONAVENTURE 115 kV
BLACK RIVER_THE RIDGES2	12.0	0.1	0.5	0.0	3.4	0.2	0.9	2.4	0	BLACK RIVER 115 kV	THE RIDGES 115 kV
BONAVENTURE_ THE RIDGES	9.1	0.1	0.5	0.0	1.7	0.4	1.6	0.0	0	BONAVENTURE 115 kV	THE RIDGES 115 kV

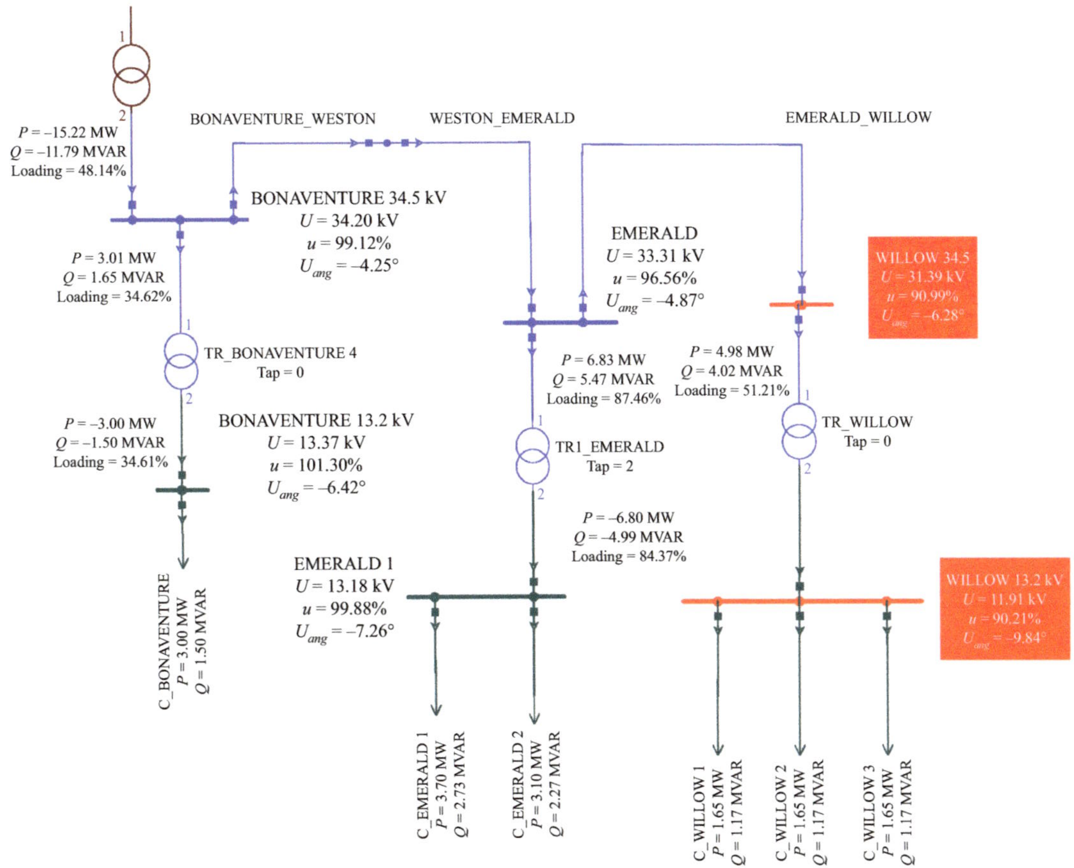

Figure 3.22 Portion of the power system of Example 3.7

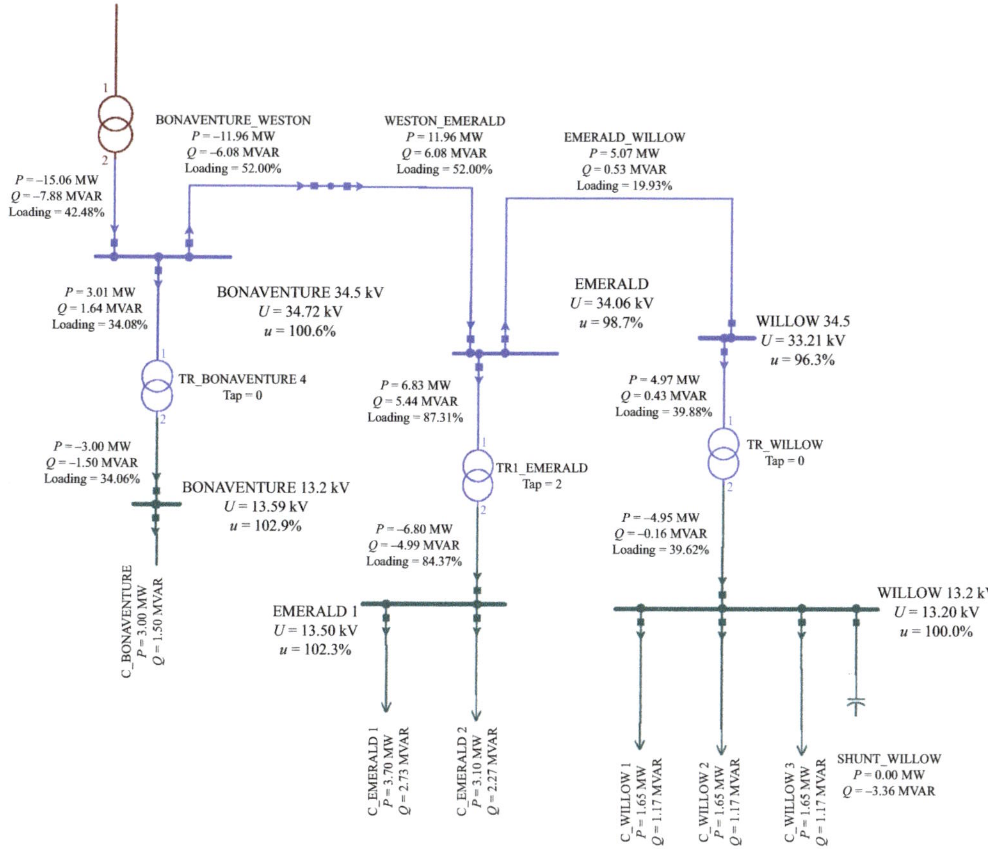

Figure 3.23 Bus voltage results considering a capacitor bank at 13.2 kV Willow bus

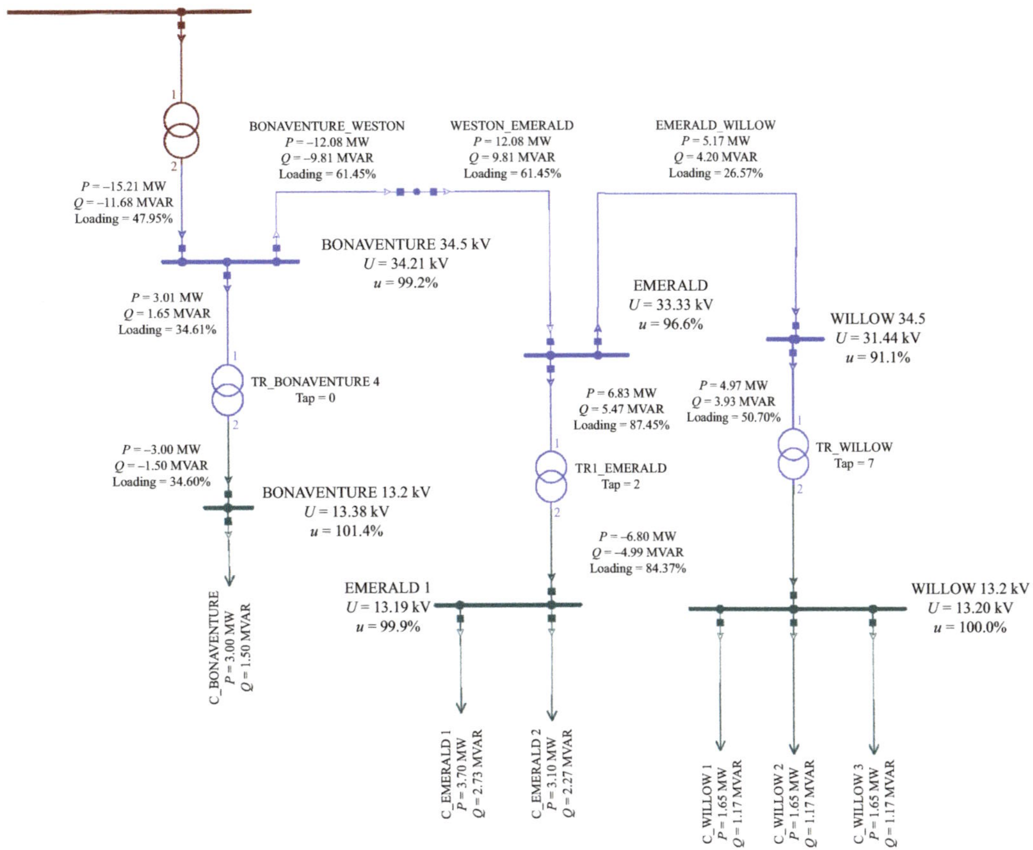

Figure 3.24 Bus voltage results of Example 3.7 considering tap changer operation

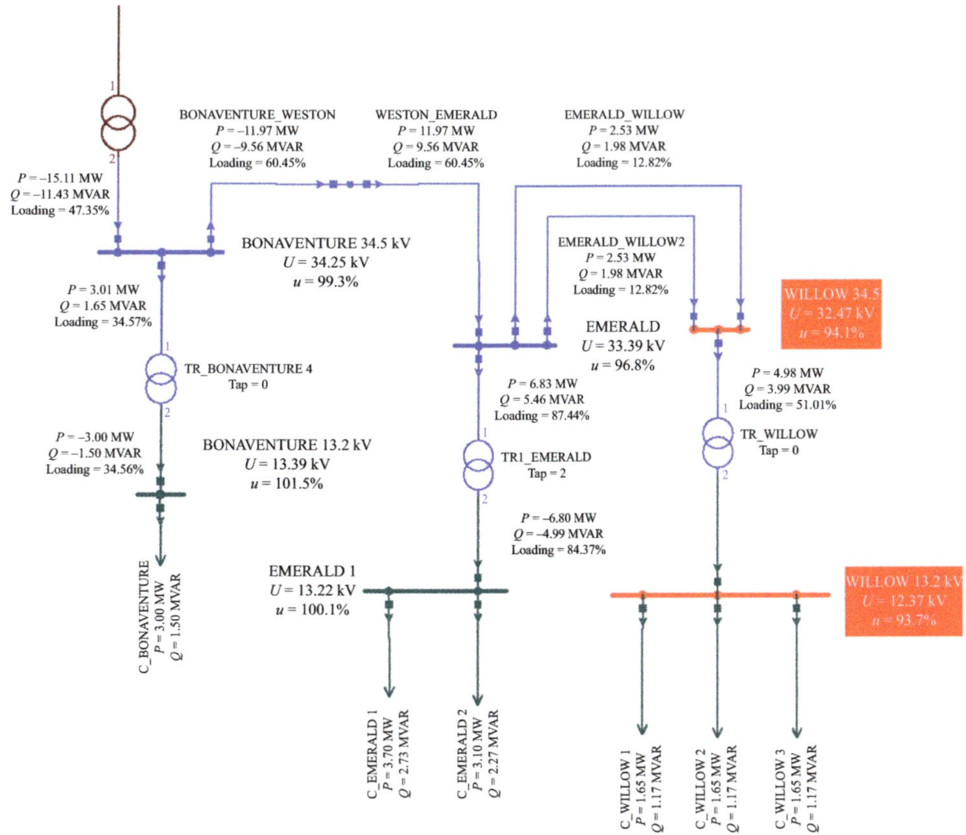

Figure 3.25 Bus voltage results of Example 3.7 considering a new line

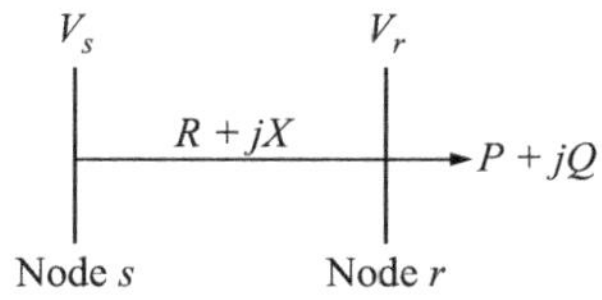

Figure 3.26 Two node systems

It is known that the vector value of current is defined by the following expression:

$$\bar{I} = \frac{P - jQ}{\bar{V}_r^*} \tag{3.33}$$

Then,

$$(V_s \angle \theta_s) - (V_r \angle \theta_r) = \frac{(P - jQ) \times (R + jX)}{(V_r \angle - \theta_r)} \tag{3.34}$$

$$(V_s \angle \theta_s) \cdot (V_r \angle - \theta_r) - V_r^2 = (PR + QX) + j(PX - QR) \tag{3.35}$$

By separating the real and imaginary parts:

$$V_s \cdot V_r \cdot cos(\theta_s - \theta_r) - V_r^2 = P \cdot R + Q \cdot X \tag{3.36}$$

$$V_s \cdot V_r \cdot sin(\theta_s - \theta_r) = P \cdot X - Q \cdot R \tag{3.37}$$

By squaring (3.36) and (3.37),

$$V_s^2 \cdot V_r^2 \cdot cos^2(\theta_s - \theta_r) = \left(V_r^2 + (P \cdot R + Q \cdot X)\right)^2 \tag{3.38}$$

$$V_s^2 \cdot V_r^2 \cdot sin^2(\theta_s - \theta_r) = (P \cdot X - Q \cdot R)^2 \tag{3.39}$$

$$cos^2(\theta_s - \theta_r) + sin^2(\theta_s - \theta_r) = 1 \tag{3.40}$$

And adding up (3.38) and (3.39), the following expression is obtained:

$$V_s^2 \cdot V_r^2 = V_r^4 + 2 \cdot V_r^2 \cdot (P \cdot R + Q \cdot X) + P^2 \cdot R^2 + 2 \cdot P \cdot X \cdot Q \cdot R + Q^2 \cdot X^2 + P^2 \cdot X^2 - 2 \cdot P \cdot X \cdot Q \cdot R + Q^2 \cdot R^2 \tag{3.41}$$

By grouping and reducing terms,

$$V_s^2 \cdot V_r^2 = V_r^4 + 2 \cdot V_r^2 \cdot (P \cdot R + Q \cdot X) + P^2 \cdot \left(R^2 + X^2\right) + Q^2 \cdot \left(R^2 + X^2\right) \tag{3.42}$$

$$V_s^2 \cdot V_r^2 = V_r^4 + 2 \cdot V_r^2 \cdot (P \cdot R + Q \cdot X) + \left(P^2 + Q^2\right) \cdot \left(R^2 + X^2\right) \tag{3.43}$$

and finally,

$$V_r^4 + \left(2 \cdot (P \cdot R + Q \cdot X) - V_s^2\right) \cdot V_r^2 + \left(P^2 + Q^2\right) \cdot \left(R^2 + X^2\right) = 0 \tag{3.44}$$

The solution for V_r at the bus load of Figure 3.19 is obtained by solving directly (3.44), which is independent of the angle. The solution is obtained easily since the equation is of quadratic form. P and Q are exact equivalents for the network connected to node r.

Active and reactive power losses can be calculated using the magnitudes for I^2R and I^2X that are obtained from the current magnitude given by the following expression:

$$I = \frac{\sqrt{P^2 + Q^2}}{V_r} \tag{3.45}$$

$$L_p = \frac{R \cdot (P^2 + Q^2)}{V_r^2} \tag{3.46}$$

$$L_q = \frac{X \cdot (P^2 + Q^2)}{V_r^2} \tag{3.47}$$

where L_p is the active losses in the branch, L_q is the reactive losses in the branch.

3.6.2 Distribution network models

3.6.2.1 Balanced three-phase models

Three-phase models are represented by their resistance and reactance in pu. In most cases the capacitance is neglected unless the line is very long.

All loads, including capacitors for reactive compensation, are represented by their active (P_0) and reactive (Q_0) components at 1.0 pu voltage. The effect of voltage variations is represented as

$$P = P_0 \cdot V^k \tag{3.48}$$

$$Q = Q_0 \cdot V^k \tag{3.49}$$

where V is the voltage magnitude, $k = 0$ for loads of constant power, $k = 1$ for loads of constant current, $k = 2$ for loads of constant impedance.

The value of k can vary according to the load characteristics.

3.6.2.2 Unbalanced three-phase network

The model is based in the following considerations: Any distribution line is made up of one, two, or three phases and the neutral conductor, which is considered to be connected to ground and therefore at ground potential. This is a valid approximation in most cases. The three-phase loads are considered to be made up by three single-phase loads connected in star.

3.6.3 Nodes and branches identification

To implement the software for radial load flow and feeder reconfiguration, it is necessary to have an adequate coding for every element to permit a proper handling of the data. Figure 3.27 shows an example of this idea.

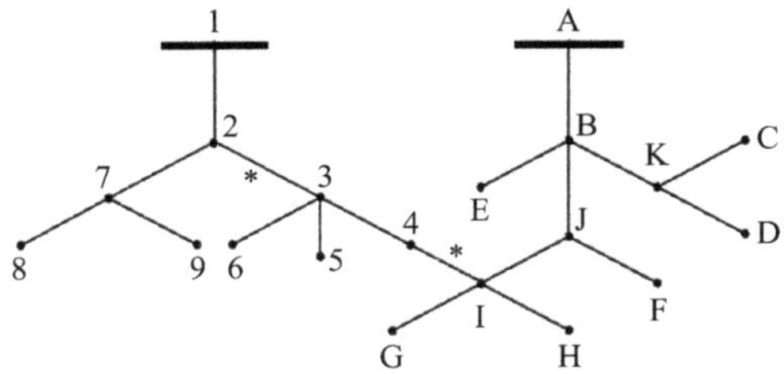

Figure 3.27 Two-feeder system with numbers and letters nomenclature

The methodology proposed considers the radiality of the feeders where any branch always has power flow in one direction. The name of each branch is always associated with the names of the respective end nodes. After any feeder is configured, it is possible to identify the path to get to the source from any node. The two columns of the configured feeder are used. This process starts by linking the node under consideration, which is in the right hand side column, to its opposite which is in the left-hand side column.

With the branches of a feeder from a database arranged in a two-column list, the process to determine the configuration of any feeder starts by picking up the node opposed the branch leaving the source node. Once this node is identified, the node or nodes at the end of the branch or branches leaving the referred node is/are in turn identified. Likewise, the rest of the nodes of the feeder can be identified until the entire feeder is configured again in a two-column arrangement. This method does not require that the elements of the initial database be ordered, which makes the process very flexible and quick.

After any feeder is configured, it is possible to identify the path to get to the source from any node. To achieve this, the two columns of the configured feeder are used. This process starts by linking the node under consideration, which is in the right-hand side column, to its opposite which is in the left-hand side column. Then a new link is established between this last node and the next one upward which has the same name located in the right-hand side column. The process continues until the source node is reached.

If any reconfiguration takes place, the method can easily identify the reconfigured feeders provided that clear indications are given in respect to the branches or links that have been opened and closed. Care must be taken to ensure that any feeder maintains its radiality and its connection to the source.

3.6.4 Illustration of nodes and branches identification

In order to illustrate the advantage of the method, Figure 3.27 shows two circuits that need to be configured, and Figure 3.28 shows the corresponding branch identification in a two-column arrangement. The branches that have operating switches are indicated with an asterisk. The initial database shows that the branches are set out in a random manner. The second database shows the configurations of the two

1) Data base

	C	K
	7	8
	B	K
	2	7
	H	I
	6	3
	E	B
	5	3
	F	J
	1	2
	9	7
	B	J
	3	4
NC *	3	2
	K	D
	A	B
	I	G
	I	J
NO *	4	I

note: SN: Sending node
RN: Receiving node
NC: Normally closed
NO: Normally opened

2) Initial configuration

i) Circuit with numbers

SN	RN
1	2
2	3
2	7
3	6
3	4
3	5
7	9
7	8

ii) Circuit with letters

A	B
B	E
B	K
B	J
K	C
K	D
J	I
J	F
I	G
I	H

3) Reconfigured circuits

Closing 4 – 1
Opening 2 – 3

i) Circuit with numbers

1	2
2	7
7	9
7	8

ii) Circuit with letters

A	B
B	E
B	K
B	J
K	C
K	D
J	I
J	F
I	4
I	H
I	G
4	3
3	5
3	6

The line shows the path between the load connected to node 5 and the source node

Figure 3.28 Two-feeder identification for the example system

circuits. The last database shows the configurations after the switches with stars have operated. The figure corresponds to the system before and after the reconfiguration.

The main advantages of this method are as follows:

- A simple and unique identification for any branch regardless of the feeder to which it belongs.
- The ability to allow the path from any node to the source node to be established easily.
- The input of branches can be carried out in any order.
- The input of branches does not need to identify which end is a source or which is a load.
- The algorithm to develop the radial load flow assumes voltages equal to 1.0 pu at every node that allows the calculation of current values entering each node.

3.6.5 Algorithm to develop radial load flow

The process is started from the furthest node from the source knowing the equivalent P and Q values. For the nodes upstream, the equivalent power loads are obtained from the P and Q values assigned to the nodes and are increased by the losses across the line that are obtained with the corresponding equations. Once the equivalent power at the source is obtained, the downstream process is carried out to calculate the voltages of the nodes. The overall process is finished when a convergence value for voltage or power is attained.

An algorithm is shown in Figure 3.29. This algorithm is built up as indicated in the following steps:

- Read network data, including line impedances, source voltages, and power values at the nodes.
- Assign voltages to the nodes for the initial start and calculate equivalent powers depending on voltage magnitudes.
- Calculate equivalent powers at each node by taking the P and Q values associated with each one and adding the power losses through the respective branches. This is the upstream routine.
- Update voltages at each node starting from the source voltage with the new power equivalents. This is the upstream routine.
- Update losses with the new voltages in the upstream process and afterward voltage levels until the convergence criteria are satisfied.
- The convergence criteria could be based on the losses value in the upstream routine or on the voltage updating in the downstream routine.

An example case is shown in Figure 3.30.

3.7 Power system analysis tool

Complex and big networks require advanced and powerful tools with the capacity to optimize and refine power system in a fast and accurate way. An equivalent mathematical model allows one to carry out analysis of any scenario on the network and develop a predictive operation. Some software developers have implemented applications for planning and operation of the electrical networks, many of which are based in load flow theory. Several simulation tools are available today and are extensively used for power system analysis, research, and education. There are a good variety of options in the market that have been developed by well-established companies.

Power system analysis tools (PSATs) are software packages designed to give the best possible performance while remaining simple to use. Some of them include a graphical interface, which should provide one-line diagrams, entering data parameters, and displaying simulation results.

PSATs should have an extensive help library and customized models and should allow functions for exporting/importing the data from different databases, including at least .txt or Excel.

In addition, warning and error messages should be displayed whenever user entries are not valid. Some of the invalid user entries and erroneous initialization should be handled by correcting to default acceptable values, displaying warning messages but without interrupting the application.

Off-line simulations in PSATs use the convergence of load flow for calculation of networks in steady state, which relates an approximation to the behavior of

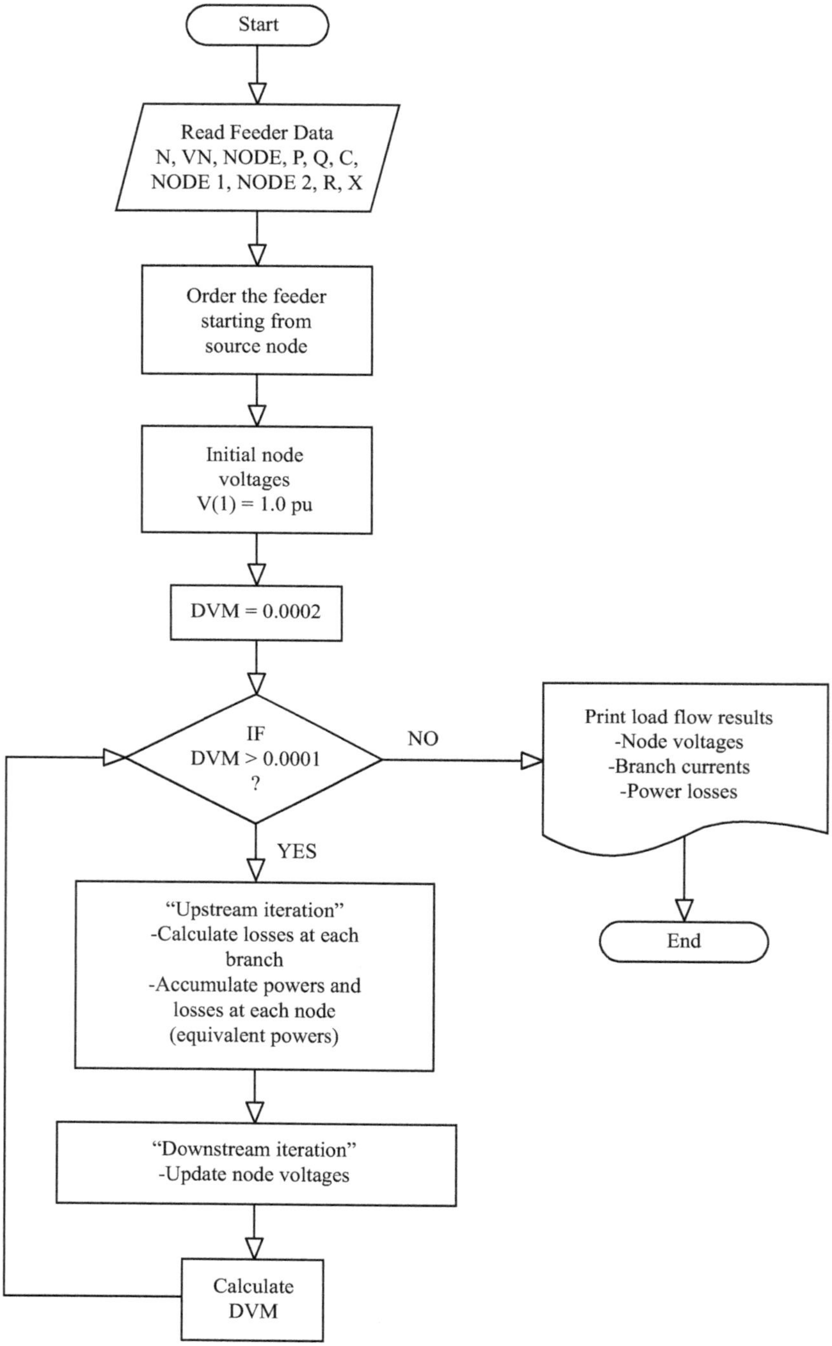

Figure 3.29 Load flow for radial systems

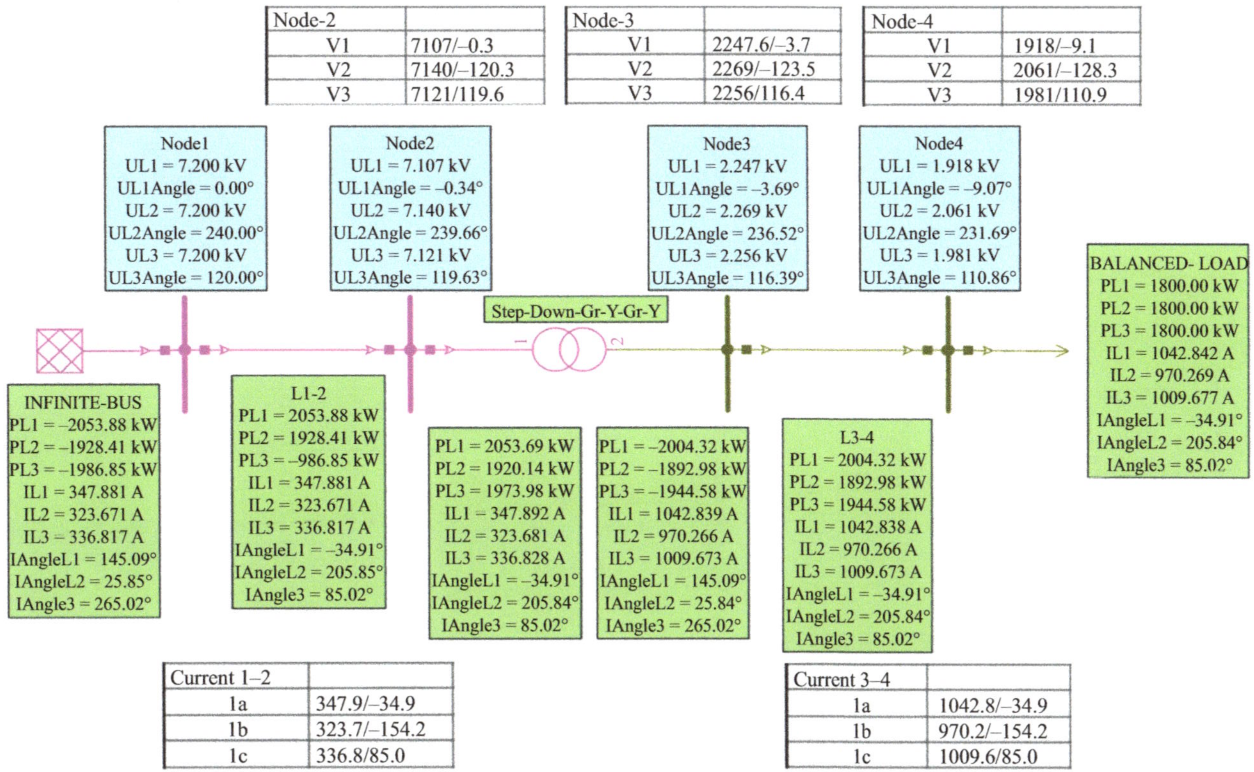

Figure 3.30 Grounded Wye–Grounded Wye step-down transformer with balanced load

power systems. These estimations have been a good solution, for planning and operation of networks, whose dispatch can be programed with minimal operational uncertainty.

Moreover, the implementation of new generation systems based in non-conventional energy sources has been notable. Information systems like SCADA play an important role for application of new solutions in monitoring and control of power systems. New technologies have made this possible to strengthen the exchanging of information through communication systems as part of monitoring and control systems in electrical networks.

Energy sources such as solar and wind are subject to a huge variability due to their nature. All the operational implications of distributed generation based on renewables lead online simulations to be closer to the real behavior of the electrical systems. New developments add an internal convergence calculation engine based on the measurements and mathematical model of the electrical system to establish control criteria for networks automation.

The state estimation concept is part of the new features in monitoring and controlling systems, where it contributes for development of automation schemes and diagnostic of information systems. It allows one to evaluate the veracity of measurements, coming from the information systems, and provides an approach to the conditions of whole networks based on measurements and mathematical convergence of the system.

3.7.1 New tendencies in PSAT applications

The evolution term reflects a process of progressive change in the attributes and applications. For power system analysis, simulation software tools offer today with add-ons and advanced modules. More complex and big networks have been implemented in recent years, thanks to the new and more advanced simulation platforms that consider all researches based on the requisitions for planning and operation of the networks.

In order to integrate PSATs with information systems in online simulations, new platforms incorporate functionalities for communication protocols application, data synchronization, information management, and data storage. A new generation of PSATs should consider applications based on real-time integrations, Cloud Computing and Big Data capabilities.

Cloud Computing is one of the most powerful techniques for Smart Grids applications and new solutions for networks operation. It allows an access to computational resources (storage, servers, and applications) immediately with a minimum effort and without desktop applications.

Cloud Computing is not only storage capability but also uses informatics resources and services. It considers an access to the information from any place and anytime. It provides solutions for most productivity, low cost, and minimum effort.

The real-time simulations can run with a timestamp based on a reference clock and hardware capabilities. Based on discrete time steps, real-time processing solves the equation systems successively; achieving an accurate representation of behavior in a dynamic system. A real-time simulator must be able to perform calculations continuously and in a minimum processing time, in order to synchronize the simulation time with a reference clock.

According to the previous idea, it is possible to acquire with a timestamp the measurements sent by measuring, protection, and/or control devices placed along the circuits. New advance in computer hardware and software technologies is enabling new ways for testing of power system controllers and digital integrations with real-time simulation concept.

3.7.2 Advanced simulations in PSATs based on load flow concept

Some special analysis in power systems can be developed with PSATs in order to optimize the network operation. In this case, the solutions for voltage profiles, losses, loading, and reactive flows are part of the criteria for implementations. Some simulation modules will be presented as examples in this section. The following chapters have more details about simulations.

3.7.2.1 Optimal capacitor placement

Inductive and resistive components on the networks contribute to the MW losses and voltage drop. The losses are proportional to the square of the current magnitude. Based on load flow results, it is possible to identify some places where the voltage level is not according to the minimal requirements for the networks operation (for instance, a regulation defined between 95% and 105%).

Load flow contributes for placement of shunt capacitors in order to reduce losses on the feeders. The reactive component of the current can be reduced by placing capacitors. Based on that, a method for maximization on the MW savings from an installation of a shunt capacitor is proposed. In the following distribution network, an operational range in which the minimum voltage level (95%) is considered. Red labels show that voltage dropped below the allowed limit.

Figure 3.31 corresponds to a distribution system whose load flow shows in red several busbars with voltages below 0.95. The addition of several capacitors bank is analyzed to increase the voltages to admissible levels.

Figure 3.32 shows the load flow results that consider four shunt capacitors after running the optimal capacitor placement simulation. Results show that all voltages are in the allowed voltage range. The capacitors are installed at busbars N1-5, N1-8, N1-14, and N1-19.

3.7.2.2 Optimal topology

Load flow simulations allow one to calculate the total losses on the networks for a specific topology. Each configuration on the network could have different results for electrical losses. It is possible to determine the topology of a distribution network, with an appropriate PSAT, in order to obtain the topology that gives the minimum losses as an objective function.

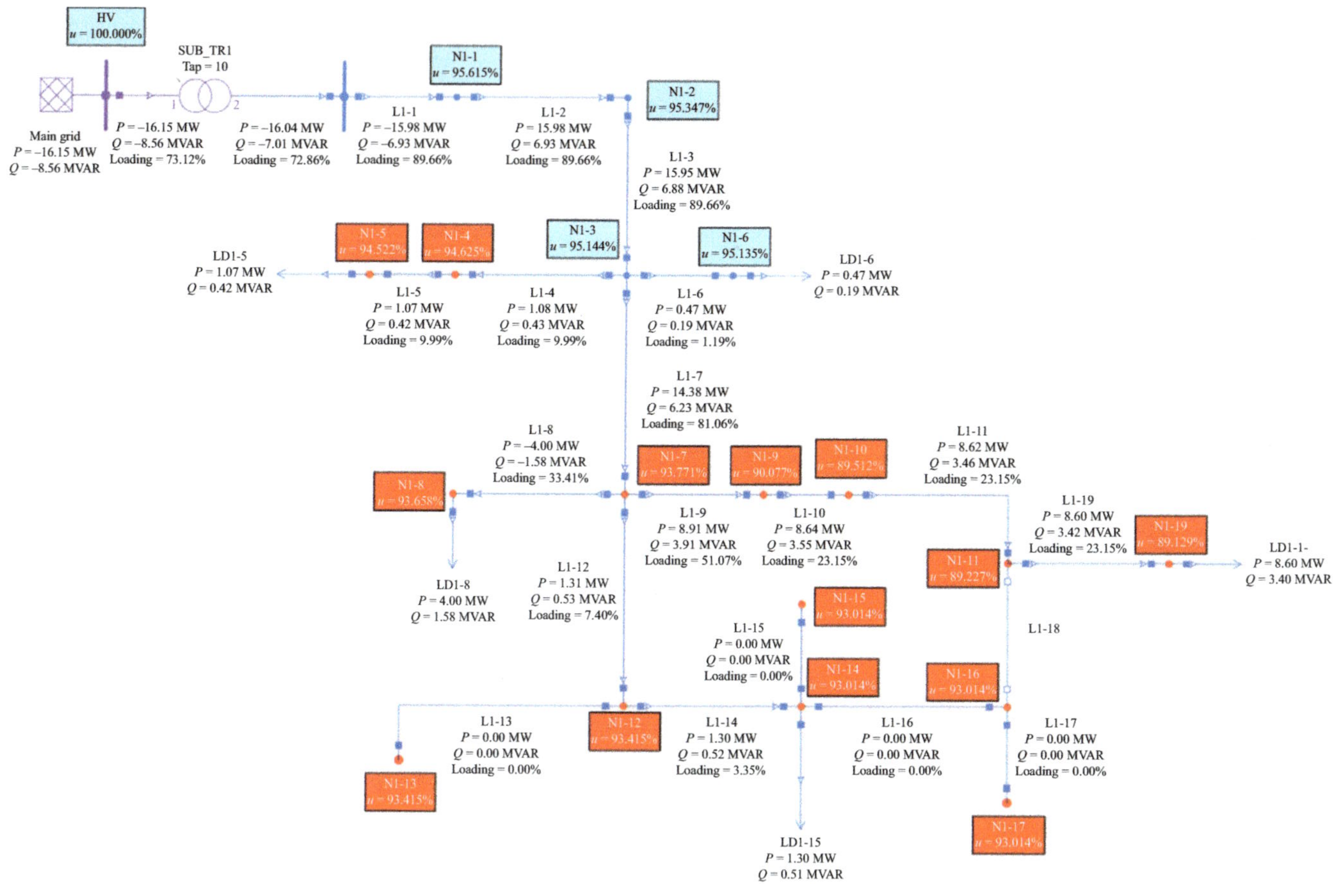

Figure 3.31 Distribution network without optimal capacitor placement

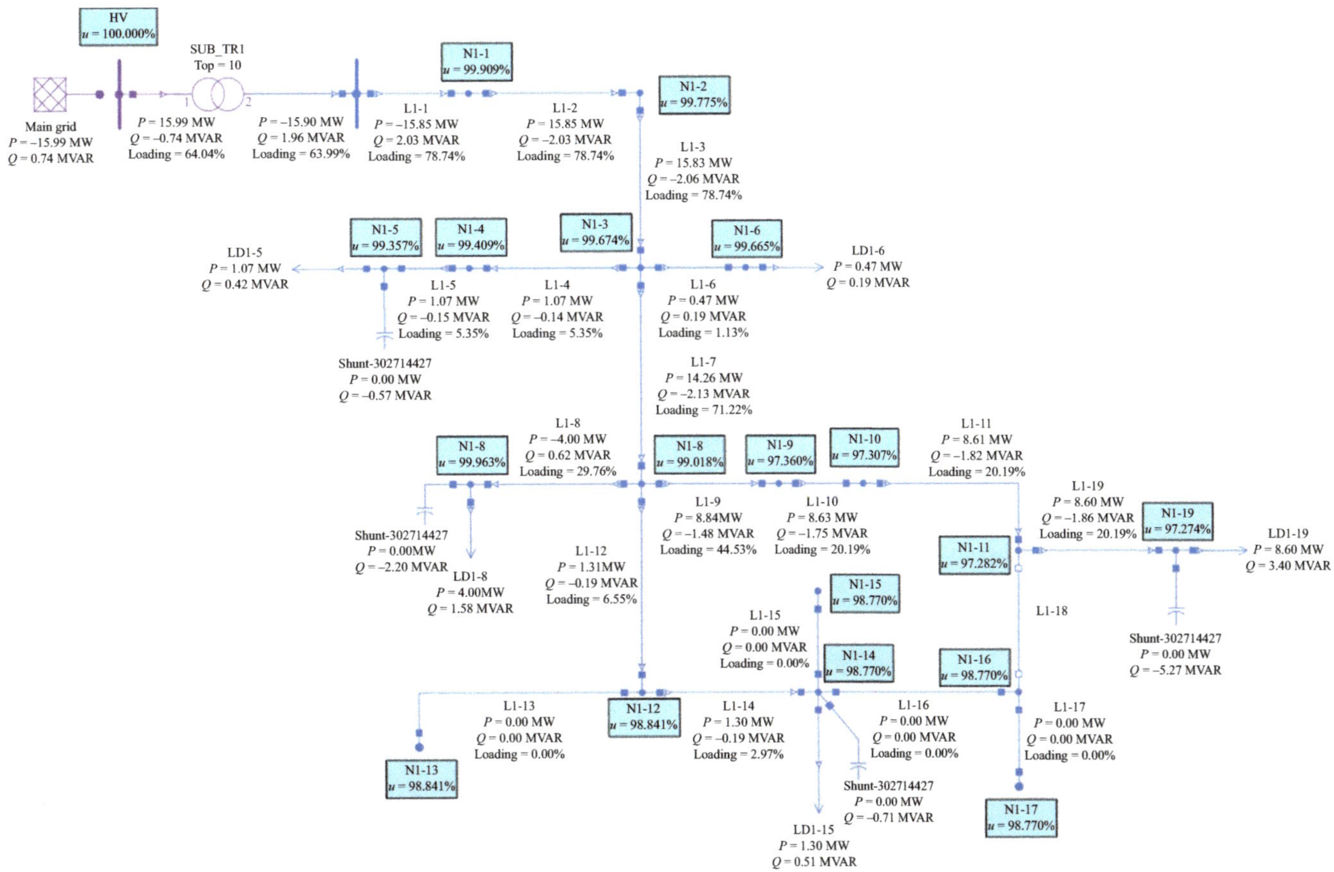

Figure 3.32 Distribution network with optimal capacitor placement simulation

Example 3.8 Figure 3.33 shows an initial feeder topology with power losses taken from a load flow analysis developed by NEPLAN. The results are presented in Table 3.11. The objective is to determine the best topology using the corresponding module of the software package. Figure 3.34 shows the final topology after completing the analysis. The results are presented in Table 3.12. It can be seen that the power losses went from 1.2766 MW to 0.7506 MW, which represent a saving in

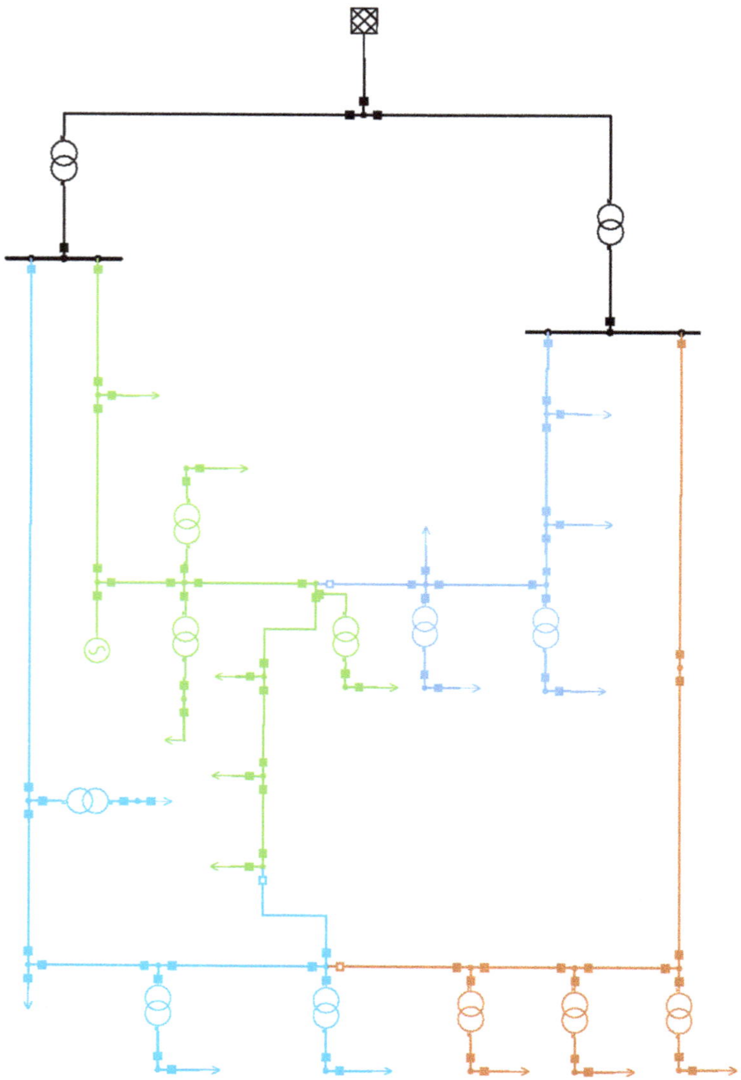

Figure 3.33 Losses on the network without optimal topology analysis

Table 3.11 Losses on the network without optimal topology analysis

Network	P_{Losses} (MW)	Q_{Losses} (MVAR)	P_{Load} (MW)	Q_{Load} (MVAR)
Total network	1.2766	1.284	10.91	7.858
FEEDER_01	0.2322	0.3983	4.512	3.351
FEEDER_02	0.0755	−0.2753	7.218	4.39
FEEDER_03	0.9212	0.9893	5.26	4.45
FEEDER_04	0.0479	0.1719	2.466	1.505

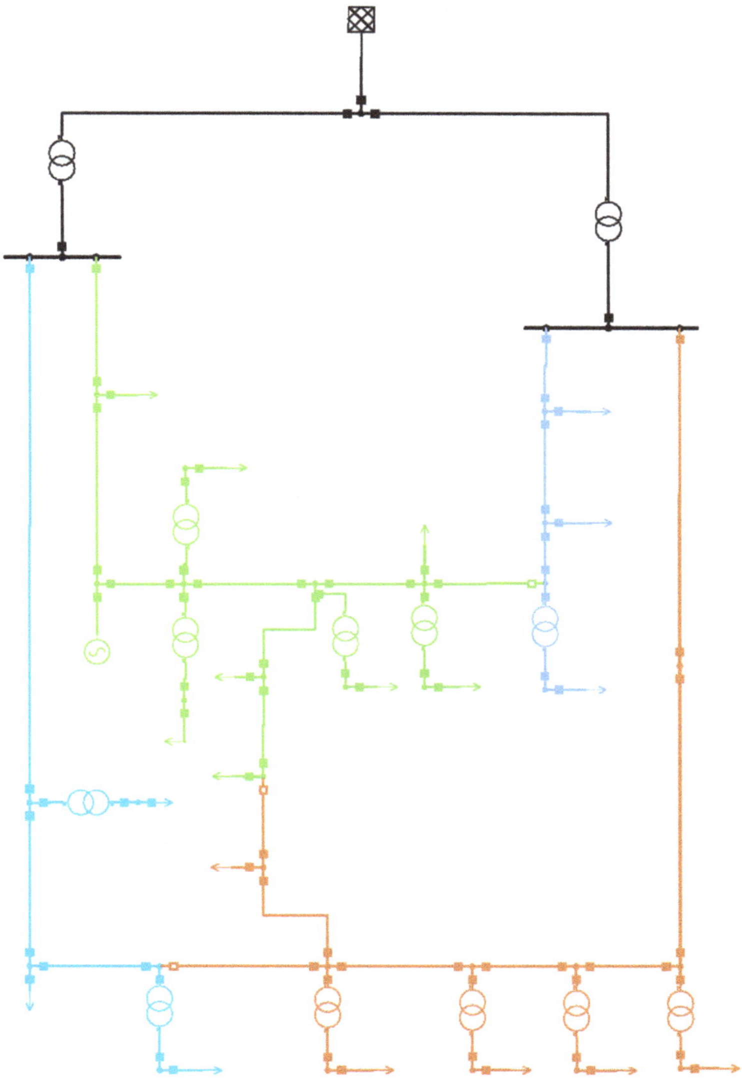

Figure 3.34 Losses on the network with optimal topology analysis

Table 3.12 Losses on the network with optimal topology analysis

Network	P_{Losses} (MW)	Q_{Losses} (MVAR)	P_{Load} (MW)	Q_{Load} (MVAR)
Total network	0.7506	0.7309	10.91	7.858
FEEDER_01	0.0787	0.1607	2.712	2.373
FEEDER_02	0.3391	0.0674	9.618	7.7182
FEEDER_03	0.0154	0.0284	1.06	0.25
FEEDER_04	0.3156	0.4729	6.066	3.3548

losses of 41% that is very significant considering the low investment required to perform the changes.

3.7.2.3 Optimal power flow

According to a specific objective function, it is possible to determine the active power injection from generators. The optimal power flow problem is formulated as a nonlinearly constrained optimization problem with a scalar, in general, nonlinear objective function.

For the network operation, it is necessary to define the operation limits, for example, voltage limits, power injection from generators, and loading. According to that, optimal load flow solutions could have different possibilities for a specific topology.

Example 3.9 For the network of Figure 3.35, the limits operation has been defined for the next limits operation:

Voltage: 95% – 105%
Loading < 100%
GEN1: 10 < MW < 200, −500 < MVAR < 500
GEN2: 10 < MW < 200, −300 < MVAR < 300

Figure 3.35 shows an initial load flow results with definition of Slack (GEN1) and *PV* (GEN2) generators both of them for voltage = 100%: The results are presented in Table 3.13.

It is possible to define a new generation scenario, an optimal power flow simulation will provide the best solution for a specific objective function. Some examples of objective functions include MW losses, MVAR losses, generation costs, MVAR flow for transformers, MVAR capacitors, among others.

The load flow solution shows the optimal solution in order to reduce the MW losses on the network. It considers voltage limits, loadings, MW, and MVAR generation limits. This is illustrated in Figure 3.36. The results are presented in Table 3.14.

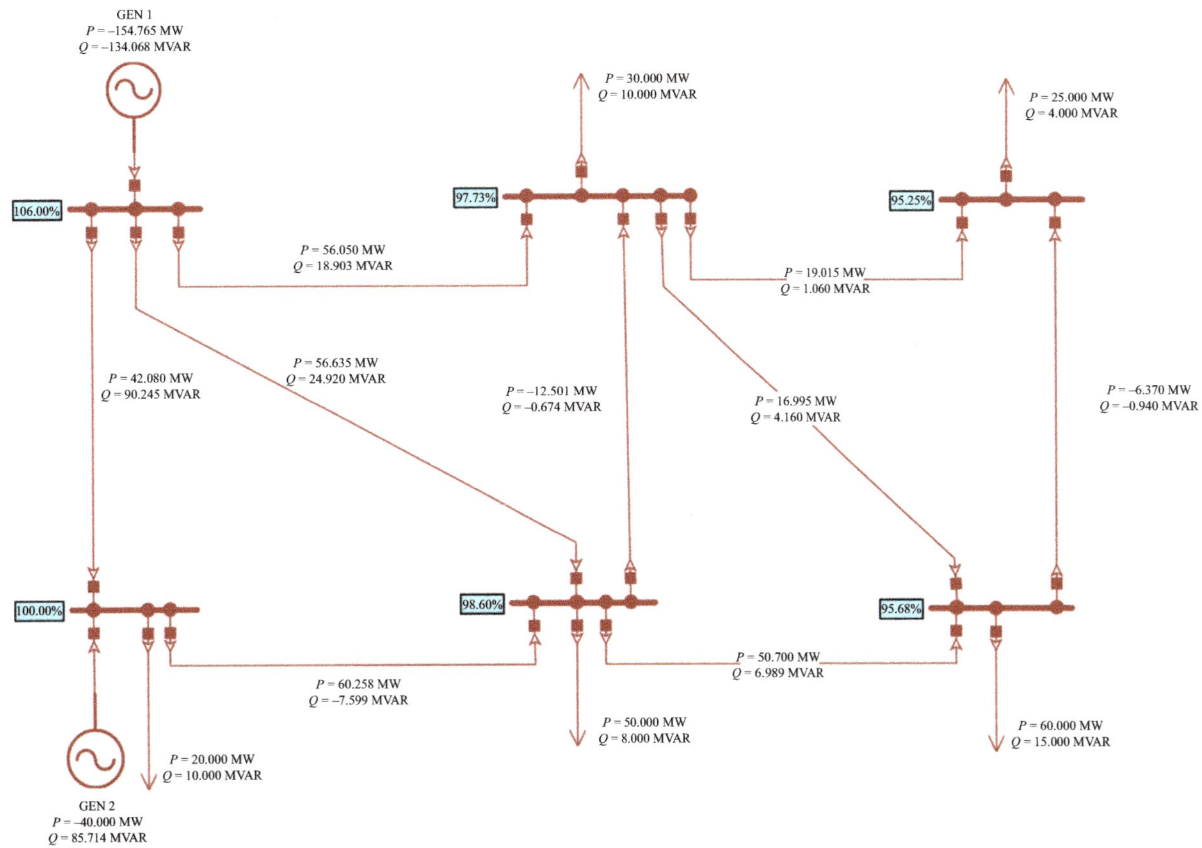

Figure 3.35 Load flow results without optimal power flow

Table 3.13 Total results on the network without optimal power flow

Network	P_{Losses} (MW)	Q_{Losses} (MVAR)	P_{Load} (MW)	Q_{Load} (MVAR)	P_{Gen} (MW)	Q_{Gen} (MVAR)
Total network results	9.7651	1.3543	185	47	194.7651	48.3543

3.7.2.4 Contingency analysis

Load flow simulation allows one to identify the impact of the elements for the operation of the network. Violation of limits and loadings are part of the results in contingency analysis and they could be obtained in distribution and transmission networks.

Contingency means that a specific element has been removed from the network and load flow simulation shows the impact in steady state.

Contingencies may occur in the power systems due to a line outage, generators outage, changes in energy demands, among others. N-1 analysis considers the remaining elements in operation within a transmission system area after the occurrence of a contingency that results on a power system element trip. The other components must be capable of accommodating to the new operational scenarios without violating any operational limits of loading, voltages, and so on. However, an N-2 analysis can be also performed in which two elements fail simultaneously.

An N-1-1 contingency is on the other hand a sequence of events that consists of the initial loss of a component, also called primary contingency with its respective system corrective actions. This contingency is followed by the loss of another power system component, known as secondary contingency.

Example 3.10 Consider the network shown in Figure 3.37 as example. A contingency N-1 will be evaluated

The results for an N-1 contingency are presented in Table 3.15 which shows that if transformers T2 and T1 get out of service, the corresponding LV voltage values go to zero when the tie switch is open. Likewise if line L3 or transformer T1 get out of service, the voltage level at busbar N3 gets below 0.95.

The results for an N-2 contingency are presented in Table 3.16 which shows that if lines L1 and L2 get out of service, the voltage levels at busbars N5, N3, and N2 get below 0.95.

3.8 Proposed exercises

1. Two voltage sources $V_1 = 450\angle -2^\circ$ and $V_2 = 440\angle 48^\circ$ are connected by a line of impedance $Z = 1.44 + j3\Omega$. Calculate the complex power for each source and the line loss.
2. Use any calculating software package (i.e., MATLAB) to elaborate the NR code for solving Example 3.6.

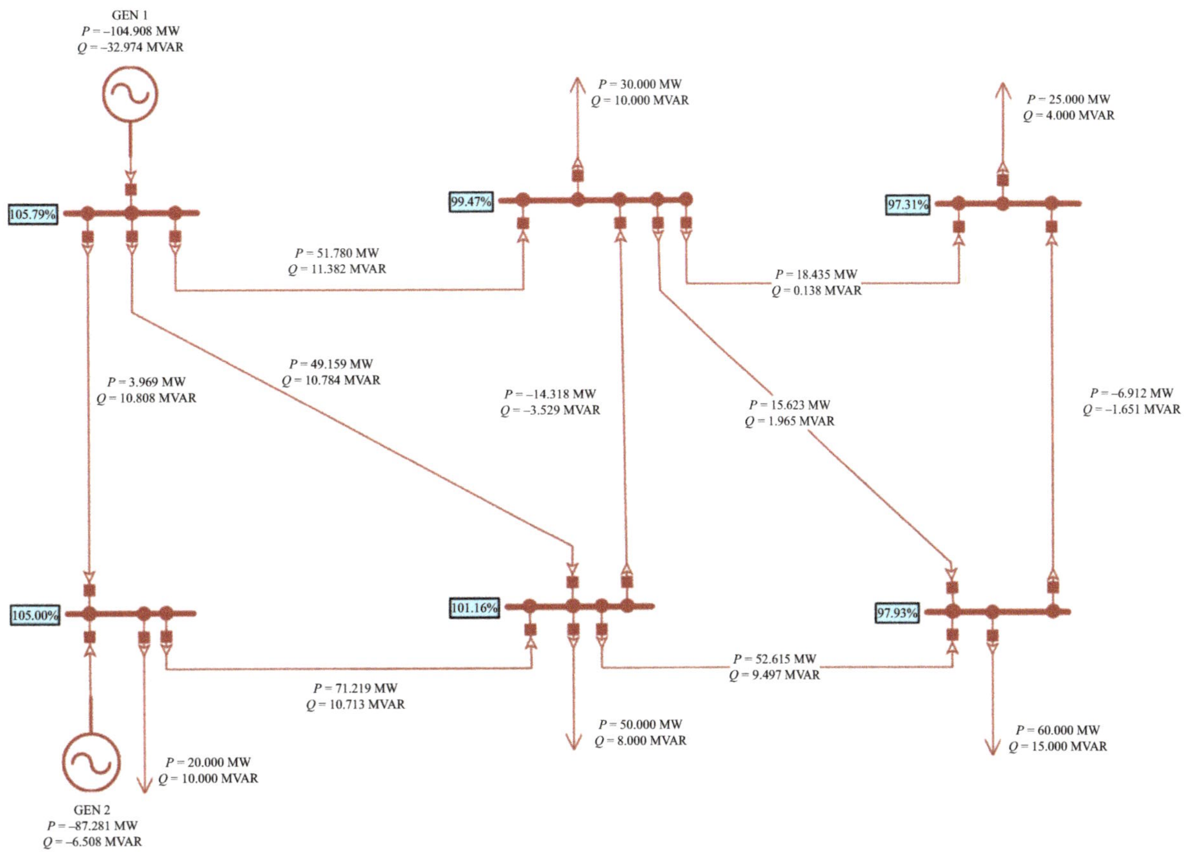

Figure 3.36 Load flow results with optimal power flow

Table 3.14 Total results on the network with optimal power flow

Network	P_{Losses} (MW)	Q_{Losses} (MVAR)	P_{Load} (MW)	Q_{Load} (MVAR)	P_{Gen} (MW)	Q_{Gen} (MVAR)
Total network results	7.189	−7.5184	185	47	192.189	39.4816

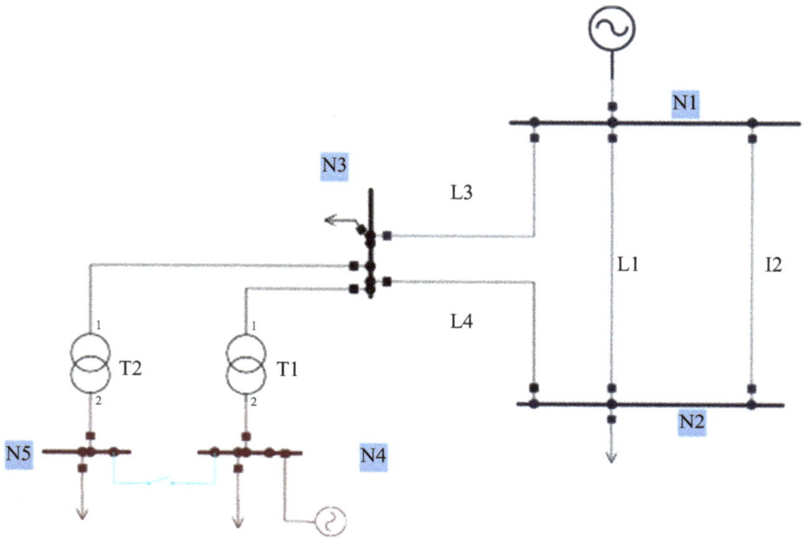

Figure 3.37 Element names on the network

Table 3.15 Contingency N-1, results for elements with limit violations

Contingency name	Violated element name	Violated element type	Voltage after N-1 contingency	Voltage before N-1 contingency
T2	N5	Busbar	0	95.0
T1	N4	Busbar	0	100
L3	N3	Busbar	94.8938	96.8665
T1	N3	Busbar	94.5947	96.8665

Table 3.16 Contingency N-2, results for L1-L2 contingency

Contingency name	Violated element name	Violated element type	Voltage after N-2 contingency	Voltage before N-2 contingency
L1-L2	N5	Busbar	89.2931	95.0
L1-L2	N3	Busbar	91.3871	96.8665
L1-L2	N2	Busbar	84.6352	95.8486

Further reading

[1] Arrillaga, J. and Arnold, C. P. Computer Analysis of Power Systems. John Wiley & Sons, New Delhi, 1990.

[2] Cespedes, R. G. "New method for the analysis of distribution networks," IEEE Transactions on Power Delivery, vol. 5, no. 1, pp. 391–396, 1990.

[3] de Vos, A., Widergren, S. E., and Zhu, J. "XML for CIM Model Exchange," PICA 2001. Innovative Computing for Power - Electric Energy Meets the Market. 22nd IEEE Power Engineering Society. International Conference on Power Industry Computer Applications (Cat. No.01CH37195), Sydney, NSW, Australia, pp. 31–37, 2001.

[4] Glover, J. D. and Sarma, M. S. Power System Analysis and Design, 3rd ed. Brooks/Cole, Pacific Grove, CA, USA, 2002.

[5] Gonen, T. Electric Power Distribution System Engineering, 2nd ed. CRC Press, Boca Raton, FL, USA, 2008.

[6] Goswami, S. K. "Distribution system planning using branch exchange technique," IEEE Transactions on Power Systems, vol. 12, no. 2, pp. 718–723, 1997.

[7] Gross, C. A. Power System Analysis, 2nd ed. John Wiley & Sons, New York, NY, USA, 1986.

[8] Harker, K. Power System Commissioning and Maintenance Practice. Peter Peregrinus, London, UK, 1998.

[9] IEEE. IEEE Std 141-1993 – IEEE Recommended Practice for Electric Power Distribution for Industrial Plants, 1993.

[10] IEEE. IEEE Std 399-1997 – IEEE Recommended Practice for Industrial and Commercial Power Systems Analysis (Brown Book), 1997.

[11] NEPLAN. Switching Optimization Module. 2018.

[12] Prabhakara, F. S., Smith, R. L., and Stratford, R. P. Industrial and Commercial Power System Handbook, 1st ed. McGraw-Hill Professional, New York, NY, USA, 1995.

[13] Short, T. A., Electric Power Distribution Handbook. CRS Press, Boca Ratón, FL, USA, 2004.

[14] Stott, B. "Decoupled Newton load flow," IEEE Transactions on Power Apparatus and Systems, vol. PAS-91, no. 5, pp. 1955–1959, 1972.

[15] Stott, B. and Alsac, O. "Fast decoupled load flow," IEEE Transactions on Power Apparatus and Systems, vol. PAS-93, no. 3, pp. 859–869, 1974.

Chapter 4
Short circuit calculation

Calculation of short circuit values is essential in power system analysis. The results are used in a number of applications like sizing of breakers and other elements of power systems, designing grounding grids, setting protective equipment, evaluating total harmonic distortion of harmonic currents, and performing arc flash analysis.

4.1 Nature of short circuit currents

When calculating short circuit currents, it is necessary to take into account two factors which could result in the currents, varying with time:

- the presence of the DC component,
- the behavior of the generator under short circuit conditions.

The circuit *L–R* in Figure 4.1 corresponds to the simplest equivalent of a power system which helps one to illustrate the concept when a short circuit is established by closing the switch:

$$e(t) = L\frac{di(t)}{dt} + Ri(t) \tag{4.1}$$

The previous expression for the short circuit current is a differential equation, whose solution has two parts:

$$i(t) = i_h(t) + i_p(t) \tag{4.2}$$

where $i_h(t)$ is the solution of the homogeneous equation corresponding to the transient period, which represents the system fed only with the energy stored in the inductor. $i_p(t)$ is the solution to the particular equation corresponding to the steady-state period, which represents the system with only the rms voltage source.

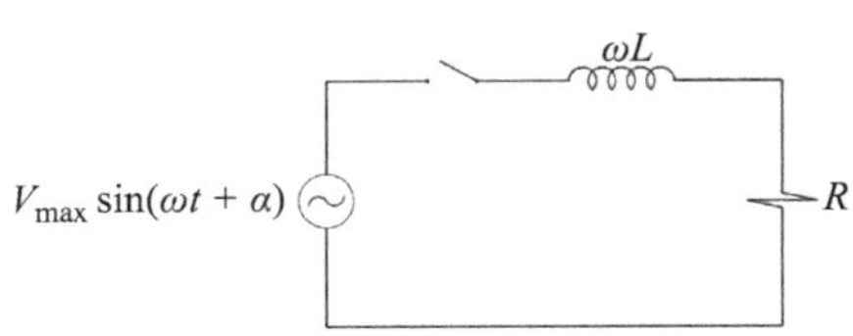

Figure 4.1 L–R circuit

The solution of this differential equation can be physically explained with the concept of the superposition theorem since there are actually two voltage sources that produce the two current components, $i_h(t)$ and $i_p(t)$. According to the theorem, each voltage source is considered one at a time while the other is short circuited. The result is the summation of both currents as the differential equation indicates.

The homogeneous equation $i_h(t)$ does not have any external source and therefore the solution has the following form:

$$Ri_h(t) + L\frac{di_h(t)}{dt} = 0 \tag{4.3}$$

If the integration is carried out, then

$$\int_{i_h(0)}^{i_h(t)} \frac{di}{dt} = -\int_0^t \frac{R}{L} dt$$

$$\int Lni_h \Big|_{i_h(0)}^{i_h(t)} = -\frac{R \cdot t}{L} \Big|_0^t$$

Then $Lni_h(t) - Ln\, i_h(0) = -(R/L) \cdot t$

$$Ln\left(\frac{i_h(t)}{i_h(0)}\right) = -\frac{R}{L} \cdot t$$

When elevating the equation to the e power, then $i_h(t)$ becomes

$$i_h(t) = i_h(0) \cdot e^{-(R/L)t} = Ke^{-(R/L)t}$$

The particular equation $i_p(t)$ considers the rms voltage source and has the following form:

$$\begin{aligned} i_p(t) &= A \sin(\omega t + \alpha) + B \cos(\omega t + \alpha) \\ \frac{di_p(t)}{dt} &= A\omega \cos(\omega t + \alpha) - B\omega \sin(\omega t + \alpha) \end{aligned} \tag{4.4}$$

The general expression for the solution is

$$\begin{aligned} e_p(t) &= L\frac{di_p(t)}{dt} + Ri_p(t) \\ V_m \sin(\omega t + \alpha) &= AR \sin(\omega t + \alpha) + BR \cos(\omega t + \alpha) + AL\omega \cos(\omega t + \alpha) \\ &\quad - BL\omega \sin(\omega t + \alpha) \end{aligned}$$

By direct comparison:

$$V_m = AR - B\omega L$$

$$0 = BR + A\omega L$$

$$B = -A\frac{\omega L}{R}$$

Replacing the value of B, V_m becomes

$$V_m = AR + A\frac{\omega^2 L^2}{R}$$
$$V_m R = AR^2 + A\omega^2 L^2$$
$$A = \frac{R}{R^2 + \omega^2 L^2} V_m$$
$$\therefore B = -\frac{\omega L}{R^2 + \omega^2 L^2} V_m$$
$$i_p(t) = V_m \frac{R}{R^2 + \omega^2 L^2} \sin(\omega t + \alpha) - V_m \frac{\omega L}{R^2 + \omega^2 L^2} \cos(\omega t + \alpha)$$

The relationship of the impedances and the angle θ can be seen in Figure 4.2 for a better illustration:

$$\cos\theta = \frac{R}{\sqrt{R^2 + \omega^2 L^2}}$$
$$\sin\theta = \frac{\omega L}{\sqrt{R^2 + \omega^2 L^2}}$$

The expression becomes

$$i_p(t) = \frac{V_m}{\sqrt{R^2 + \omega^2 L^2}} \left[\frac{R}{\sqrt{R^2 + \omega^2 L^2}} \sin(\omega t + \alpha) - \frac{\omega L}{\sqrt{R^2 + \omega^2 L^2}} \cos(\omega t + \alpha) \right]$$
$$i_p(t) = \frac{V_m}{\sqrt{R^2 + \omega^2 L^2}} [\sin(\omega t + \alpha) \cos\theta - \cos(\omega t + \alpha) \sin\theta]$$
$$i_p(t) = \frac{V_m}{\sqrt{R^2 + \omega^2 L^2}} [\sin(\omega t + \alpha - \theta)]$$

Since $i(t) = i_h(t) + i_p(t)$ from (4.2), then

$$i(t) = Ke^{-(R/L)t} + \frac{V_m}{\sqrt{R^2 + \omega^2 L^2}} [\sin(\omega t + \alpha - \theta)]$$

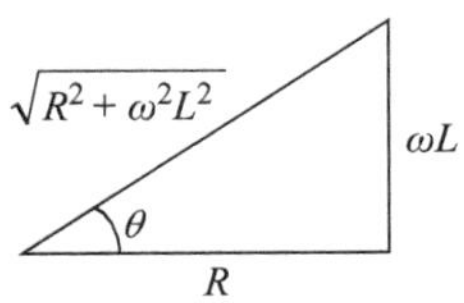

Figure 4.2 Impedance components

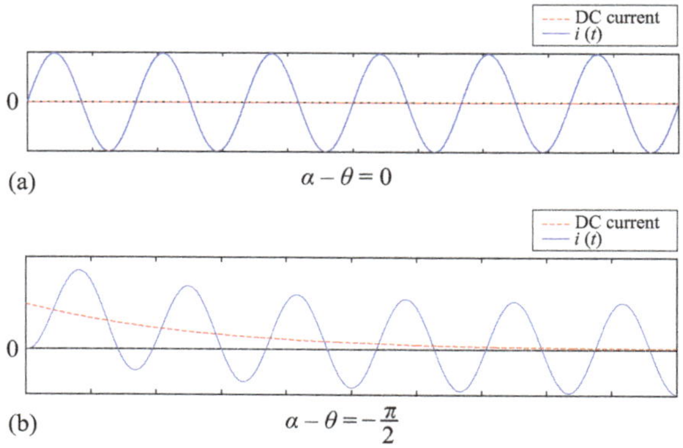

Figure 4.3 Variation of fault current due to the DC component when (a) $\alpha - \theta = 0$; *(b)* $\alpha - \theta = \frac{-\pi}{2}$

For the initial condition, when $t = 0^+ \Rightarrow i(t) = 0$

$$0 = K + \frac{V_m}{\sqrt{R^2 + \omega^2 L^2}}[\sin(\alpha - \theta)]$$

$$K = -\frac{V_m}{\sqrt{R^2 + \omega^2 L^2}}[\sin(\alpha - \theta)]$$

Replacing the K value, the total current $i(t)$ considering $i_p(t)$ and $i_h(t)$ becomes

$$i(t) = -\frac{V_m}{\sqrt{R^2 + \omega^2 L^2}}[\sin(\alpha - \theta)]e^{-(R/L)t} + \frac{V_m}{\sqrt{R^2 + \omega^2 L^2}}[\sin(\omega t + \alpha - \theta)]$$

Rearranging the terms, the final value for $i(t)$ is

$$i(t) = \frac{V_m}{\sqrt{R^2 + \omega^2 L^2}}[\sin(\omega t + \alpha - \theta) - \sin(\alpha - \theta)e^{-(R/L)t}] \quad (4.5)$$

where α is the closing angle which defines the point on which the fault occurs and θ is defined as $\tan^{-1}(\omega L/R)$.

The first term varies sinusoidally and is called the steady state or AC component. The second term decreases exponentially and is called the transient or DC component.

Figure 4.3 shows how the current can vary with two different values between α and θ.

Example 4.1 Write a code in MATLAB® to calculate the fault current during an event for the first 0.05 s, and plot the AC and DC components of the fault current, and the total current $i(t)$, for the circuit shown in Figure 4.1. Run the case considering a nominal voltage of 13,200 V, a frequency of 60 Hz, and the following parameters for the circuit: $R = 1.5\ \Omega$, $L = 0.032$, $\alpha = 30°$.

The MATLAB code for solving the problem can be as follows:

```
V=input('Write the Voltage pre-fault (in V): ');
R=input('Write the equivalent resistance of the system (in ohms): ');
L=input('Write the equivalent inductance of the system (in hernies):
');
f=input('Write the nominal frequency (in hertz): ');
alpha=input('Write the angle when the fault occurred (in degree): ');

Vm=V/sqrt(3);
angle=alpha*(pi()/180);
w=2*pi()*f;
theta=atan(w*L/R);
Z=sqrt((R^2)+((w*L)^2));

t=[0:0.0001:0.05]';

i_ac=(Vm/Z).*(sin(w.*t+angle-theta));
i_dc=(-Vm/Z)*(sin(angle-theta).*exp(-(R/L).*t));
i=i_ac+i_dc;

result=[t, i_ac, i_dc, i];
disp(' ')
disp('    t[s]     iAC(t)[A] iDC(t)[A]    i(t)')
disp(result)

plot (t,i_ac,'-g')
hold on
plot(t,i_dc,'-r')
hold on
plot(t,i,'-b')
xlabel('Time (sec)')
ylabel('Current (Amp)')
title('Asymmetrical Fault Current')
hleg1 = legend('AC Current','DC Current','Total
Current','Location','NorthEastOutside');
```

Figure 4.4 shows the plot of the AC, DC, and total currents for Example 4.1.

When a fault occurs close to the terminals of rotating machinery, a decaying AC current is produced, similar in pattern to that flowing when an AC voltage is applied to an *L–R* circuit as discussed in the previous section. Here, the decaying pattern is due to the fact that the magnetic flux in the windings of rotating machinery cannot change instantaneously because of the nature of the magnetic circuits involved. The reduction in current from its value at the onset, due to the gradual decrease in the magnetic flux caused by the reduction of the mmf of the induction current, can be seen in Figure 4.5. This effect is known as armature reaction.

The physical situation that is presented to a generator, which makes the calculations quite difficult, can be interpreted as a reactance which varies with time. Notwithstanding this, in the majority of practical applications it is possible to take account of the variation of reactance in only three stages without producing significant errors. In Figure 4.6, it will be noted that the variation of current with time, $I(t)$, comes close to the three discrete levels of current, I'', I', and I, the subtransient, transient, and steady-state currents, respectively. The corresponding values of

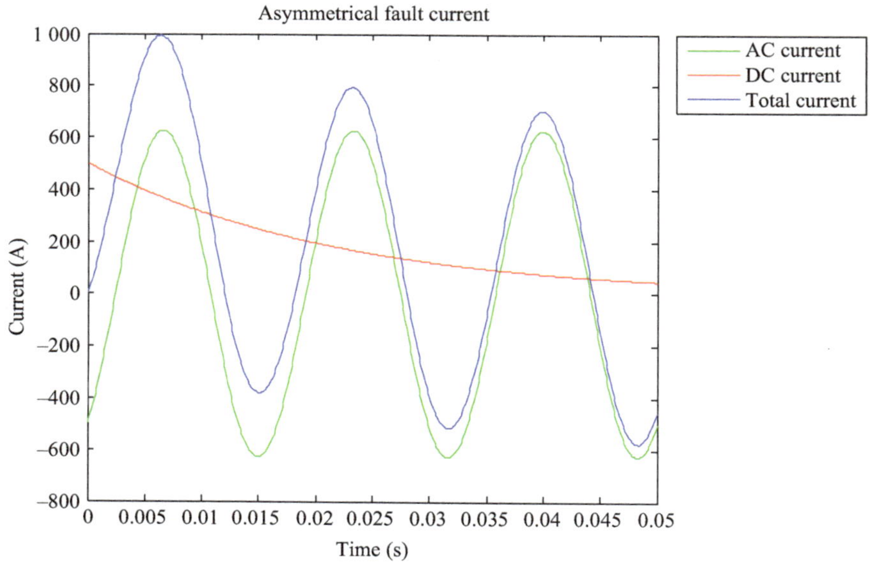

Figure 4.4 Results for Example 4.1

direct axis reactance are denoted by X_d'', X_d', and X_d, and the typical variation with time for each of these is illustrated in Figure 4.7.

Short circuit levels vary considerably during a fault, taking into account the rapid drop of the current due to the armature reaction of the synchronous machines and the fact that extinction of an electrical arc is never achieved instantaneously. Therefore, short circuit currents have to be calculated carefully in order to obtain the correct value for the respective applications.

The following paragraphs refer to the short circuit currents which are specifically used for the selection of interrupting equipment and protection relay settings—the so-called normal duty rating. ANSI/IEEE Standards Std. C37.010-2016 and IEC 60909 refer to four duty types defined as first cycle or momentary, peak, interrupting or breaking, and time-delayed or steady-state currents [1,2].

First cycle currents, also called momentary currents, are the currents that present one-half of a cycle after fault initiation. In European Standards these values are indicated by I_k''. These are the currents that are sensed by circuit breaker protection equipment when a fault occurs and are therefore also called close and latch currents. They are calculated with DC offset but no AC decrement in the sources, and using the machine subtransient reactances. Peak currents correspond to the maximum currents during the first cycle after the fault occurs and differ from the first cycle currents which are totally asymmetrical rms currents.

Interrupting currents, also known as contact parting currents, are the values that have to be cleared by interrupting equipment. In European Standards, these values are called breaking currents and typically are calculated in the range from

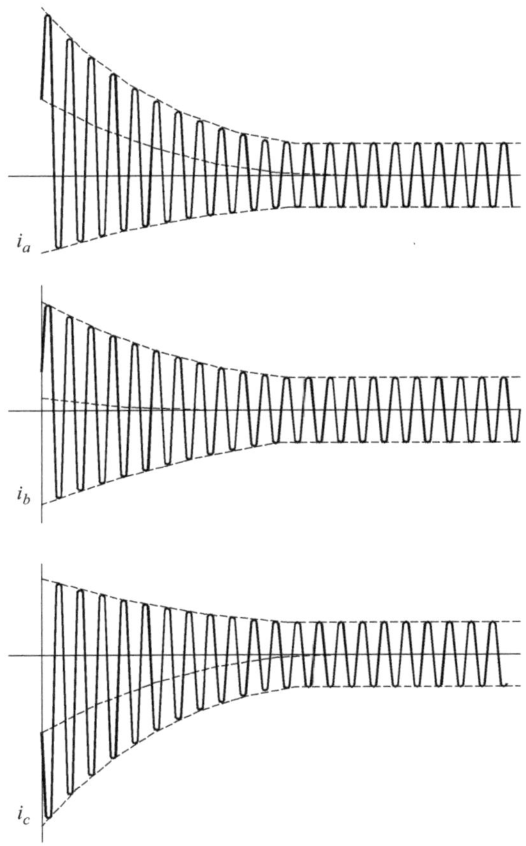

Figure 4.5 Transient SC current at generator terminals

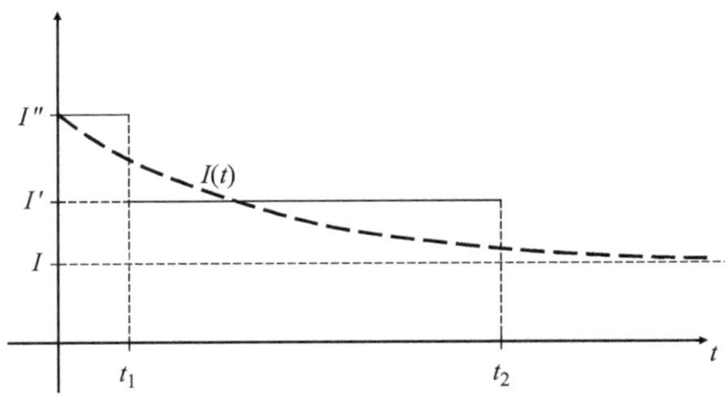

Figure 4.6 Variation of current with time during a fault at generator terminals

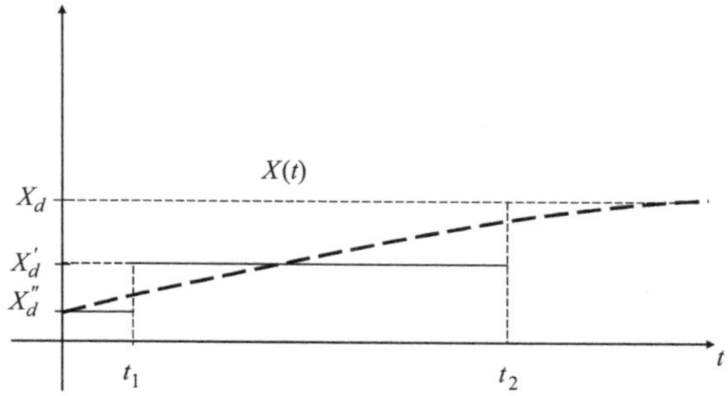

Figure 4.7 Variation of generator reactance during a fault

three to five cycles. These currents contain DC offset and some decrement of the AC current. Time-delayed or steady-state short circuit currents correspond to the values obtained between 6 and 30 cycles. These currents should not contain DC offset, and synchronous and induction contributions should be neglected and transient reactances or higher values should be used in calculating the currents.

Reactance values to be used for the different duties are reproduced in Figure 4.8, based on IEEE Standard 399-1990 [3]. For each case, asymmetrical or symmetrical rms values can be defined depending on whether the DC component is included or not. The peak values are obtained by multiplying the rms values by 2.

An approximate approach accepted is that the asymmetrical values are calculated as the square root of the sum of the squares of the DC component and the rms. value of the AC current, i.e.,

$$I_{\text{rms}} = \sqrt{I_{DC}^2 + I_{AC}^2} \tag{4.6}$$

4.2 Calculation of fault duty values

The momentary current is used when specifying the closing current of switchgear. Typically, the AC and DC components decay to 90% of their initial values after the first half cycle. From this, the value of the rms current would then be

$$\begin{aligned} I_{rms.asym.closing} &= \sqrt{I_{DC}^2 + I_{AC.rms.sym}^2} \\ &= \sqrt{\left(0.9\sqrt{2}V/X_d''\right)^2 + \left(0.9V/X_d''\right)^2} \\ &= 1.56V/X_d'' = 1.56I_{rms.sym} \end{aligned} \tag{4.7}$$

Usually a factor of 1.6 is used by manufacturers and in international standards so that, in general, this value should be used when carrying out similar calculations.

The peak value is obtained by arithmetically adding together the AC and DC components. It should be noted that, in this case, the AC component is multiplied

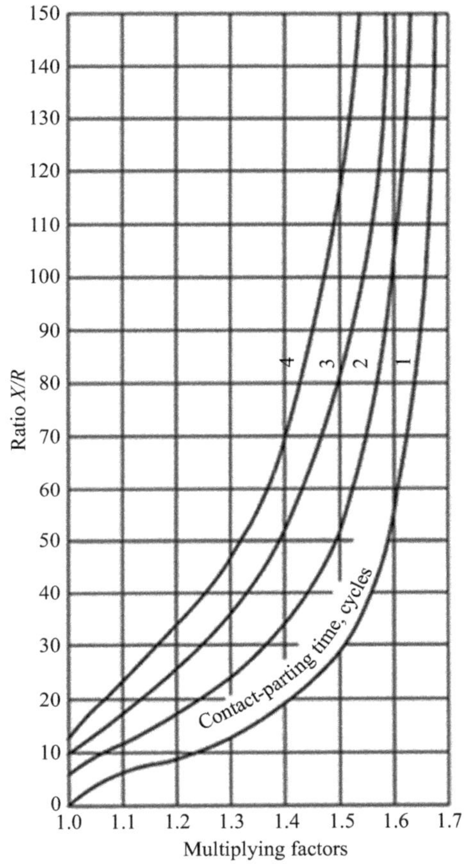

Figure 4.8 Multiplying factors for 3 Ph and L–L ground faults [4]

by a factor of $\sqrt{2}$. Thus,

$$\begin{aligned} I_{peak} &= I_{DC} + I_{AC} \\ &= \left(0.9\sqrt{2}V/X_d''\right) + \left(0.9\sqrt{2}V/X_d''\right) \\ &= 2.55 I_{rms.sym} \end{aligned} \tag{4.8}$$

When considering the specification for the switchgear opening current, the so-called rms value of interrupting current is used in which, again, the AC and DC components are taken into account, and therefore,

$$I_{rms.asym.int} = \sqrt{I_{DC}^2 + I_{AC.rms.sym.int}^2}$$

Replacing the DC component by its exponential expression gives

$$\begin{aligned} I_{rms.asym.int} &= \sqrt{\left(\sqrt{2} I_{rms.sym.int} e^{-(R/L)^t}\right)^2 + I_{rms.sym.int}^2} \\ &= I_{rms.sym.int}\sqrt{2e^{-2(R/L)^t} + 1} \end{aligned} \tag{4.9}$$

The expression ($I_{rms.asym.int}/I_{rms.sym.int}$) has been drawn for different values of X/R, and for different switchgear contact separation times, in ANSI Standard C37.5-1979 [5]. The multiplying factor graphs are reproduced in Figure 4.8.

As an illustration of the validity of the curves for any situation, consider a circuit breaker with a total contact separation time of two cycles—one cycle due to the relay and one related to the operation of the circuit breaker mechanism.

If the frequency, f, is 60 Hz and the ratio X/R is given as 50, with $t = 2$ cycles $= 0.033$ s, then $(X/R) = (\omega L/R) = 50$.

Thus $(L/R) = (50/\omega) = (50/2\pi f) = 0.132$. Therefore,

$$\frac{I_{asym}}{I_{sym}} = \sqrt{2e^{2(-R/L)t} + 1} = \sqrt{2e^{2(-(0.033/0.132))} + 1} = 1.49$$

Example 4.2 A three-phase, 500 MVA, 13.8 kV synchronous generator has the following parameters:

$$\begin{array}{lll} X_d = 1.20 \text{ pu} & X_d' = 0.15 \text{ pu} & X_d'' = 0.09 \text{ pu} \\ T_d' = 0.60 \text{ s} & T_d'' = 0.035 \text{ s} & T_a = 0.13 \text{ s} \end{array}$$

A three-phase short circuit occurs at the terminals of the machine while it is operating at rated voltage and no-load. Assume that the air-gap voltage is 100% of the rated voltage; that is, $E_{air\text{-}gap} = 1.10V_{rated}$. Determine the instantaneous value of current for a completely displaced wave.

The total AC component of the short circuit current may be expressed as follows:

$$i_{AC} = \left[\left(i_d'' - i_d'\right)e^{-t/T_d''} + \left(i_d' - i_d\right)e^{-t/T_d'} + i_d \right] \cos \omega t$$

Or

$$\begin{aligned} i_{AC} = & \left[\sqrt{2}\left(\frac{V_{rated}}{X_d''} - \frac{V_{rated}}{X_d'}\right)e^{-t/T_d''} \right. \\ & \left. + \sqrt{2}\left(\frac{V_{rated}}{X_d'} - \frac{E_{air.gap}}{X_d}\right)e^{-t/T_d'} + \sqrt{2}\left(\frac{E_{air.gap}}{X_d}\right) \right] \cos \omega t \end{aligned}$$

The total DC component of the short circuit current may be expressed as follows:

$$\begin{aligned} i_{DC} &= i_{DC,\mathrm{MAX}} e^{-t/T_a} \\ i_{DC,\mathrm{MAX}} &= -i_d'' = -\sqrt{2}\left(\frac{V_{rated}}{X_d''}\right) \\ i_a(t) &= [6.28e^{-28.57t} + 8.13e^{-1.667t} + 1.3] \cos 120\pi t - 15.71e^{-7.69t} \end{aligned}$$

The expression obtained is plotted in Figure 4.9.

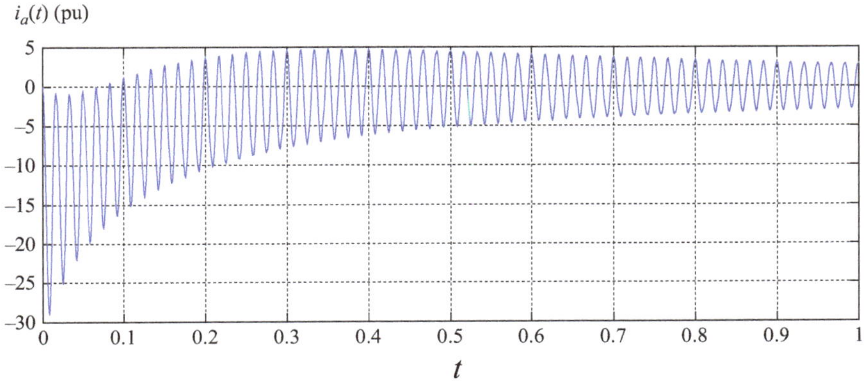

Figure 4.9 Short circuit total current

Table 4.1 Short circuit currents nomenclature for standards ANSI/IEEE and IEC

ANSI/IEEE SC currents		IEC SC currents	
First cycle	($I_{1/2c}$)	Initial	(I''_k)
Closing-latching	(I_{cres})	Peak	(I_p)
Interrupting	(I_{int})	Breaking	(I_b)
Time delayed	(I_{30c})	Steady state	(I_k)

It is interesting to note that the nomenclature to refer to short circuit currents by the Standards IEC and IEEE is not the same which might cause confusion. Table 4.1 presents the equivalences for both standards.

4.3 Fault calculation for symmetrical faults

The calculation of short circuit values for symmetrical faults is the simplest. The system has to be modeled by using only single phase values.

It is important to emphasize that symmetrical faults produce the largest short circuit values for power and sometimes for currents when systems are not solidly grounded.

Example 4.3 Write a code in MATLAB to calculate the short circuit current for a three-phase fault on the system below considering line voltages of 13.8 kV, 22 kV, and 34.5 kV. Plot the short circuit currents for fault distances from the substation to 20 km away in steps of 1 km (Figure 4.10).

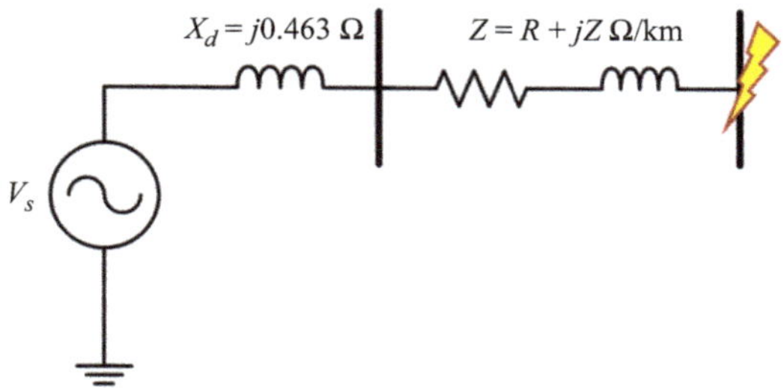

Figure 4.10 System for Example 4.3

For a line impedance of $0.124 + j0.625\ \Omega$/km, the code for solving the problem can be the following:

```
V_1_mag=input('Write the line voltage magnitude of V1 [kV] = ');
V_2_mag=input('.Write the line voltage magnitude of V2 [kV] = ');
V_3_mag=input('Write the line voltage magnitude of V3 [kV] = ');
R= input ('Write the line resistance [ohm/km] = ');
X= input ('Write the line reactance [ohm/km] = ');

Zl=R+i*X;
line_var=[0:1:20];
Zl_var=Zl*line_var;
Zs=i*0.463;
Z=abs(Zs+Zl_var);

I1=V_1_mag./(sqrt(3)*Z);
I2=V_2_mag./(sqrt(3)*Z);
I3=V_3_mag./(sqrt(3)*Z);

plot(line_var, I1,line_var, I2,line_var, I3)
legend('13.8 kV','22 kV','34.5 kV')

ylabel('3-phase fault current [kA]')
xlabel('Distance from substation, km')
```

Figure 4.11 shows the plot of the short circuit currents and the variation with the distance from the substation for three different voltage levels.

4.4 Symmetrical components

Symmetrical faults, that is, three-phase faults and three-phase-to-earth faults, with symmetrical impedances to the fault, leave the electrical system balanced and therefore can be treated by using a single-phase representation. This symmetry is lost during asymmetric faults—line-to-earth, line-to-line, and line-to-line-to-earth—and in these cases a method of analyzing the fault that provides a convenient means of dealing with the asymmetry is required.

On June 28, 1918, Mr. Charles L. Fortescue presented at the 34th Annual Convention of the AIEE, in Atlantic City, NJ a very famous paper entitled "Method of Symmetrical Co-ordinates Applied to the Solution of Poly-phase Networks" [6]. The excerpt of this paper is shown in Figure 4.12. This method allowed the

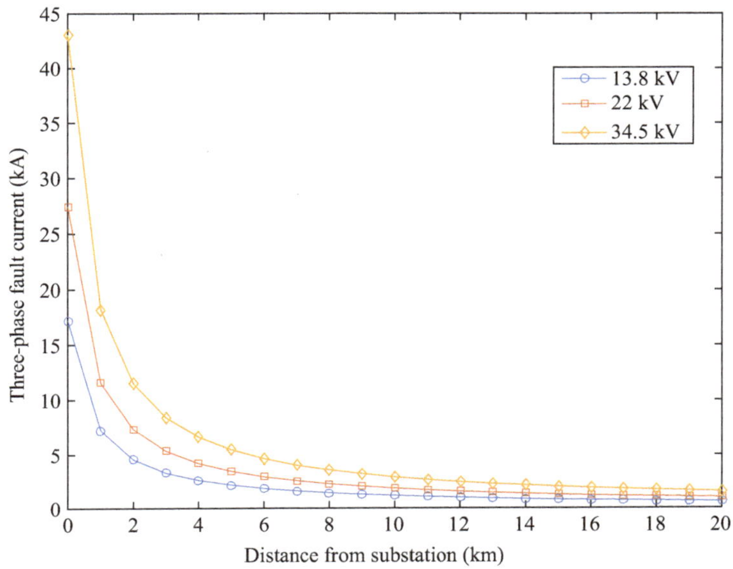

Figure 4.11 Short circuit variation for Example 4.3

representation of any unbalanced system with *n* phases, into *n* systems of balanced networks. The method was later called Symmetrical Components. The main use of this method was of course in the analysis of unbalanced three-phase systems.

While the method can be applied to any unbalanced polyphase system, the theory is summarized here for the case of an unbalanced three-phase system.

When considering a three-phase system, each vector quantity, voltage or current, is replaced by three components so that a total of nine vectors uniquely represent the values of the three phases. As is shown in Figure 4.13, the three-system balanced phasors are designated as

1. Positive-sequence components, which consist of three phasors of equal magnitude, spaced 120° apart, and rotating in the same direction as the phasors in the power system under consideration, i.e., the positive direction.
2. Negative-sequence components, which consist of three phasors of equal magnitude, spaced 120° apart, rotating in the opposite direction of the positive-sequence phasors, i.e., the negative direction.
3. Zero-sequence components, which consist of three phasors equal in magnitude and in phase with each other, rotating in the same direction as the positive sequence phasors.

It is possible to define all the phases in terms of phase *a*, since the method allows one to obtain three balanced systems, as shown in Figure 4.14.

Currents values of any three-phase system, I_a, I_b, and I_c, can be represented as

$$
\begin{aligned}
I_a &= I_{a0} + I_{a1} + I_{a2} \\
I_b &= I_{b0} + I_{b1} + I_{b2} \\
I_c &= I_{c0} + I_{c1} + I_{c2}
\end{aligned}
$$

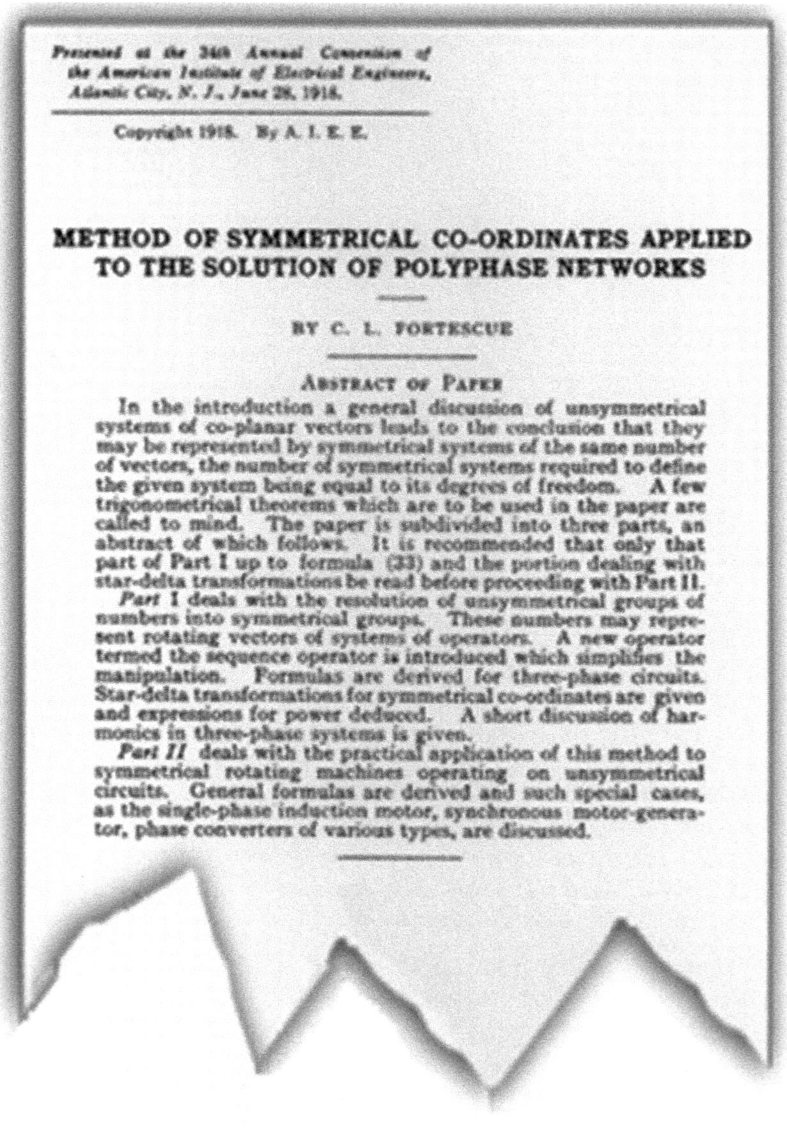

Presented at the 34th Annual Convention of the American Institute of Electrical Engineers, Atlantic City, N. J., June 28, 1918.

Copyright 1918. By A. I. E. E.

METHOD OF SYMMETRICAL CO-ORDINATES APPLIED TO THE SOLUTION OF POLYPHASE NETWORKS

BY C. L. FORTESCUE

ABSTRACT OF PAPER

In the introduction a general discussion of unsymmetrical systems of co-planar vectors leads to the conclusion that they may be represented by symmetrical systems of the same number of vectors, the number of symmetrical systems required to define the given system being equal to its degrees of freedom. A few trigonometrical theorems which are to be used in the paper are called to mind. The paper is subdivided into three parts, an abstract of which follows. It is recommended that only that part of Part I up to formula (33) and the portion dealing with star-delta transformations be read before proceeding with Part II.

Part I deals with the resolution of unsymmetrical groups of numbers into symmetrical groups. These numbers may represent rotating vectors of systems of operators. A new operator termed the sequence operator is introduced which simplifies the manipulation. Formulas are derived for three-phase circuits. Star-delta transformations for symmetrical co-ordinates are given and expressions for power deduced. A short discussion of harmonics in three-phase systems is given.

Part II deals with the practical application of this method to symmetrical rotating machines operating on unsymmetrical circuits. General formulas are derived and such special cases, as the single-phase induction motor, synchronous motor-generator, phase converters of various types, are discussed.

Figure 4.12 Excerpt of the famous paper by C.L. Fortescue [6]

Replacing the sequence component values, the following equations are obtained:

$$I_a = I_{a0} + I_{a1} + I_{a2}$$
$$I_b = I_{a0} + a^2 I_{a1} + a I_{a2}$$
$$I_c = I_{a0} + a I_{a1} + a^2 I_{a2}$$

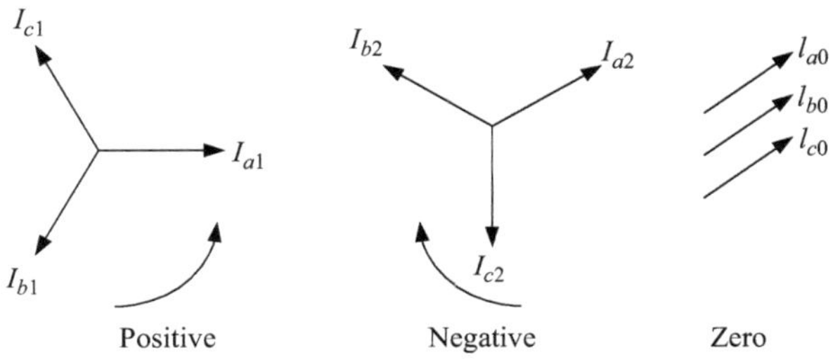

Figure 4.13 Illustration of symmetrical components

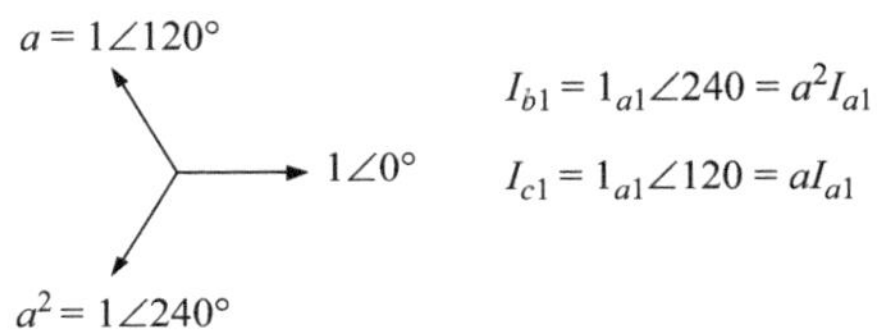

Figure 4.14 Magnitudes of the positive sequence network

Therefore, the following matrix relationship can be established:

$$\begin{bmatrix} I_a \\ I_b \\ I_c \end{bmatrix} = \begin{bmatrix} 1 & 1 & 1 \\ 1 & a^2 & a \\ 1 & a & a^2 \end{bmatrix} \cdot \begin{bmatrix} I_{a0} \\ I_{a1} \\ I_{a2} \end{bmatrix}$$

Inverting the matrix of coefficients:

$$\begin{bmatrix} I_{a0} \\ I_{a1} \\ I_{a2} \end{bmatrix} = \frac{1}{3} \begin{bmatrix} 1 & 1 & 1 \\ 1 & a & a^2 \\ 1 & a^2 & a \end{bmatrix} \cdot \begin{bmatrix} I_a \\ I_b \\ I_c \end{bmatrix}$$

From the above matrix, it can be deduced that

$$\begin{aligned} I_{a0} &= \frac{1}{3}(I_a + I_b + I_c) \\ I_{a1} &= \frac{1}{3}\left(I_a + aI_b + a^2 I_c\right) \\ I_{a2} &= \frac{1}{3}\left(I_a + a^2 I_b + aI_c\right) \end{aligned}$$

In three-phase systems the neutral current is equal to $I_n = (I_{a0} + I_{b0} + I_{c0})$ and, therefore, $I_n = 3I_{a0}$.

Voltage values of any three-phase system, V_a, V_b, and V_c, can be represented as

$$\begin{aligned} V_a &= V_{a0} + V_{a1} + V_{a2} \\ V_b &= V_{b0} + V_{b1} + V_{b2} \\ V_c &= V_{c0} + V_{c1} + V_{c2} \end{aligned}$$

Replacing the sequence component values, the following equations are obtained:

$$V_a = V_{a0} + V_{a1} + V_{a2}$$
$$V_b = V_{a0} + a^2V_{a1} + aV_{a2}$$
$$V_c = V_{a0} + aV_{a1} + a^2V_{a2}$$

Therefore, the following matrix relationship can be established:

$$\begin{bmatrix} V_a \\ V_b \\ V_c \end{bmatrix} = \begin{bmatrix} 1 & 1 & 1 \\ 1 & a^2 & a \\ 1 & a & a^2 \end{bmatrix} \cdot \begin{bmatrix} V_{a0} \\ V_{a1} \\ V_{a2} \end{bmatrix}$$

Inverting the matrix of coefficients,

$$\begin{bmatrix} V_{a0} \\ V_{a1} \\ V_{a2} \end{bmatrix} = \frac{1}{3}\begin{bmatrix} 1 & 1 & 1 \\ 1 & a & a^2 \\ 1 & a^2 & a \end{bmatrix} \cdot \begin{bmatrix} V_a \\ V_b \\ V_c \end{bmatrix}$$

From the above matrix it can be deduced that

$$V_{a0} = \frac{1}{3}(V_a + V_b + V_c)$$
$$V_{a1} = \frac{1}{3}(V_a + aV_b + a^2V_c)$$
$$V_{a2} = \frac{1}{3}(V_a + a^2V_b + aV_c)$$

In three-phase systems the neutral current is equal to $V_n = (V_{a0} + V_{b0} + V_{c0})$ and, therefore, $V_n = 3V_{a0}$.

An example of this is shown in Figure 4.15, for an arbitrary unbalanced three-phase system.

4.4.1 Importance and construction of sequence networks

The impedance of a circuit in which only positive-sequence currents are circulating is called the positive sequence impedance and, similarly, those in which only negative- and zero-sequence currents flow are called the negative- and zero-sequence impedances. These sequence impedances are designated Z_1, Z_2, and Z_0, respectively, and are used in calculations involving symmetrical components. Since generators are designed to supply balanced voltages, the generated voltages are of positive sequence only. Therefore, the positive-sequence network is composed of an emf source in series with the positive-sequence impedance. The negative- and zero-sequence networks do not contain emfs but only include impedances to the flow of negative- and zero-sequence currents, respectively.

The positive- and negative-sequence impedances of overhead line circuits are identical, as are those of cables, being independent of the phase if the applied voltages are balanced. The zero-sequence impedances of lines differ from the positive- and negative-sequence impedances since the magnetic field creating the positive- and negative-sequence currents is different from that for the zero-sequence currents. The following ratios may be used in the absence of detailed information. For a single

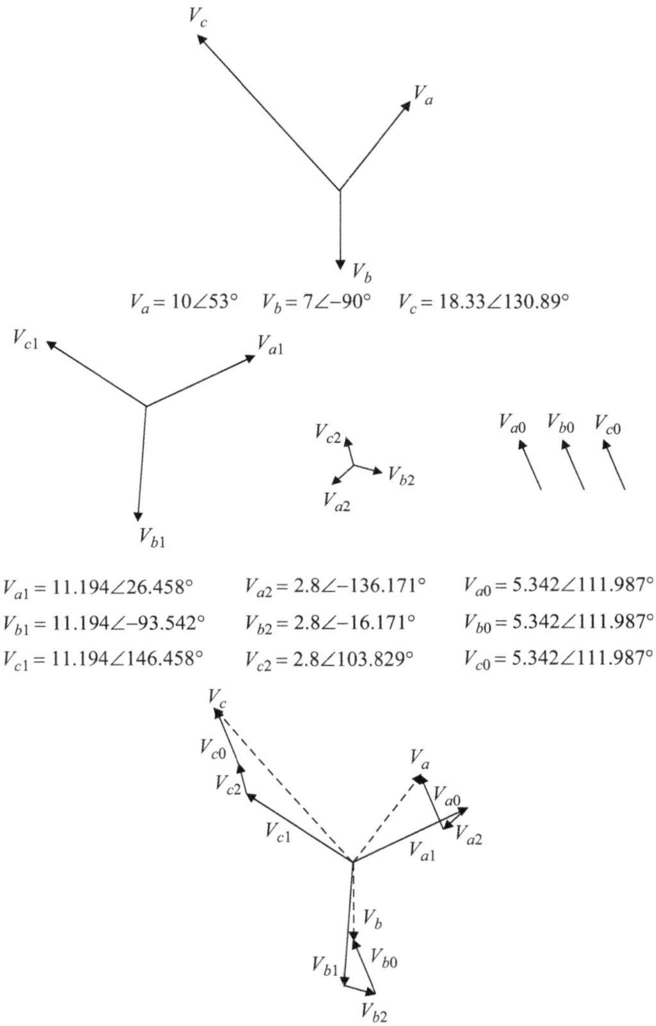

Figure 4.15 Symmetrical components of an unbalanced three-phase system

circuit line, $Z_0/Z_1 = 2$ when no earth wire is present, and 3.5, with an earth wire. For a double circuit line $Z_0/Z_1 = 5.5$. For underground cables Z_0/Z_1 can be taken as 1–1.25 for single core, and 3–5 for three-core cables.

For transformers, the positive- and negative-sequence impedances are equal because in static circuits these impedances are independent of the phase order, provided that the applied voltages are balanced. The zero-sequence impedance is either the same as the other two impedances, or infinite, depending on the transformer connections. The resistance of the windings is much smaller and can generally be neglected in short circuit calculations.

When modeling small generators and motors, it may be necessary to take resistance into account. However, for most studies only the reactances of synchronous machines are used. Three values of positive-reactance are normally quoted—subtransient, transient, and synchronous reactances, denoted by X_d'', X_d', and X_d. In fault studies the subtransient and transient reactances of generators and motors must be included as appropriate, depending on the machine characteristics and fault clearance time. The subtransient reactance is the reactance applicable at the onset of the fault occurrence. Within 0.1 s the fault level falls to a value determined by the transient reactance and then decays exponentially to a steady-state value determined by the synchronous reactance.

In connecting sequence networks together, the reference busbar for the positive- and negative-sequence networks is the generator neutral which, in these networks, is at earth potential, so only zero-sequence currents flow through the impedances between neutral and earth. The reference busbar for zero-sequence networks is the earth point of the generator. The current that flows in the impedance Z_n between the neutral and earth is three times the zero-sequence current. Figure 4.16 illustrates the sequence networks for a generator. The zero-sequence network carries only zero-sequence current in one phase which has an impedance of $Z_0 = 3Z_n + Z_{e0}$.

The voltage and current components for each phase are obtained from the equations given for the sequence networks. The equations for the components of voltage, corresponding to the a phase of the system, are obtained from the point a on phase a relative to the reference busbar and can be deduced from Figure 4.16 as follows:

$$V_{a1} = E_a - I_{a1}Z_1$$

$$V_{a2} = -I_{a2}Z_2$$

$$V_{a0} = -I_{a0}Z_0$$

where E_a is the no-load voltage to earth of the positive-sequence network, Z_1 is the positive-sequence impedance of the generator, Z_2 is the negative-sequence impedance of the generator, Z_0 is the zero-sequence impedance of the generator (Z_{g0}) plus three times the impedance to earth.

The above equations can be applied to any generator that carries unbalanced currents and are the starting point for calculations for any type of fault. The same approach can be used with equivalent power systems or to loaded generators, E_a then being the voltage behind the reactance before the fault occurs.

4.4.2 Calculation of asymmetrical faults using symmetrical components

The positive-, negative-, and zero-sequence networks, carrying currents I_1, I_2, and I_0, respectively, are connected together in a particular arrangement to represent a given unbalanced fault condition. Consequently, in order to calculate fault levels using the method of symmetrical components, it is essential to determine the individual sequence impedances and combine these to make up the correct sequence networks. Then, for each type of fault, the appropriate combination of

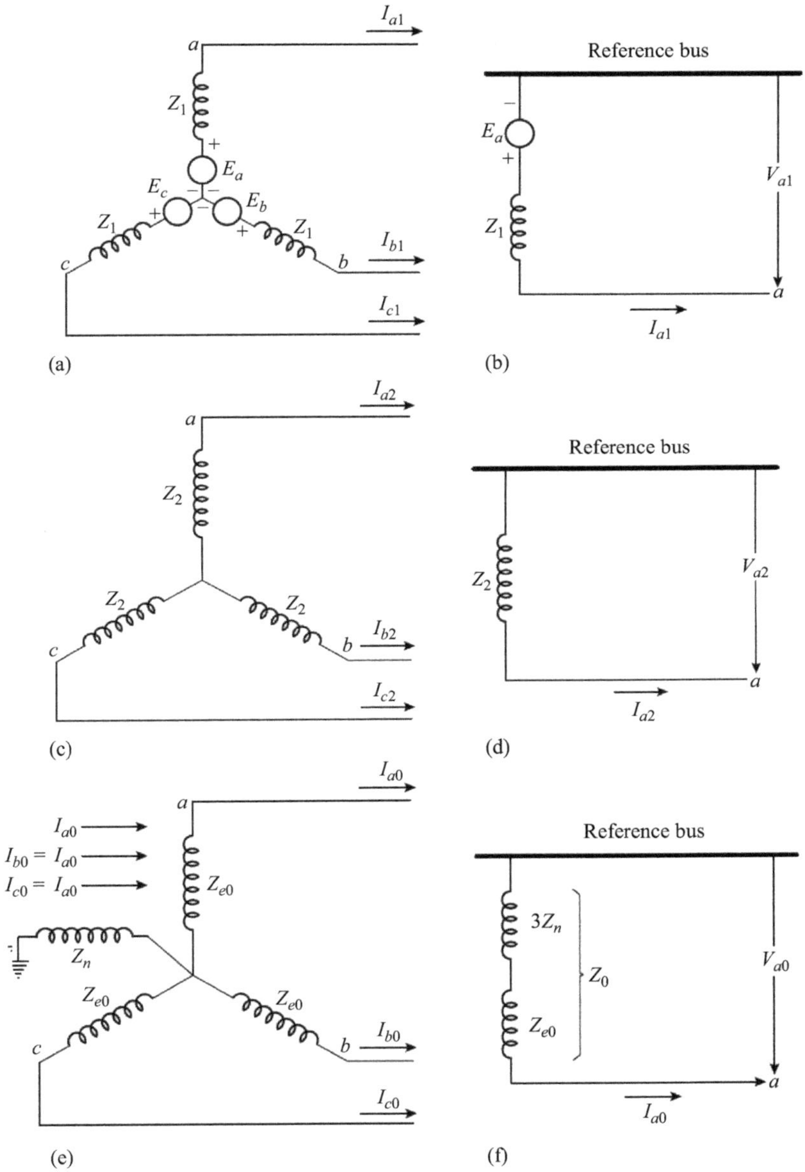

Figure 4.16 Representation of sequence impedances: (a) 3-phase positive-sequence; (b) single phase positive-sequence equivalent; (c) 3-phase negative-sequence; (d) single phase negative-sequence equivalent; (e) 3-phase zero-sequence; and (f) single phase zero-sequence equivalent

sequence networks is formed in order to obtain the relationships between fault currents and voltages.

4.4.2.1 Line-to-earth fault

The conditions for a solid fault from line a to earth are represented by the equations $I_b = 0$, $I_c = 0$, and $V_a = 0$. As in the previous equations, it can easily be deduced that $I_{a1} = I_{a2} = I_{a0} = E_a/(Z_1 + Z_2 + Z_0)$. Therefore, the sequence networks will be connected in series, as indicated in Figure 4.17(a). The current and voltage conditions are the same when considering an open-circuit fault in phases b and c, and thus the treatment and connection of the sequence networks will be similar.

4.4.2.2 Line-to-line fault

The conditions for a solid fault between lines b and c are represented by the equations $I_a = 0$, $I_b = -I_c$, and $V_b = V_c$. Equally it can be shown that $I_{a0} = 0$ and $I_{a1} = E_a/(Z_1 + Z_2) = -I_{a2}$. For this case, with no zero-sequence current, the zero-sequence network is not involved and the overall sequence network is composed of the positive- and negative-sequence networks in parallel as indicated in Figure 4.17(b).

4.4.2.3 Line-to-line-to-earth fault

The conditions for a fault between lines b and c and earth are represented by the equations $I_a = 0$ and $V_b = V_c = 0$. From these equations it can be proved that

$$I_{a1} = \frac{E_a}{Z_1 + (Z_0 Z_2/(Z_0 + Z_2))}$$

The three sequence networks are connected in parallel as shown in Figure 4.17(c).

4.4.3 *Equivalent impedances for a power system*

When it is necessary to study the effect of any change on the power system, the system must first of all be represented by its corresponding sequence impedances. The equivalent positive- and negative-sequence impedances can be calculated directly from

$$Z = V^2/P$$

where Z is the equivalent positive- and negative-sequence impedances, V is the nominal phase-to-phase voltage, P is the three-phase short circuit power.

The equivalent zero-sequence of a system can be derived from the expressions of sequence components referred to for a single-phase fault, i.e., $I_{a1} = I_{a2} = I_{a0} = V_{LN}/(Z_1 + Z_2 + Z_0)$, where V_{LN} = the line-to-neutral voltage.

For lines and cables the positive- and negative-sequence impedances are equal. Thus, on the basis that the generator impedances are not significant in most distribution network fault studies, it may be assumed that overall $Z_2 = Z_1$, which simplifies the calculations. Thus, the above formula reduces to $I_a = 3I_{a0} = 3V_{LN}/(2Z_1 + Z_0)$, where V_{LN} = line to neutral voltage and $Z_0 = (3V_{LN}/I_a) - 2Z_1$.

4.4.4 *Supplying the current and voltage signals to protection systems*

In the presence of a fault the current transformers circulate current proportional to the fault current to the protection equipment without distinguishing between the

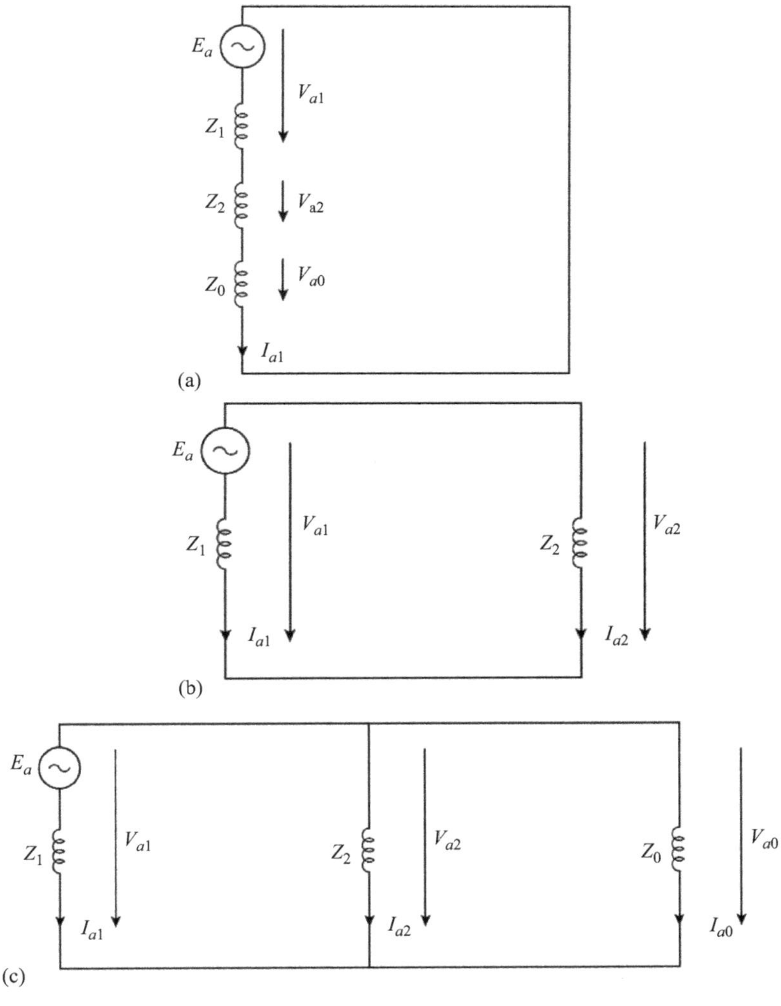

Figure 4.17 Representation of sequence networks for (a) Line-to-earth fault; (b) Line-to-line fault; and (c) Line-to-line-to-earth fault

vectorial magnitudes of the sequence components. Therefore, in the majority of cases, the relays operate on the basis of the corresponding values of fault current and/or voltages, regardless of the values of the sequence components. It is very important to emphasize that, given this, the advantage of using symmetrical components is that they facilitate the calculation of fault levels even though the relays in the majority of cases do not distinguish between the various values of the symmetrical components.

In Figure 4.18, the positive- and negative-sequence values of current and voltage for different faults are shown together with the summated values of current

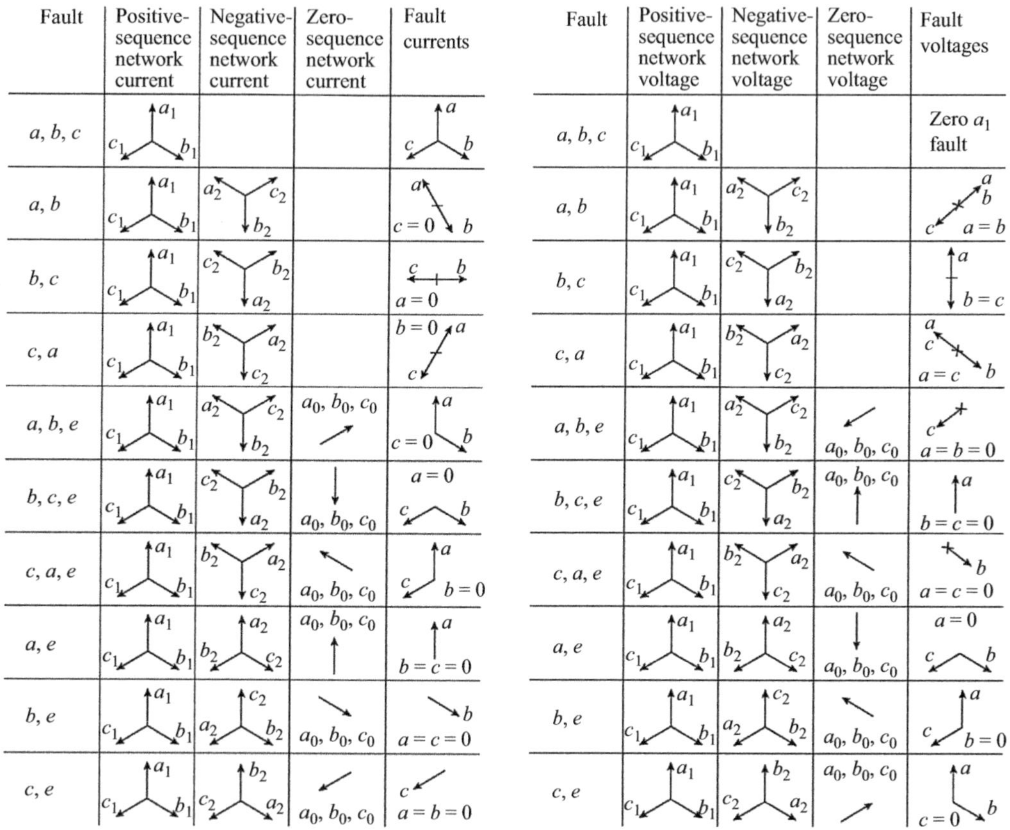

(a) (b)

Figure 4.18 Sequence currents and voltages for different types of faults and voltage. Relays usually only operate using the summated values in the right-hand columns. However, relays are available, which can operate with specific values of some of the sequence components. In these cases there must be methods for obtaining these components, and this is achieved by using filters which produce the mathematical operations of the resultant equations to resolve the matrix for voltages and for currents. Although these filters can be constructed for electromagnetic elements, the growth of electronics has led to their being used increasingly in logic circuits. Amongst the relays that require this type of filter in order to operate are those used in negative-sequence and earth-fault protection.

Example 4.4 For the power system shown in Figure 4.19 perform the following:

- Calculate the fault currents for (i) symmetrical three-phase fault and (ii) single line-to-ground (SLG) fault at BUS C.
- For the three-phase fault, calculate the positive sequence contributions from G_1 and G_2, and for the SLG fault, calculate zero-sequence contributions from G_1 and G_2.

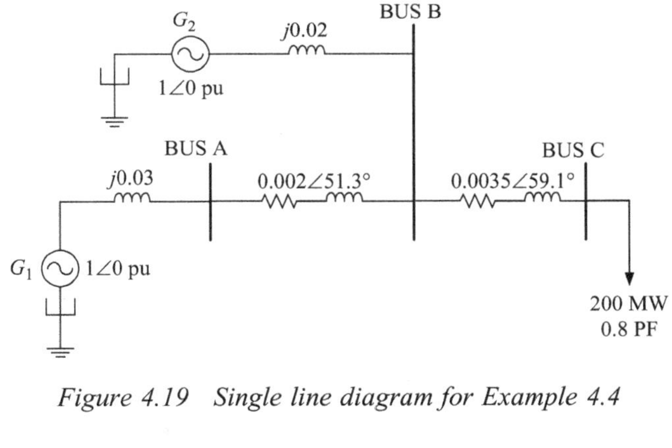

Figure 4.19 Single line diagram for Example 4.4

Notes:

- All the impedances shown in the above figure are in per unit (pu) values expressed in the bases of 230 kV and 100 MVA.
- For lines (branches between BUS A and BUS B and also between BUS B and BUS C), the zero-sequence impedance (Z_0) is 2.5 times the positive sequence impedance (Z_1).
- For generators (source impedances behind BUSes A and B), both the zero-sequence impedance (Z_0) and positive-sequence impedance (Z_1) are equal.
- G_1 and G_2 are perfect voltage sources generating 1.0 pu voltage with zero source impedances. These voltage sources and their series impedances are Thévenin equivalent sources behind BUSes A and B, respectively.

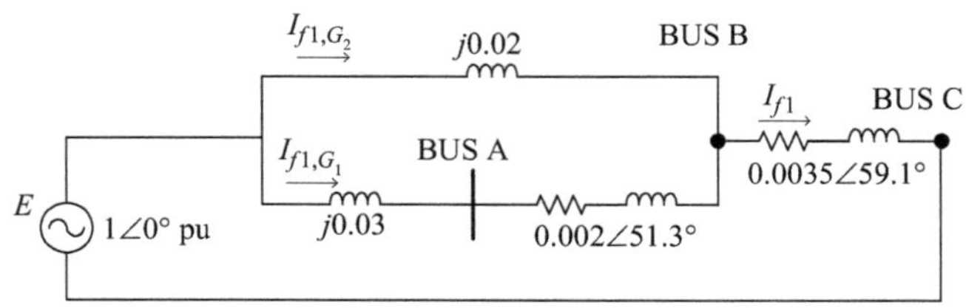

Figure 4.20 Positive sequence diagram for Example 4.4

1. Calculations for symmetrical *L–L–L* fault at BUS C

   The fault impedance is neglected here. The positive sequence network for *L–L–L* fault calculations at BUS C is shown in Figure 4.20.

   Positive sequence fault current for symmetrical *L–L–L* fault would be

$$I_{f1} = \frac{E}{Z_{eq}} = \frac{1\angle 0}{0.01538\angle 89.5822^\circ} = 65.0251\angle -82.5822 \text{ pu.}$$

Using current divider rule, positive sequence fault current contributions for G_1 and G_2 would be

$$\begin{aligned} I_{f1,G_1} &= \frac{j0.02}{j0.02 + 0.03158\angle 87.73} \cdot 65.0251\angle -82.5822 \text{ pu} \\ &= 25.2153\angle -81.1929 \text{ pu} \end{aligned}$$

$$I_{f1,G_2} = I_{f1} - I_{f1,G_1} = 39.822\angle -83.4619 \text{ pu}$$

Fault currents in three phases,

$$\begin{aligned} I_{base} &= \frac{100 \times 10^6}{\sqrt{3} \cdot 230 \times 10^3} = 251.022 \text{ A} \\ I_{f,A} &= (65.0251\angle -82.5822 \text{ pu})(251.022) \\ &= 16322.7\angle -82.5822 \text{ A} \end{aligned}$$

Similarly, for the other two phases,

$$\begin{aligned} I_{f,B} &= 16{,}322.7\angle -202.582 \text{ A} \\ I_{f,C} &= 16{,}322.7\angle -37.4178 \text{ A} \end{aligned}$$

2. The calculations for *S–L–G* fault at BUS C are done by connecting the three sequence diagrams in series as indicated in Figure 4.21.

$$I_{f1} = I_{f2} = I_{f0} = \frac{E}{Z_{eq1} + Z_{eq2} + Z_{eq0}}$$

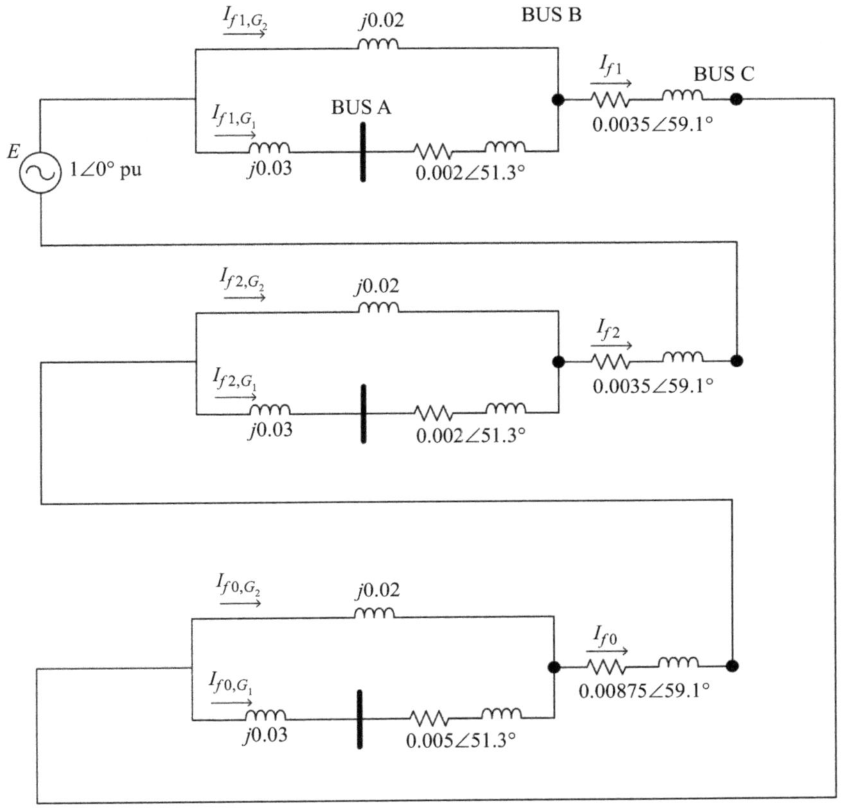

Figure 4.21 Sequence network connection for a single-phase fault of Example 4.4

Note:

For transmission lines, the zero-sequence impedance (Z_0) is 2.5 times the positive-sequence impedance (Z_1), whereas, for equivalent generators, both the zero-sequence impedance (Z_0) and positive-sequence impedance (Z_1) are equal.

$$\begin{aligned}
Z_{eq1} &= Z_{eq2} = [(j0.03 + 0.002\angle 51.3)\|j0.02] + 0.0035\angle 59.1 \\
&= 0.01538\angle 82.5822^\circ \text{ pu} \\
Z_{eq0} &= [(j0.03 + 0.005\angle 51.3)\|j0.02] + 0.00875\angle 59.1 \\
&= 0.02071\angle 76.2472^\circ \text{ pu}
\end{aligned}$$

Sequence fault currents for *S–L–G* fault would be

$$\begin{aligned}
I_{f1} = I_{f2} = I_{f0} &= \frac{E}{Z_{eq1} + Z_{eq2} + Z_{eq0}} \\
&= \frac{1\angle 0}{2(0.01538\angle 82.5822^\circ) + 0.02071\angle 76.2472^\circ} \\
&= 19.4574\angle -80.0338^\circ \text{ pu}
\end{aligned}$$

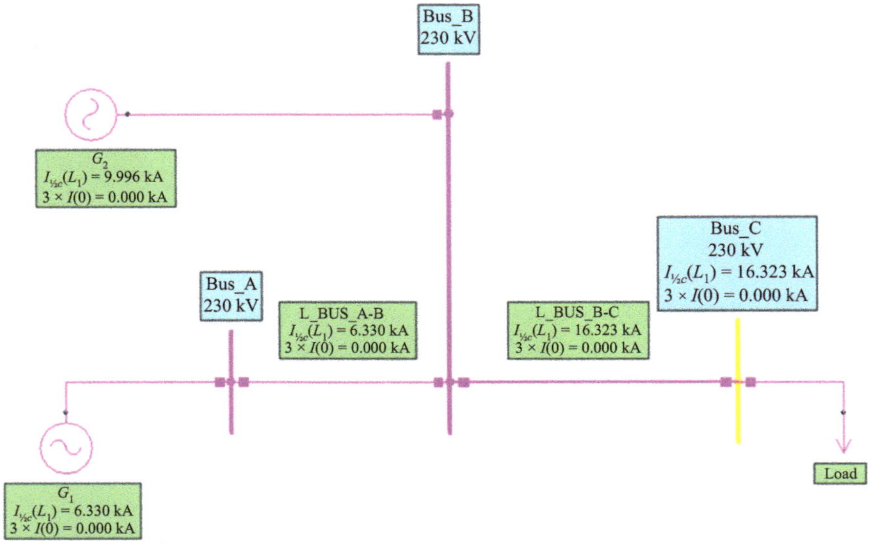

Figure 4.22 Solution of Example 4.4 by using a PSAT for a three-phase fault

Using current divider rule, zero-sequence fault current contributions for G_1 and G_2 would be

$$
\begin{aligned}
I_{f0,G_1} &= \frac{j0.02}{j0.02 + 0.03405\angle 84.7315} \cdot 19.4574\angle -80.0338^\circ \text{ pu} \\
&= 7.2074\angle -76.7145^\circ \text{ pu} \\
I_{f0,G_2} &= I_{f0} - I_{f0,G_1} = (19.4574\angle -80.0338^\circ) \\
&\quad - (7.2074\angle -76.7145^\circ) = 12.2692\angle -81.983 \text{ pu}
\end{aligned}
$$

Fault currents in three phases,

$$
\begin{aligned}
I_{base} &= \frac{100 \times 10^6}{\sqrt{3} \cdot 230 \times 10^3} = 251.022 \text{ A} \\
I_{f,A} &= I_{f1} + I_{f2} + I_{f0} = 3(19.4574\angle -80.0338^\circ \text{ pu}) \\
&= 58.3722\angle -80.0338^\circ \text{ pu} \\
I_{f,A} &= (58.3722\angle -80.0338^\circ \text{ pu}) \cdot (251.022) \\
&= 14{,}652.71\angle -80.0338 \text{ A} \\
I_{f,B} &= I_{f,C} = 0 \text{ A}
\end{aligned}
$$

The results for the three-phase fault by using a PSAT are shown in Figure 4.22.

Example 4.5 For Figure 4.23, calculate the short circuit currents for a three-phase faults at busbar 115 kV The Ridges. Notice that the plant connected at South Post is considered out for maintenance.

Figure 4.23 shows the reduced diagram when the three-phase fault occurred. Only the equivalent system before bus Black River 230 kV, the three-winding transformers between Black River 230 kV bus and Black River 115 kV bus, and the lines between Black River 115 kV bus and The Ridges 115 kV bus have effect on this fault.

Considering that $S_{base} = 100$ kV and $V_{base} = 230$ kV, for Thévenin equivalent of the system:

$$S_{cc} = 2{,}514.9 \text{ MVA}$$

$$Z_{cc} = \frac{{V_{BR230}}^2}{S_{cc}} = \frac{(230 \text{ kV})^2}{2{,}514.9 \text{ MVA}} = 21.035\ \Omega$$

$$Z_{Thevenin\_pu} = \frac{Z_{cc}}{Z_{Thevenin\_base}} = \frac{21.035\ \Omega}{(230 \text{ kV}^2/100 \text{ MVA})} = 0.0398 \text{ pu}$$

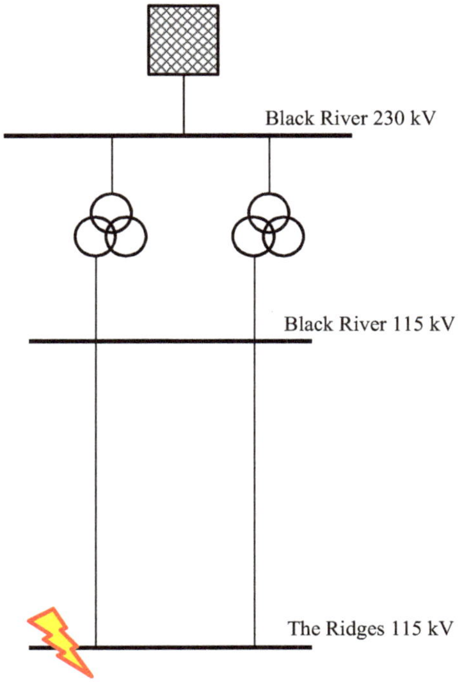

Figure 4.23 Reduced diagrams for the fault analysis in Example 4.5

For the transformers,

$$Z_{Tr1} = Z_{Tr2} = Z_{Trpu}\frac{S_{new}}{S_{old}} = 0.1\text{ pu}\frac{100\text{ MVA}}{90\text{ MVA}} = 0.111\text{ pu}$$

$$Z_{Tr} = Z_{Tr1}||Z_{Tr2} = \frac{0.111\text{ pu}}{2} = 0.0556\text{ pu}$$

For the lines,

$$Z_{L_1} = Z_{L_2} = 0.4654\ \Omega/\text{km}$$

$$L_{L_1} = L_{L_2} = 12\text{ km}$$

$$Z_L = Z_{L_1}||Z_{L_2} = \frac{0.4654\ \Omega/\text{km}\cdot 12\text{ km}}{2} = 2.7924\ \Omega$$

$$Z_{L\text{pu}} = \frac{Z_L}{Z_{Lbase}} = \frac{2.7924\ \Omega}{115\text{ kV}^2/100\text{ MV A}} = 0.0211\text{ pu}$$

Figure 4.24 shows the diagram with all the elements involved in the three-phase fault on the bus The Ridges 115 kV.

The three-phase fault current in this case is

$$I_{3\varphi\text{pu}} = \frac{V_{3\varphi\text{pu}}}{Z_{Thevenin\_\text{pu}} + Z_{Tr1} + Z_{L\text{pu}}} = \frac{1}{0.0398 + 0.0556 + 0.0211} = 8.6\text{ pu}$$

$$I_{base} = \frac{S_{base}}{V_{base}} = \frac{100\text{ MV A}}{\sqrt{3}\cdot 115\text{ kV}} = 502.04\text{ A}$$

$$I_{3\varphi\text{pu}}\cdot I_{base} = 8.6\cdot 502.04\text{ A} = 4,309.2\text{ A}$$

The results are shown in Figure 4.25 using a software package. It can be seen the fault current distribution throughout the system.

4.5 Proposed exercises

1. A short transmission line of inductance 0.04 H and resistance 1.2 Ω is suddenly short-circuited at the receiving end, while the source voltage is: $V = 311.13$

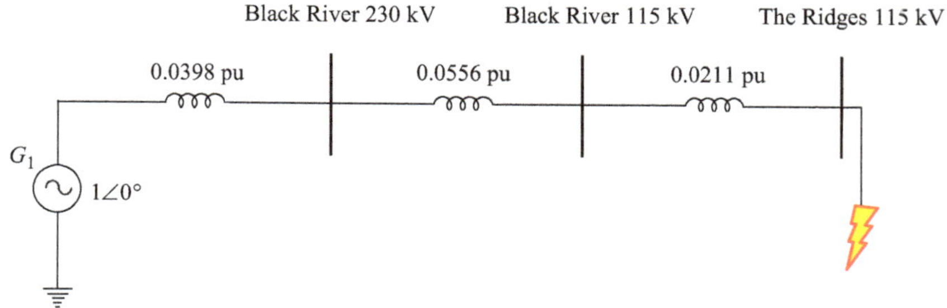

Figure 4.24 Impedance diagram for the three-phase fault in node The Ridges 115 kV

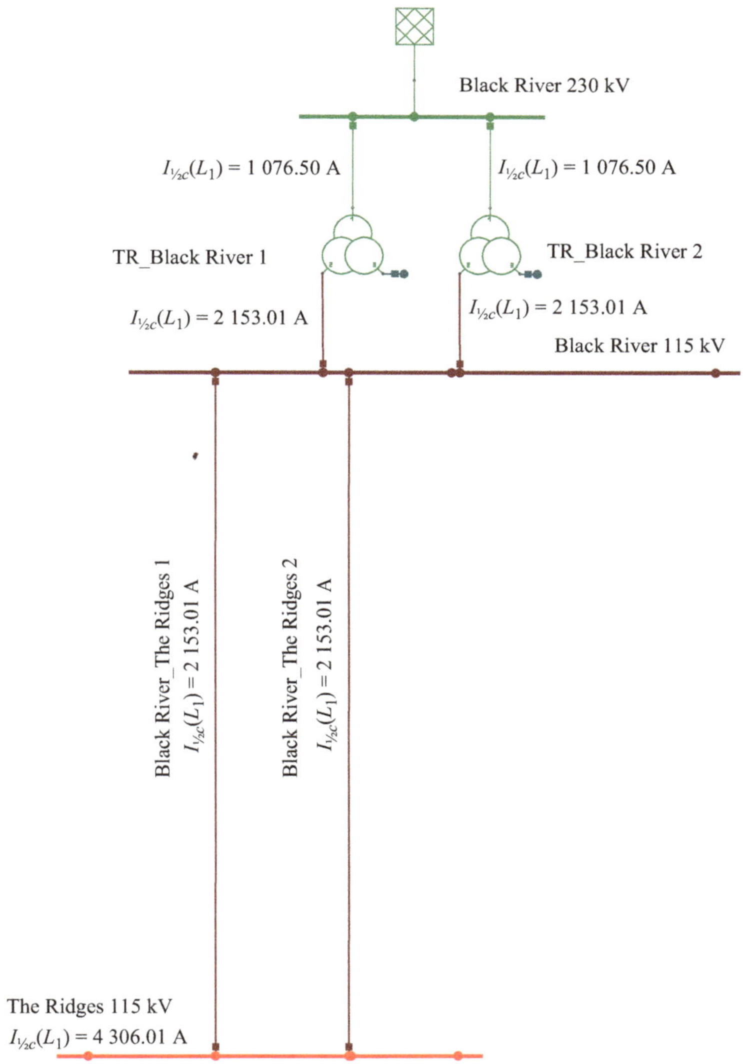

Figure 4.25 Solution of Example 4.5 by using a software package

sin(377*t* − 30°). Calculate the instant of the short circuit fault where the dc offset will be zero and maximum.

2. For the power system in Figure 4.26, form sequence impedance networks (including the zero-sequence network). Assume that loads do not contribute to the short circuit currents and use a common base of 120 MV A.
3. For Figure 4.27 convert the sequence impedance networks to single impedances as seen from the fault point. Use the following numerical values on a

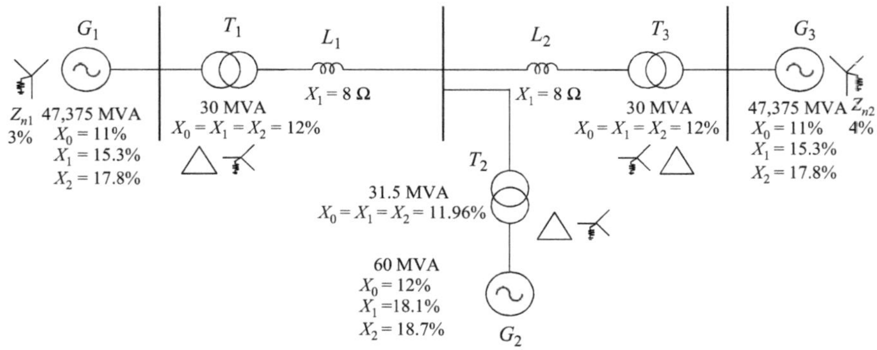

Figure 4.26 Diagram for exercise 2

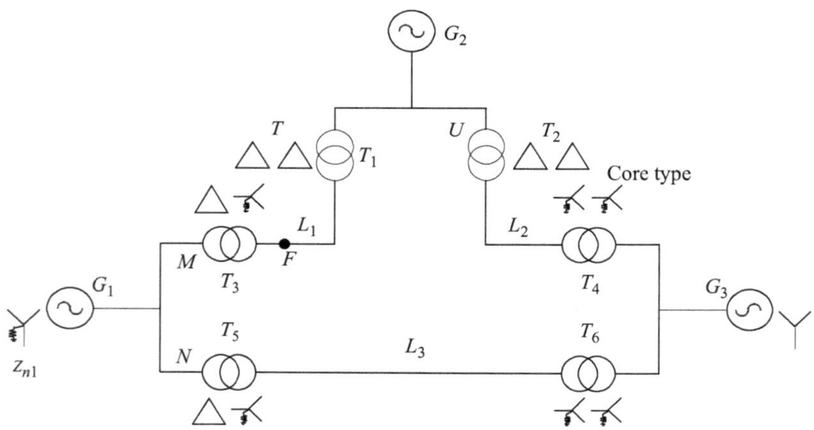

Figure 4.27 Diagram for exercise 3

per unit basis (all on a common MV A base). Neglect resistances and ignore the load currents.

Generators G_1, G_2, and G_3: $Z_1 = 0.158$, $Z_2 = 0.161$, $Z_0 = 0.06$, Z_n (neutral grounding impedance) $= 0.22$;

Transmission lines L_1, L_2, and L_3: $Z_1 = 0.15$, $Z_2 = 0.15$;

Transformers T_1, T_2, T_3, T_4, T_5, and T_6: $Z_1 = Z_2 = 0.0563$, transformer T_3: $Z_0 = 0.0563$

4. For the industrial system of Figure 4.28 specify the sequence network when a SLG fault occurs at the secondary of the transformer terminals, and calculate the fault current. Consider a load operating power factor of 0.879 and an overall efficiency of 93%.

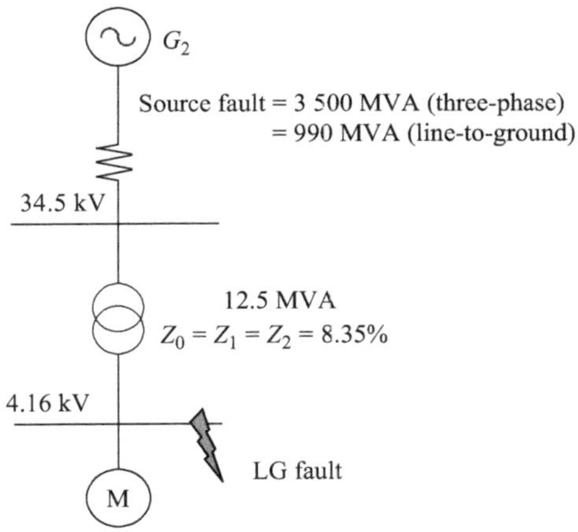

Figure 4.28 Diagram for exercise 4

References

[1] IEEE. IEEE Std. C37.010-2016 – Application Guide for AC High Voltage Circuit Breakers > 1000 Vac Rated on a Symmetrical Current Basis (Revision of IEEE Std C37.010-1999). 2016.

[2] IEC. IEC 60909 Series: Short-Circuit Currents in Three-Phase a.c. Systems, 2016.

[3] IEEE. IEEE Std. 399-1997 – IEEE Recommended Practice for Industrial and Commercial Power Systems Analysis (Brown Book), 1997.

[4] IEEE. IEEE Std. 551-2006 – IEEE Recommended Practice for Calculating AC Short-Circuit Currents in Industrial and Commercial Power Systems (Violet Book), 2006.

[5] ANSI/IEEE. IEEE Std. C37.5-1979 – IEEE Guide for Calculation of Fault Currents for Application of AC High-Voltage Circuit Breakers Rated on a Total Current Basis, 1979.

[6] Fortescue, C. L. "Method of symmetrical co-ordinates applied to the solution of poly-phase networks (with discussion)," in Presented at the 34th Annual Convention of the AIEE (American Institute of Electrical Engineers), Atlantic City, NJ, USA, 28 June 1918; vol. 37, pp. 1027–1140.

Further reading

[1] Anderson, P. M. Analysis of Faulted Power Systems. The Iowa State University Press, Ames, IA, USA, 1995.

[2] Blackburn, J. I. Protective Relaying Principles and Applications, 3rd ed. CRC Press, Boca Raton, FL, 2008.

[3] Burnett, J. "IDMT relay tripping of main incoming circuit breakers," Power Engineering Journal, vol. 4, pp. 51–56, 1990.
[4] Cooper Power Systems. Electrical Distribution System Protection Manual, 3rd ed., Cooper Power Systems, 1990.
[5] Elmore, W. A. Protective Relaying Theory and Applications. ABB, New York, NY, USA, 1994.
[6] Glover, J. D. and Sarma, M. S. Power System Analysis and Design, 3rd ed. Brooks/Cole, Pacific Grove, CA, USA, 2002.
[7] Gonen, T. Electric Power Distribution System Engineering, 2nd ed. CRC Press, Boca Raton, FL, 2008.
[8] Gross, C. A. Power System Analysis, 2nd ed. John Wiley & Sons, New York, NY, USA, 1986.
[9] Short, T. A. Electric Power Distribution Handbook. CRC Press, Boca Raton, FL, USA, 2004.

Chapter 5
Reliability of distribution systems

Knowing that the most important benefit of a Smart Grids is the increase in the reliability indices, a proper examination of this concept is important. The term "reliability" refers to the notion that the system performs its specified task correctly for a certain duration of time. The term "availability" refers to the readiness of a system or component to be immediately ready to perform its task. Both terms have precise definitions within the reliability engineering field and have specified equations and methods to provide quantitative metrics for them.

Availability is the probability of something being energized. It is the most basic aspect of reliability and is typically measured in percent or per-unit. The complement of availability is unavailability.

Unavailability can be computed directly from interruption duration information. If a customer experiences 24 h of interrupted power in a year, unavailability is equal to $24/8{,}760 = 0.00274 = 0.274\%$. Availability is equal to $100\% - 0.2739\% = 99.7260\%$.

Each distribution system component can be described by a set of reliability parameters. Simple reliability models are based on component failure rates and component repair times, but sophisticated models make use of many other reliability parameters. Some of the most common reliability parameters are the following:

- Failure rate, indicated by the symbol λ, describes the number of times per year that a component can expect to experience a failure.
- Mean time to repair (MTTR), indicated by the letter r, represents the expected time it will take for a failure to be repaired (measured from the time that the failure occurs). A single value of MTTR is typically used for each component, but separate values can be used for different failure modes.
- Probability of operational failure (POF) is the conditional probability that a device will not operate if it is supposed to operate. For example, if an automated switch fails to function properly five times out of every 100 attempted operations, it has a POF of 5%. This reliability parameter is typically associated with switching devices and protection devices.

5.1 Network modeling

The analysis of reliability in physical networks is done by using the concepts of series and parallel circuits of electrical systems, which are clear and flexible tools to handle. The solutions involve simple mathematical methods that represent accurately the models under analysis.

Network modeling is a component-based technique rather than a state-based technique. Each component is described by a probability of being available, P, and a probability of not being available, Q. Since components are assumed to be either available or not available, Q and P are arithmetic complements: $Q = 1 - P$. Care has to be exercised in reliability studies since the symbol P represents the non-supplied power, especially in software packages.

If a component is described by an annual failure rate λ and an MTTR r in hours, the probability of being available can be computed as follows:

$$P = \frac{8{,}760 - \lambda r}{8{,}760} \tag{5.1}$$

Therefore,

$$Q = \frac{\lambda r}{8{,}760} \tag{5.2}$$

As mentioned earlier, networks are composed of series and parallel components. Two components are in series if it is required that both are available. Two components are in parallel if only one of the components is required.

For a series network, the probability of an available path is equal to the product of the individual component availabilities. For a parallel network, the probability of an unavailable path is equal to the product of the individual component's unavailability. Components in series reduce availability and components in parallel improve availability:

$$P_{series} = \prod P_{component} \tag{5.3}$$

$$Q_{parallel} = \prod Q_{component} \tag{5.4}$$

The failure rate λ is a measure of unreliability. The product λr (failure rate $\times$ average downtime per failure) is equal to the forced downtime hours per year and it can be considered as a measure of forced unavailability since a scale factor of 8,760 converts one value to the other. The average downtime per failure r could be called restorability. The procedure and formulas for calculating the reliability indices are given in IEEE Std. 493-2007 [1] and are reproduced in (5.5)–(5.10). A schematic demonstrating these formulas is shown in Figure 5.1 for two components numbered "1" and "2" connected in series and in Figure 5.2 for two components "3" and "4" connected in parallel.

In these schematics, scheduled outages are assumed to be zero and the units for f, λ, and r are, respectively, frequency of failures for the whole system, failure rate

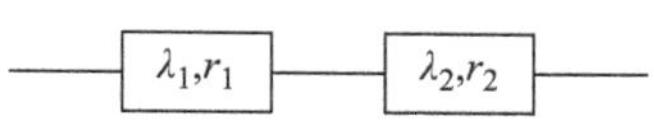

Figure 5.1 Repairable components in series—both must work for success [1]

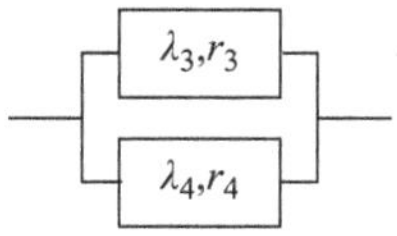

Figure 5.2 Repairable components in parallel—one or both must work for success [1]

per year of each component, and hours of downtime per failure. The equations in Figures 5.1 and 5.2 assume the following:

1. The component failure rate is constant with age.
2. The outage time after a failure has an exponential distribution.
3. Each failure event is independent of any other failure event.

$$f_s = \lambda_1 + \lambda_2 \tag{5.5}$$

$$f_s r_s = \lambda_1 r_1 + \lambda_2 r_2 \tag{5.6}$$

$$r_s \cong \frac{\lambda_1 r_1 + \lambda_2 r_2}{\lambda_1 + \lambda_2} \tag{5.7}$$

$$f_p = \frac{\lambda_3 \lambda_4 (r_3 + r_4)}{8{,}760} \tag{5.8}$$

$$f_p r_p = \frac{\lambda_3 r_3 \lambda_4 r_4}{8{,}760} \tag{5.9}$$

$$r_p = \frac{r_3 r_4}{r_3 + r_4} \tag{5.10}$$

Example 5.1 Figure 5.3 shows a power system with two lines, each one with three failures per year and 0.75 h downtime per failure. It has a load/total load/load per line of 20 MW. Calculate the frequency of failures, the minutes per year of unreliability, and the non-supplied power. First assume that each line can support the whole load; then calculate again, assuming that each line cannot support the total load.

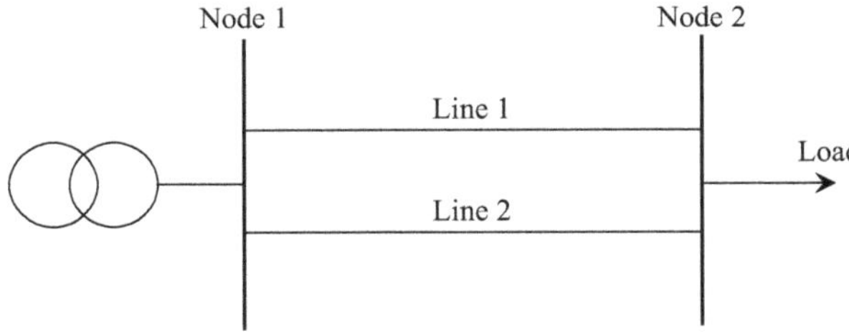

Figure 5.3 Diagram system for Example 5.1

---

1. For the first situation, where each line can support the whole load, it is valid to assume a parallel connection.

$$f_p = \frac{3 \times 3 \times (0.75 + 0.75)}{8{,}760} = 0.0015411/\text{year}$$
$$r_p = \frac{0.75 \times 0.75}{0.75 + 0.75} = 0.375\ \text{h}$$
$$Q_p = f_p \times r_p = 0.0015411 \times 0.375 = 0.000578\ \text{h/year}$$
$$= 0.03467\ \text{min/year}$$

or

$$Q_p = Q_1 \times Q_2 = (Q_1)^2 = (135\ \text{min/year})^2 = (0.0002568)^2$$
$$= 0.03467\ \text{min/year}$$

The non-supplied power per year is

$$p_p = p_{load} \times f_p = 0.031\ \text{MW/year}$$

These results were validated with power system calculation software. The result is shown in Figure 5.4.

2. For the second case where it is assumed that each individual line cannot support the total load, the reliability analysis is developed by treating the system as a series connection, although both lines are physically connected in parallel.

$$f_s = 3 + 3 = 6/\text{year}$$
$$r_s \cong \frac{3 \times 0.75 + 3 \times 0.75}{3 + 3} = 0.75\ \text{h}$$
$$Q_s = f_s \times r_s = 6 \times 0.75 = 4.5\ \text{h/year} = 270\ \text{min/year}$$

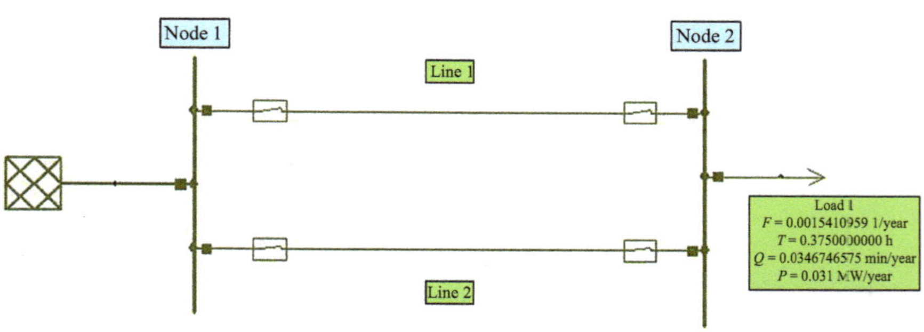

Figure 5.4 Diagram system results for the first part of Example 5.1

or

$$1 - Q_s = P_s = P_1 \times P_2 = {P_1}^2 = (1 - Q_1)^2$$

$$Q_s = 1 - (1 - Q_1)^2 = 1 - (1 - 3 \times 0.75/8{,}760)^2 = 1 - (1 - 0.0002568)^2$$

$$= 270 \text{ min/year}$$

The non-supplied power is

$$p_s = p_{load} \times f_s = 20 \times 6 = 120 \text{ MW/year}$$

These results were validated with power system calculation software. The result is shown in Figure 5.5.

5.2 Network reduction

Complex systems have a high number of series and parallel connections. To process them, a network reduction is required. This is accomplished by simplifying parallel and series components into equivalent network components until a single component is obtained. The availability of the last component is equal to the availability of the original system. An example of network reduction is shown in Figure 5.6.

In this system, a line section (component 1) provides power to a distribution transformer (component 2). The distribution transformer can also receive power from an alternate path consisting of three line sections in series (components 3, 4, and 5). Three applications of network reduction are required to compute the availability of the system.

A good approach for network reduction is the so-called minimal cut set method. In this method, it is required to determine the minimal number of elements needed to keep the system running.

Figure 5.7 shows an example of a portion of a network that has five switches. There are five scenarios, each corresponding to a minimal cut set. Each one requires only three switches at a time for the system to keep running. It is clear that

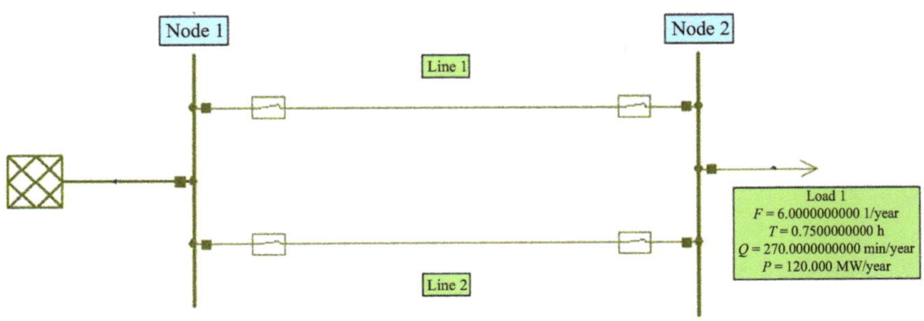

Figure 5.5 Diagram system results for the second part of Example 5.1

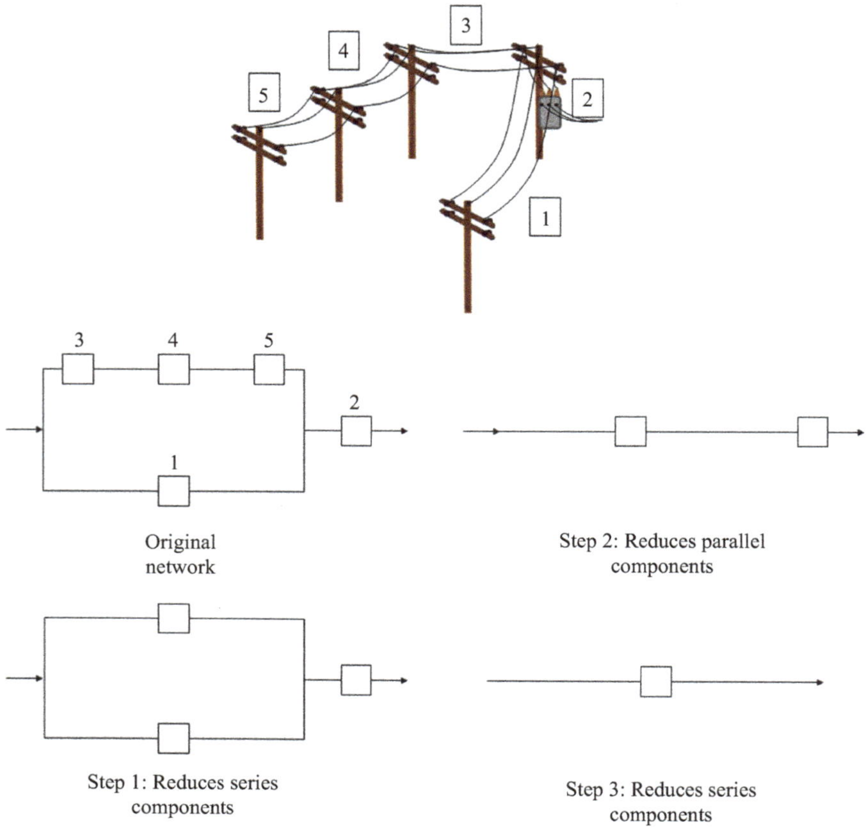

Figure 5.6 An example of network reduction

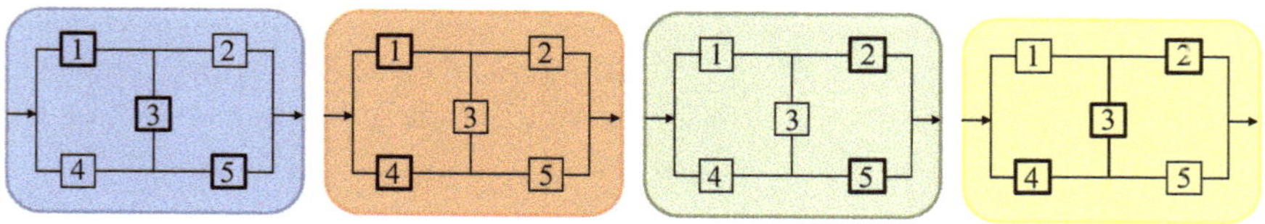

Figure 5.7 Minimal cut sets of a simple system

the unavailability of the system is equal to the sum of the unavailability of these minimal cut sets.

5.3 Quality indices

Quality indices, also called reliability indices, measure the performance of a power system. The most important benefit of Smart Grids is that they improve reliability performance of power systems. Given the importance of these indices to enhance a Smart Grid, a proper consideration to their meaning and application is explained in this section.

Several key definitions relating to distribution reliability include the following:

- Fault: An abnormal operating condition of an electrical system that normally develops a short circuit. It can be caused by natural events, rough weather conditions, animal presence, equipment failure, and even vandalism. Faults can be categorized as self-clearing, temporary and permanent. A self-clearing fault will extinguish itself without any external intervention. A temporary fault will clear if de-energized and then reenergized. A permanent fault lasts until repaired by human intervention.
- Contingency: An unexpected event such as a fault or an open circuit. Another term for a contingency is an unscheduled event.
- Outage: An outage occurs when a piece of equipment is de-energized. Outages can be either scheduled or unscheduled. Scheduled outages are known in advance (e.g., outages for periodic maintenance). Unscheduled outages result from contingencies. Different utilities have different criteria to define an outage. Some utilities consider outages those interruptions that exceed 1 min. Others use 2 min and other up to 5 min.
- Open circuit: A point in a circuit that interrupts load current without causing fault current to flow. An example of an open circuit is the false tripping of a circuit breaker.
- Momentary interruption: A momentary interruption occurs when a customer is de-energized for just a few minutes. Most momentary interruptions result from reclosing or automated switching. Multiple reclosing operations result in multiple momentary interruptions.
- Sustained interruption: A sustained interruption occurs when a customer is de-energized for more than a few minutes. Most sustained interruptions result from open circuits and faults.

Poor reliability on the part of the electrical utility is penalized or rewarded based on quantification by reliability indices. Some utilities also pay bonuses to utility personnel, based in part on indices. Commercial and industrial customers enquire about reliability indices when locating a new facility. Most regulatory bodies have established targets for quality indices. If utilities do not fulfill them (have figures higher than those defined), they can be penalized. The two main indices are the system average interruption duration index (SAIDI) and system average interruption frequency index (SAIFI) which are explained in the following paragraphs.

SAIDI (system average interruption duration index)

SAIDI is defined as the average interruption duration for customers served during a specified time period. It is calculated by summing the customer-minutes off for each interruption during a specified time period and dividing the sum by the average number of customers served during the period. The unit is minutes.

The index enables the utility to report the time (normally in minutes) customers would have been out of service if all customers were out at one time. The benchmarking survey SAIDI average for the USA is around 90 min:

$$SAIDI = \frac{\sum \text{Customer interruption durations}}{\text{Total number of customers served}} \Big\} \text{min/year} \quad (5.11)$$

SAIFI (system average interruption frequency index)

SAIFI is defined as the average number of times that a customer is interrupted during a specified time period. It is calculated by dividing the total number of customers interrupted in a time period by the average number of customers served. (This seems like outages per customer, not the number of interruptions.)

The index is calculated for both interruptions of more than 1 min (SAIFI-long) and less than 1 min (SAIFI-short). The resulting unit is "interruptions per customer." The benchmarking survey SAIFI average for the USA is around 1.2 interruptions per customer per year:

$$SAIFI = \frac{\text{Total number of customer interruptions}}{\text{Total number of customers served}} \Big/ \text{year} \quad (5.12)$$

CAIDI (customer average interruption duration index)

Customer average interruption duration index (CAIDI) is the ratio of the SAIDI over the SAIFI and is defined with the following expression:

$$CAIDI = \frac{\sum \text{Customer interruption durations}}{\text{Total number of customer interruptions}} \text{h} \quad (5.13)$$

Example 5.2 Table 5.1 shows an excerpt from one utility's customer information system database for feeder 25, which serves 1,500 customers with a total load of 3 MW. In this example, feeder 25 constitutes the "system" for which the indices are calculated. More typically the "system" combines all circuits together in a region or for a whole company.

Table 5.1 Example of outage data

Date	**Time**	**Time on**	**Total time (min)**	**Circuit**	**Event code**	**Number of customers**	**Load (kV A)**	**Interruption type**
2/15	14:10:17	14:25:39	15.37	25	51	250	431	S
4/20	17:20:39	17:38:41	18.03	25	306	368	800	S
5/1	06:33:36	07:14:49	41.22	25	468	23	150	S
6/2	23:18:10	23:18:57	0.78	25	522	590	1,200	M
6/8	02:39:52	03:55:34	15.7	25	634	87	200	S
9/29	09:29:05	09:30:02	0.95	25	811	1,500	3,000	M
11/14	17:15:49	17:16:18	0.48	25	963	700	1,500	M
12/4	12:16:32	13:01:44	45.2	25	1,021	1,000	1,800	S

"S" indicates a sustained interruption, and "M," a momentary interruption. Applying the previous equation, it is possible to find these quality indexes.

$$SAIFI = \frac{250 + 368 + 23 + 87 + 1{,}000}{1{,}500} = 1.152/\text{year}$$

$$SAIDI = \frac{(15.37 \times 250) + (18.03 \times 368) + (41.22 \times 23) + (15.7 \times 87) + (45.2 \times 1{,}000)}{1{,}500} = 38.66\ \text{min/year}$$

$$CAIDI = \frac{SAIDI}{SAIFI} = \frac{38.66}{1.152} = 33.56\ \text{min}$$

Momentary average interruption frequency index (MAIFI) and momentary average interruption event frequency index (MAIFIE)

The momentary average interruption frequency index (MAIFI), as defined by IEEE Std 1366-2003 [2], indicates the average frequency of momentary interruptions. Mathematically, this is given in the following equation:

$$MAIFI = \frac{\sum \text{Total number of customer momentary interruptions}}{\text{Total number of customers served}} \tag{5.14}$$

The momentary average interruption event frequency index (MAIFIE) as defined by the same IEEE Standard indicates the average frequency of momentary interruption events, not including the events immediately preceding a lockout. Mathematically, this is given the following equation:

$$MAIFIE = \frac{\sum \text{Total number of customer momentary interruption events}}{\text{Total number of customers served}} \tag{5.15}$$

To calculate these indices, the following equations can be used:

$$MAIFI = \frac{\sum IM_i N_{mi}}{N_T} \tag{5.16}$$

$$MAIFIE = \frac{\sum IM_E N_{mi}}{N_T} \tag{5.17}$$

where IM_i is the number of momentary interruptions, IM_E is the number of momentary interruption events, N_{mi} is the number of interrupted customers for each momentary interruption event during the reporting period, N_T is the total number of customers served for the areas.

Example 5.3 To better illustrate the concepts of momentary interruptions, sustained interruptions, and the associated indices, consider Figure 5.8. The figure illustrates a feeder composed by a recloser and a sectionalizer.

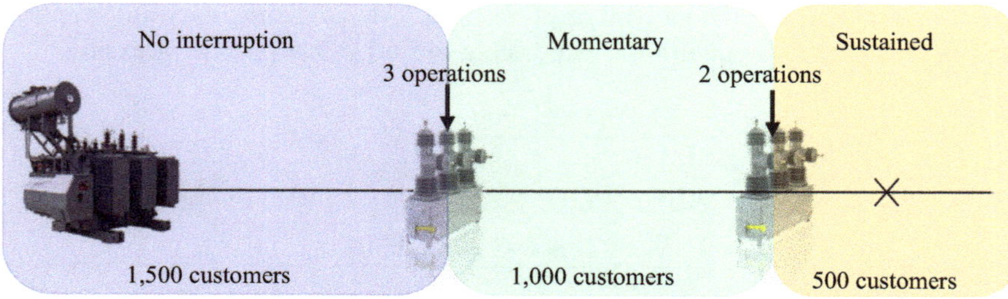

Figure 5.8 Representation of events used in calculating indices

For this scenario, 1,000 customers would experience a momentary interruption, and 500 customers would experience a sustained interruption. Calculations for SAIFI, MAIFI, and MAIFIE on a feeder basis are shown in the following equations. Notice that the numerator of MAIFI is multiplied by 2 because the recloser took two shots. However, MAIFIE is multiplied by 1 because it only recognizes that a series of momentary events occurred.

$$SAIFI = \frac{500}{3,000} = 0.167$$
$$MAIFI = \frac{2 \cdot 1,500}{3,000} = 1$$
$$MAIFIE = \frac{1 \cdot 1,500}{3,000} = 0.5$$

A good comparison of reliability indices has been worked out by the Council of European Energy Regulators (CEER) in the report "CEER Benchmarking Report 6.1 on the Continuity of Electricity and Gas Supply, data update 2015/2016" [3]. Tables 5.2 and 5.3 present the SAIDI and SAIFI data for different European countries from 2002 to 2016.

5.4 Proposed exercises

1. Determine a general expression for the availability, the reliability, the frequency, the MTTF, and the hazard rate for series and parallel connections having three components.
2. Calculate the equivalent reliability for each system in Figure 5.9, considering each component's respective reliability value.
3. A substation transformer has a constant hazard rate of 0.007 per day. Calculate the probabilities that it will fail during the next 7 years, and also the probability that it will not fail during the next 7 years. Calculate the MTTF for this transformer and determine the probability of having at least three of four transformers out of service at any given time.
4. Table 5.4 shows database for feeder 43, which serves 3,500 customers with a total load of 5 MW. Complete the table and calculate the quality indices SAIFI, SAIDI, and CAIDI. The indices must be calculated for sustained interruptions, which are those that last more than one minute.

Table 5.2 Planned and unplanned SAIDI, without exceptional events [3]

Country	2002	2003	2004	2005	2006	2007	2008	2009	2010	2011	2012	2013	2014	2015	2016
Austria											48.67	47.96	49.79	42.31	37.91
Croatia													416.49	404.42	325.25
Czech Republic			250.83	286.69	247.20	274.46	252.52	243.30	265.64	261.86	257.36	257.69	246.71	253.32	233.0
Denmark									20.37	21.04	19.44	15.90	16.65	19.55	19.34
France	46.00	56.30	57.30	60.20	79.40	68.50	82.00	90.40	86.90	71.50	75.70	84.00	66.00	67.90	66.60
Germany					36.63	33.10	30.06	26.16	24.56	25.43	27.74	22.55	19.84	19.73	23.09
Great Britain					69.61	82.99	79.92	79.91	76.74	74.63	62.14	60.39	58.94	49.89	42.30
Greece											250.00	252.00	228.00	199.00	208.00
Hungary														250.00	232.00
Ireland														158.70	144.90
Italy	186.85	177.50	139.14	124.51	107.63	98.63	102.45	93.03	103.48	105.44	111.42	97.55	100.92	112.02	115.96
Latvia											520.00	472.00	409.00	332.00	260.00
Lithuania				208.35	199.63	167.18	145.33	124.29	211.09	217.43	234.90	264.16	266.89	238.76	220.98
Luxembourg												33.64	25.09	28.81	21.72
Poland							503.56	456.77	445.96	462.15	401.32	393.97	311.17	340.04	260.36
Portugal	386.75	366.14	197.97	181.98	170.78	111.64	135.15	187.62	174.55	99.30	80.16	90.16	77.48	69.20	65.99
Romania							1,025.0	959.00	963.00	880.00	876.00	697.00	591.00	519.00	474.00
Slovenia										128.44	119.68	121.44	122.96	133.43	121.77
Spain											63.24	61.26	64.68	69.12	65.82
Sweden											100.70	101.06	92.66	100.25	87.90
Switzerland									28.00	29.00	34.00	25.00	22.00	21.00	19.00

Table 5.3 Planned and unplanned SAIFI, without exceptional events [3]

Country	2002	2003	2004	2005	2006	2007	2008	2009	2010	2011	2012	2013	2014	2015	2016
Austria											0.69	0.78	0.77	0.70	0.72
Croatia													3.46	3.60	2.95
Czech Republic			2.68	2.49	2.42	2.91	2.32	2.17	2.23	2.19	2.31	2.21	2.11	2.09	1.95
Denmark									0.44	0.44	0.44	0.36	0.35	0.41	0.42
France	0.11	0.18	0.16	0.13	0.15	0.19	0.32	0.31	0.27	0.19	0.19	0.23	0.20	0.22	0.22
Germany						0.45	0.43	0.39	0.35	0.41	0.39	0.55	0.42	0.88	0.59
Great Britain					0.73	0.80	0.76	0.74	0.72	0.70	0.63	0.61	0.62	0.56	0.50
Greece											2.43	2.50	2.40	2.10	2.07
Hungary														1.64	1.52
Ireland														1.31	1.30
Italy	3.23	3.17	2.82	2.71	2.52	2.40	2.27	2.24	2.19	2.04	2.15	2.00	2.00	2.12	1.91
Latvia											4.34	3.86	3.37	2.97	2.88
Lithuania				1.18	1.17	1.38	1.20	0.96	1.22	1.29	1.32	1.26	1.23	1.25	1.20
Luxembourg												0.40	0.38	0.43	0.28
Poland							4.82	4.46	4.42	4.96	4.12	3.64	3.51	4.10	3.45
Portugal	6.22	5.11	2.92	2.90	2.82	2.10	2.38	2.78	3.15	1.95	1.63	1.76	1.57	1.46	1.46
Romania							8.32	7.87	7.38	6.90	6.48	5.74	5.15	4.96	4.48
Slovenia										1.04	1.07	1.00	1.11	1.05	0.92
Spain											3.34	1.43	1.29	1.30	1.18
Sweden											1.47	1.44	1.46	1.34	1.33
Switzerland									0.40	0.40	0.45	0.37	0.30	0.32	0.30

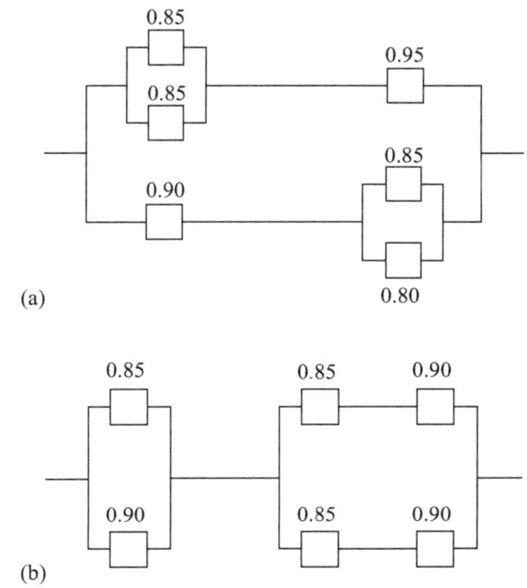

Figure 5.9 Elements of exercise: (a) case a and (b) case b

Table 5.4 Database for exercise 3

Date	Time	Time on	Total time (min)	Number of customers	Load (kV A)	Interruption type
1/23	13:24:28	14:45:43		343	567	
2/28	12:15:46	12:23:58		576	776	
3/11	04:06:09	04:07:02		102	120	
3/11	18:23:24	19:01:01		102	120	
3/18	01:03:43	01:53:24		432	683	
4/10	12:13:56	12:13:59		1,500	2,480	
4/30	08:20:18	08:20:56		700	1,315	
5/02	05:40:11	05:44:23		1,400	2,269	
6/29	03:56:32	04:38:12		1,973	2,984	
8/12	17:48:39	17:59:04		2,534	3,542	
8/13	13:08:02	14:09:01		28	35	
8/13	13:25:36	13:56:26		28	35	
9/22	23:49:27	23:54:24		1,769	2,647	
10/16	20:09:37	20:44:55		2,046	3,018	
12/05	19:31:15	19:31:46		378	588	

References

[1] IEEE. IEEE Std 493-2007 – IEEE Recommended Practice for the Design of Reliable Industrial and Commercial Power Systems (Gold Book), 2007.

[2] IEEE. IEEE Std 1366-2012 (Revision of IEEE Std 1366-2003) – IEEE Guide for Electric Power Distribution Reliability Indices.
[3] Council of European Energy Regulators. CEER Benchmarking Report 6.1 on the Continuity of Electricity and Gas Supply, data update 2015/2016. Energy Quality of Supply Work Stream (EQS WS), 2018.

Further reading

[1] Eto, J. H., Lacommare, K., Sohn, M. D., and Caswell, H. C. "Evaluating the performance of the IEEE Standard 1366 method for identifying major event days," IEEE Transactions on Power Systems, vol. 32, no. 2, pp. 1327–1333, 2017.
[2] Ndawula, M. B., Zhao, P., and Hernando-Gil, I. "Smart Application of Energy Management Systems for Distribution Network Reliability Enhancement," in Proceedings – 2018 IEEE International Conference on Environment and Electrical Engineering and 2018 IEEE Industrial and Commercial Power Systems Europe, EEEIC/I and CPS Europe 2018, 2018, vol. 6, pp. 1–5.
[3] Karacelebi, M., Gol, M., and Ercin, O. "A New Index for Voltage Quality Assessment at Distribution and Systems," in 2018 53rd International Universities Power Engineering Conference (UPEC), 2018, pp. 1–6.
[4] Short, T. A. Electric Power Distribution Handbook, 2nd ed. CRC Press, Boca Raton, FL, USA, 2004.
[5] Agalgaonkar, Y. P., and Hammerstrom, D. J. "Evaluation of smart grid technologies employed for system reliability improvement: Pacific Northwest smart grid demonstration experience," IEEE Power and Energy Technology Systems Journal, vol. 4, no. 2, pp. 24–31, 2017.
[6] Brown, R. E. Electric Power Distribution Reliability, 2nd ed. CRC Press, Boca Raton, FL, USA, 2009.

Chapter 6

Reconfiguration and restoration of distribution systems

For decades, distribution systems were designed with rigid topologies. In order to determine topologies that allow better operating conditions, a number of algorithms have been developed. The best topology could be used as the normal condition to run the system and to determine the location of flexibility equipment in order to have temporary changes in topology, best referred as reconfiguration.

Improvements achieved by the manufacturers in recent years have prompted the use of feeder reconfiguration to modify the operating condition of distribution networks. These improvements include remotely controlled switches or breakers for pole installation, numerical protection, and appropriate communication systems.

Under normal operating conditions, feeder reconfiguration aims for a more efficient operating condition of the network. Under faulty conditions, feeder reconfiguration aims to restore the service to the maximum number of users in the shortest time. Prior to determining the location of switches to allow changes in configuration, it is highly recommended to find the best topology for a distribution system.

6.1 Optimal topology

The optimal topology of a distribution system is that one where operating conditions are satisfied with the lowest possible value of system losses. The conditions include all loads operating within acceptable ranges, no overloaded elements, and all loads served.

To determine the optimal topology, all the poles in overhead feeders and, in particular, the double dead (DD)–end poles, or switching boxes in underground feeders, are potential open points. This allows us to determine the best boundaries among feeders to reduce the overall losses. Once the optimal topology is achieved, switches, breakers, or even reclosers can be installed instead the open points selected, to allow for system reconfiguration. This will improve remarkably SAIFI, SAIDI, and CAIDI indices.

The optimal topology can be found by running load flows, one for each case. This would result in a huge number of runs especially for large systems, thus making this approach unfeasible. That is why several methods have been developed to determine the best topology for a distribution system. These include heuristic methods, linear programming, neural networks, expert systems, fuzzy logic, simulated annealing, and genetic algorithms, among others.

The heuristic method is the most commonly used since it can produce fast results with good accuracy. Linear programming is used mainly in planning applications to minimize capital investment.

Among the heuristic methods, the reconfiguration of feeders according to Civanlar *et al.* [1] is very popular. The method considers the closing of a normally open (NO) link and the opening of another normally closed (NC) in order to transfer load from a feeder with a higher voltage drop to another with a higher voltage level so as to reduce the active losses produced in the system. The reduction in losses could be calculated easily from the results of load flows run for the system configurations before and after the reconfiguration takes place. The method can be illustrated with the three feeder-distribution system shown in Figure 6.1.

The branches A, B, and C represent links among the feeders with a switch NO each. On the other hand, it is assumed that each branch has a section switch.

The objective function according to Civanlar *et al.* method to determine the losses change as a result of transferring loads from Feeder 2 to Feeder 1 of a system is given by the following expression:

$$\Delta P = \mathrm{Re}\left\{2\left(\sum_{i\in D} I_i\right)(E_m - E_n)^*\right\} + R_{path}\left|\sum_{i\in D} I_i\right|^2 \tag{6.1}$$

where D is the group of nodes disconnected from Feeder 2 and connected to Feeder 1; m is the node of Feeder 1 to which the loads coming from Feeder 2 will be connected;

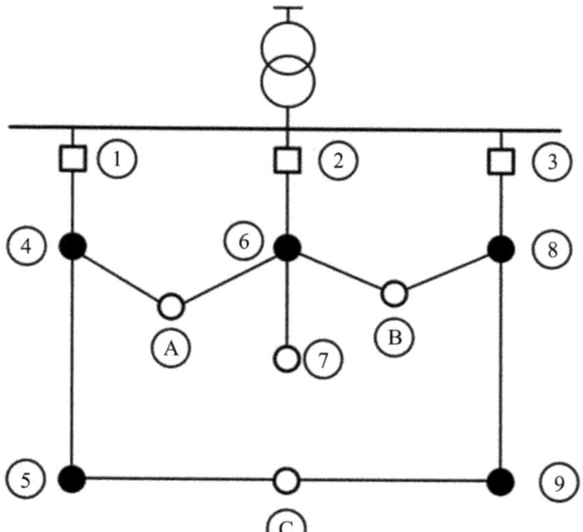

Figure 6.1 Three feeder-distribution system

n is the node of Feeder 2 that will be connected to node m through a link switch; I_i is the complex current of node i; R_{path} is the series resistance of the path that connects the nodes of Feeders 1 and 2 through a link switch; E_m is the component of $E = R_{BUS}I_{BUS}$ corresponding to node m of Feeder 1; E_n is similar to E_m but defined for node n of Feeder 2.

E_m and E_n are calculated with the node currents I_i of the base case before the load transfer. Under these circumstances, the first term becomes negative if $|E_m| < |E_n|$. The effect of the capacitors in the node currents should be incorporated to have a convenient modeling. The second term of the right part of (6.1) is always positive. Therefore, it is possible to achieve a reduction in losses unless the first term is significantly negative.

Shirmohammadi and Hong proposed an algorithm called "optimum flow pattern" (OFP) [2] to determine the minimum real power losses for radial networks. The OFP is computed on a purely resistive distribution network, neglecting the reactive components of the branch impedances, after closing all the connection points. Initially, the switches are closed resulting in a meshed system. Then the switches are opened one at a time according to the OFP. The connection link to be opened in a certain loop is the one located in the branch with the lowest current flow. The method provides good results for small systems but it is not the ideal alternative in general, since the meshed system does not represent the real operation state of the network.

In order to illustrate the previous idea, a small distribution network of Figure 6.2 is analyzed (adapted from [1]). This consists of three radial feeders connected at the station transformer and a group of switches in their normal, open or closed, positions. Network loads, assumed constant, are represented by voltage-dependent current injections at the network nodes. The objective of many

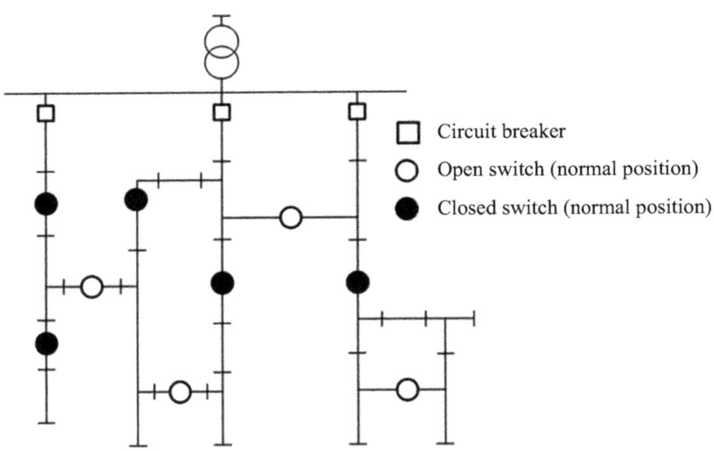

Figure 6.2 Radial distribution network with normally closed and open switches [1]

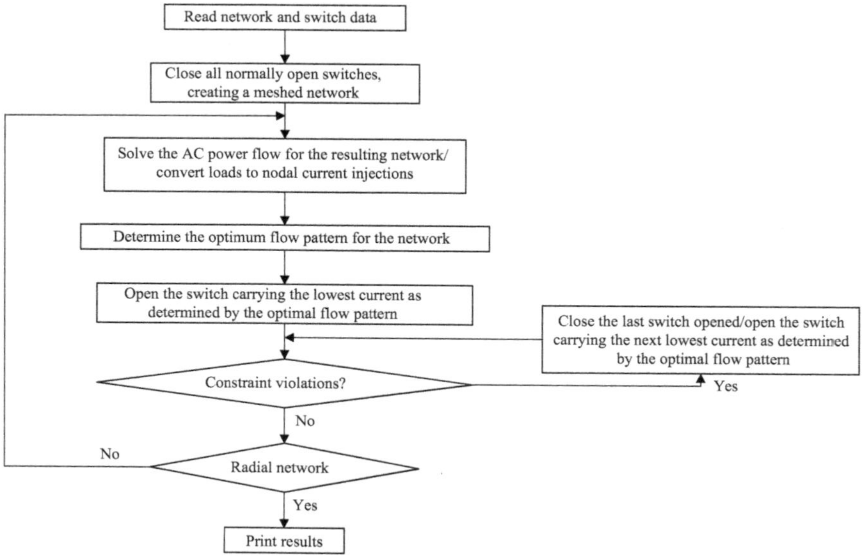

Figure 6.3 Basic algorithm for the reconfiguration of distribution networks [2]

topology-definition methodologies is to configure this distribution network, by defining open and close links, in order to minimize its resistive line losses. The final network configuration must be radial and all the loads must remain connected.

The solution algorithm first closes all the network switches, converting it from a radial configuration to a meshed one. An AC power-flow solution for the meshed network is then performed, and network loads are converted to nodal current injections, using the values of the node voltages.

The entire process, starting with the AC power-flow solution of the resulting network up to the opening of the connection link with the lowest current for the optimal flow pattern, is repeated until the network becomes radial. Figure 6.3 shows the flow chart for this algorithm.

The best approach for determining the basic topology is by considering that all the poles in overhead lines, or in the connection boxes of underground feeders, can feasibly be opening points or feeder ends. This is illustrated in Figure 6.4 that has a simple system of two feeders.

Several specialized software packages have been developed to determine the optimal topology that some software developers refer to as "optimal separation points." The packages can have different optimization criteria, such as minimum losses, elimination of overloads, and voltage control. The most common criterion is minimum losses.

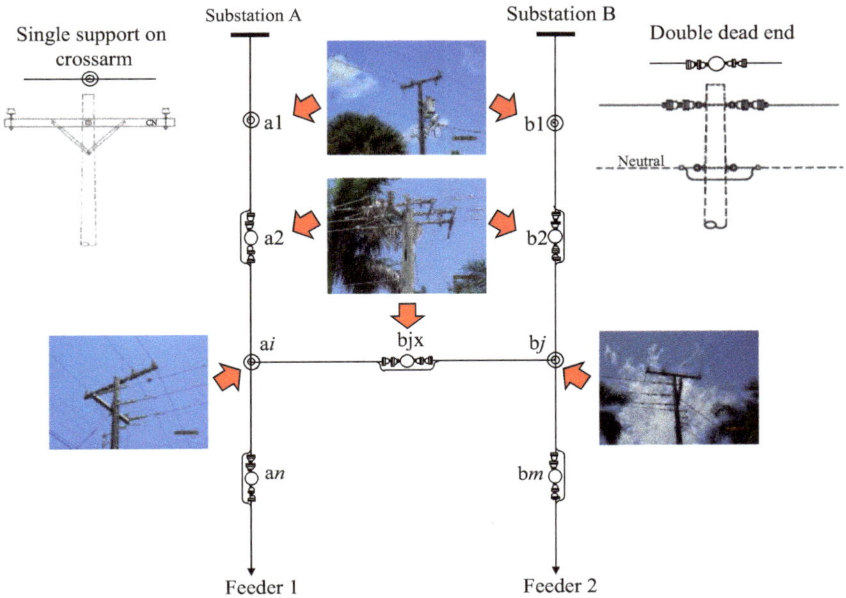

Figure 6.4 Two-feeder system showing location of potential opening points

---

Example 6.1 A pilot project was developed on the basis of the Civanlar approach, which considered a distribution system that has approximately 110 feeders at 13.2 kV with a total loss figure above 15%. An optimized configuration was required to reduce losses.

A sample of around 5% of the distribution network was studied. This was a good representation of the entire system.

The prototype was selected such that the characteristics of the feeders were typical and the area included several social strata, hospitals, sport arenas, and other key locations of the city. The analysis was carried out by running a load flow for each case. Table 6.1 presents the results obtained after applying the Civanlar algorithm for feeder reconfiguration.

The first columns of Table 6.2 show the identification of the poles and the feeders they correspond to. The next columns present the initial and final loss levels for the circuits receiving loads (CRL) and circuits transferring loads (CTL). The loss reduction is shown in two columns. The first corresponds to the results calculated with load flow runs and the second with the optimal topology methods. It is interesting to see the excellent accuracy of the later. The last column shows the voltage profile to make sure that all nodes are within the established range. Table 6.3 is similar and presents the best three options for reconfiguring the system at the poles that offer the highest loss reduction.

Table 6.1 Data of feeders encompassed within the prototype

Substation	**Circuit name**	**Circuit code**	**Nodes**	**Transf.**	**Installed capacity (kVA)**	**Load factor**	**Links**	**Total losses**	
								(kW)	**(kVAR)**
St. Anthony	Crystals	0106	283	152	16,122	0.312	10	61.08	234.53
St. Anthony	10th Street	0109	100	47	9,927	0.253	2	10.99	32.60
St. Anthony	St. Ferdinand	0110	152	81	7,750	0.324	7	9.11	37.86
South	Britain	0513	469	251	27,582	0.390	6	226.72	802.16
South	Lido	0517	405	222	27,147	0.333	9	292.36	809.14
South	Cedar	0518	417	230	30,760	0.283	8	99.98	323.07

Table 6.2 Result of reconfiguration analysis

Link no.	Location	Initial losses (kW)			Final losses (kW)			Reduction (kW)		Volt min (pu)
	Circuits	CLR	CTL	Total	CRL	CTL	Total	Opt. topology	Load flow	
1	0110	9.11	292.36	301.47	46.58	163.26	209.84	90.45	91.63	0.970302
2	0110	9.11	61.08	70.19	35.87	23.21	59.08	11.05	11.11	0.973206
3	0110	9.11	61.08	70.19	43.51	16.66	60.20	9.88	9.99	0.970410
4	0110	9.11	61.08	70.19	50.48	14.45	64.93	5.18	5.26	0.967839
5	0110	9.11	61.08	70.19	9.37	60.48	69.85	0.33	0.34	0.989122
6	0110	9.11	61.08	70.19	9.11	61.08	70.19	0.00	0.00	0.989240
7	0518	99.98	292.36	392.34	101.8	289.43	391.26	1.10	1.08	0.955976
8	0518	99.98	292.36	392.34	100.6	291.12	391.76	0.61	0.58	0.956378
9	0518	99.98	292.36	392.34	130.1	252.65	382.80	9.11	9.54	0.947586
10	0518	99.98	292.36	392.34	99.98	292.36	392.34	0.04	0.00	0.956510
11	0106	61.08	99.98	161.06	74.64	84.55	159.19	1.90	1.87	0.958511
12	0106	61.08	99.98	161.06	82.34	73.40	155.74	5.37	5.32	0.958584
13	0106	61.08	99.98	161.06	89.36	69.60	158.96	2.14	2.10	0.957062
14	0513	226.7	292.36	519.08	229.2	288.57	517.85	1.20	1.23	0.932889
15	0109	10.99	226.72	237.71	56.66	118.02	174.68	62.63	63.03	0.968333
16	0109	10.99	226.72	237.71	64.87	111.53	176.40	60.63	61.31	0.963412
17	0106	61.08	292.36	353.44	127.5	189.47	317.01	35.62	36.43	0.945557
18	0110	9.11	226.72	235.83	45.35	138.54	183.89	51.67	51.94	0.968140
19	0518	99.98	61.08	161.06						
20	0513	226.7	292.36	519.08	255.7	255.93	511.63	7.32	7.45	0.928727
21	0513	226.7	292.36	519.08	250.1	266.32	516.51	2.41	2.57	0.928589

Table 6.3 Options where the highest losses reductions are obtained

Link no.	Location	Initial losses (kW)			Final losses (kW)			Reduction (kW)		Volt min (pu)
	Circuits	CLR	CTL	Total	CRL	CTL	Total	Opt. topology	Load flow	
1	0110	9.11	292.36	301.47	46.58	163.26	209.84	90.45	91.63	0.970302
12	0106	61.08	99.98	161.06	82.34	73.40	155.74	5.37	5.32	0.958584
15	0109	10.99	226.72	237.71	56.66	118.02	174.68	62.63	63.03	0.968333
Total		81.18	619.06	700.24	185.58	354.68	540.26	158.45	159.98	

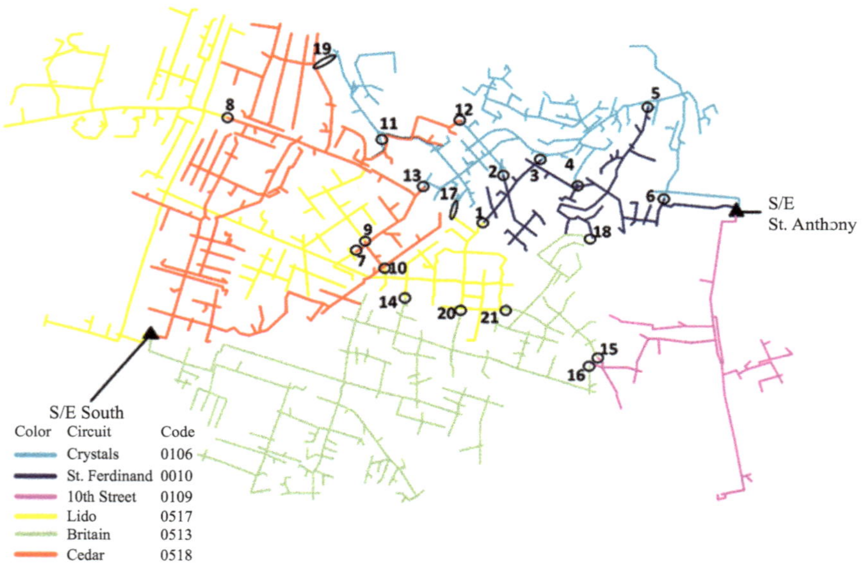

Figure 6.5 Initial topology of case study

Figure 6.5 shows the initial topology of the system, whereas Figure 6.6 shows the topology of the system after the topology change was implemented to attain losses reduction. The numbers indicate the existing boundaries among the feeders, which are defined by DD ends.

A simple cost-saving evaluation data considering a load factor that is equal to 1 gives the following results:

kWh cost: \$0.10 \$/kWh
Loss savings: 160 kW
Annual saving: 160 kW $\times$ 8,760 h $\times$ 0.10 \$/kWh = \$140,160

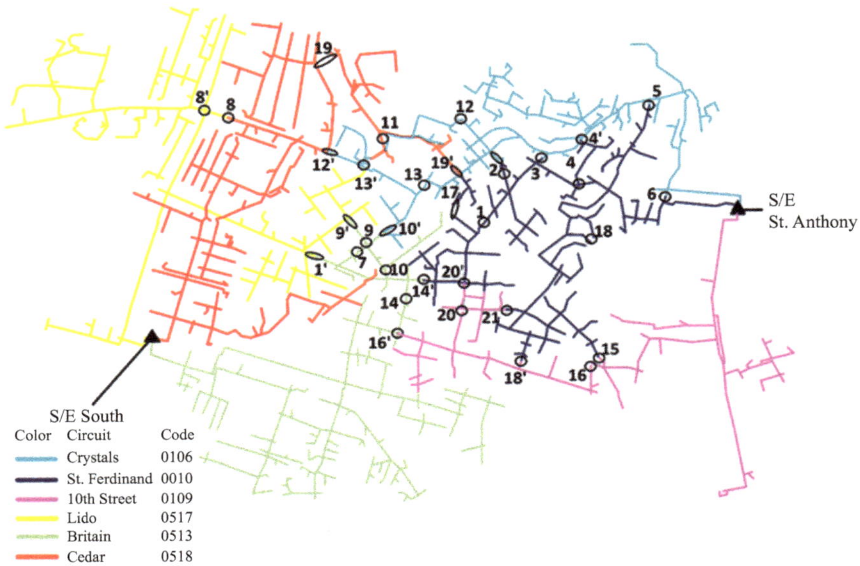

Figure 6.6 Optimal topology considering loss reduction

If the prototype represents 5% of the overall system, the total savings amount is $2,803,200/year.

6.2 Location of switches controlled remotely

It was shown the impact of defining an optimal topology, for example, to reduce losses. However, this is just merely for defining a fixed topology that remains working in a steady-state operation. Additional elements like switches in the electric grid are required to allow operators defining new topologies when a fault occurs, which respond to this new operational condition in a fast and reliable way. Switches can be breakers, reclosers, or sectionalizers and should have the means for remote operation to guarantee a fast reconfiguration when required. Location of them on the grid should be carefully studied.

The placement of switches is carried out in such a way that the reliability and flexibility criteria are considered. The reliability of distribution networks can be greatly improved by adding switches along the feeders. The benefit of adding the switches can be measured by examining the improvement in reliability performance. This is explained in the following sections.

6.2.1 Considerations to increase reliability

The basic concepts of reliability were treated in Chapter 5. Those concepts will be now used to show how the reliability of distribution networks can be greatly improved by adding switches along the feeders. Faults downstream a load cause the tripping of the corresponding protection.

If breakers with the corresponding protective equipment are installed on feeder sections downstream important loads, faults on those sections will not affect the loads upstream, improving then the reliability indices. The benefit of adding the switches can be measured by examining the improvement in reliability performance.

Example 6.2 Figure 6.7 shows a distribution system with three feeders modeled with a PSAT. Table 6.4 contains the data for each load (active and reactive power), and Table 6.5 shows the reliability characteristics for each line. Calculate the frequency of failures F, the hours per year of unavailability T, the probability to being unavailable Q, and the nonsupplied power and the nonsupplied energy W at Node 11, for the following scenarios:

(a) Normal operation
(b) Removing the breaker at substation N11 associated with line N11–N12
(c) Removing the breaker at substation N10 associated with line N2–N10
(d) Removing the circuit breaker at substation N2 associated with line N2–N8
(e) Removing the circuit breaker at substation N4 associated with line N4–N13

Compare the effect from each scenario on the whole system using the reliability values.

Considering the possibility of multiple faults in this system, the calculation becomes difficult, since the analysis for each node results from a combination of scenarios/combinatory selection of relationships between different distribution system elements. For that reason, a PSAT is used.

The results for the base normal operating scenario, Case (a), are presented in Figure 6.8.

In Case (b) (see Figure 6.9), the circuit breaker just after node N11 was removed. It can be seen that the reliability decreases. This is explained because the circuit breaker removed could have cleared all the faults beyond node N11. Then faults beyond this point can only be cleared when a crew identifies isolates and restores the service at this point. This increases the probability that node N11 is impaired by a fault for a particular amount of time.

For Case (c) (see Figure 6.10), the circuit breaker just before node N10 was removed. Here, the reliability does not change compared to the first scenario. Removing this element from the system does not change the ability to clear a fault beyond this node.

For Case (d) (see Figure 6.11), the circuit breaker removed is near node N2, between the nodes N2 and N8. Removing this element from the system impairs the reliability of node N11 in comparison with the first scenario.

Changes in other feeders can also impair reliability in neighboring feeders. In Case (e) (see Figure 6.12), a circuit breaker from the green feeder was removed.

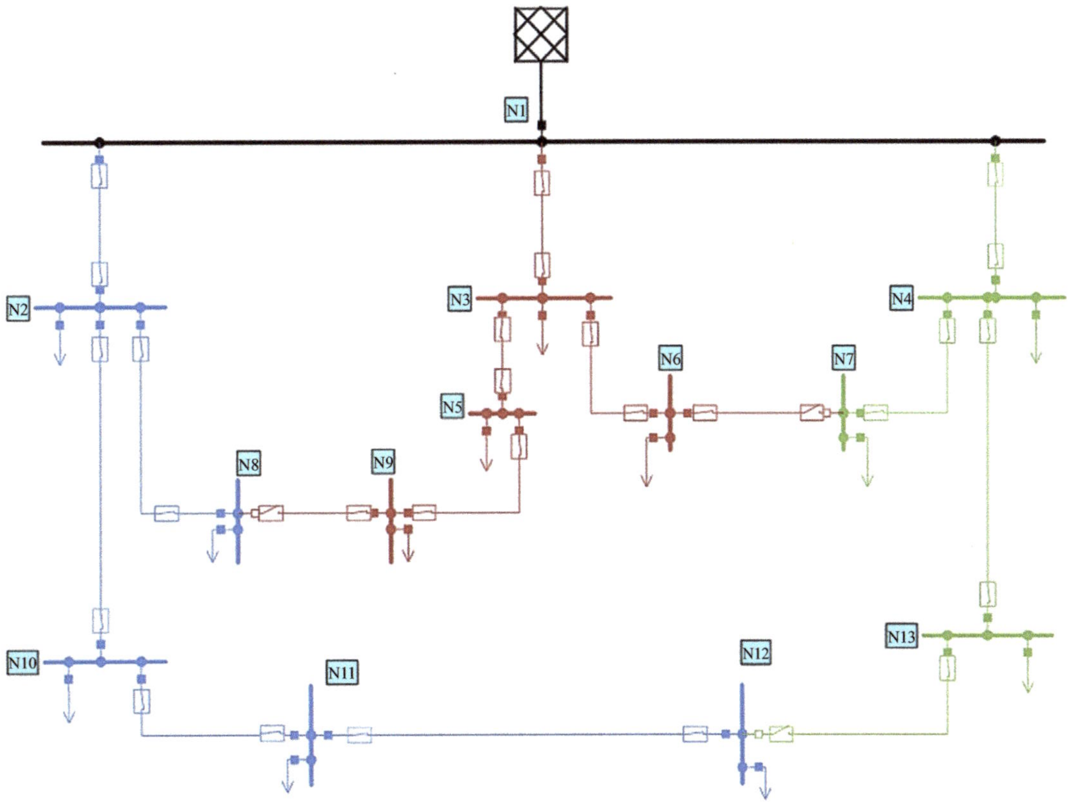

Figure 6.7 System diagram for Example 6.2

Table 6.4 Load data for Example 6.2

Name	*P* (MW)	*Q* (MVAR)
LN_{02}	2	1.0
LN_{03}	2	1.0
LN_{04}	2	1.0
LN_{05}	5	2.5
LN_{06}	5	2.5
LN_{07}	5	2.5
LN_{08}	5	2.5
LN_{09}	5	2.5
LN_{10}	7	3.5
LN_{11}	5	2.0
LN_{12}	2	1.0
LN_{13}	7	3.5

Table 6.5 Line data for Example 6.2

Name	Length (km)	*T* (1/year km)	*H* (*h*)
L_{01-03}	1.0	1.4	0.5
L_{06-07}	1.2	1.4	0.5
L_{03-06}	1.7	1.4	0.5
L_{04-13}	1.9	1.0	0.75
L_{01-04}	0.8	1.2	0.45
L_{02-08}	2.6	1.2	0.45
L_{01-02}	2.0	1.0	0.75
L_{09-08}	2.2	1.4	0.5
L_{10-11}	3.0	1.4	0.5
L_{11-12}	1.0	1.2	0.45
L_{03-04}	1.5	1.4	0.5
L_{05-09}	2.0	1.2	0.45
L_{03-05}	1.4	1.0	0.75
L_{02-10}	2.1	1.4	0.5
L_{13-12}	1.5	1.4	0.5

This affects the reliability of node N11, since any fault that takes place at this point disconnects the other feeders (Table 6.6).

6.2.2 Considerations to increase flexibility

Distribution networks should no longer be considered to have rigid configurations, since remotely controlled switches are available and allow better configuration. It is convenient that the feeders are sectionalized into equally loaded portions as far as it is possible (Figure 6.13). Figure 6.14 evaluates the flexibility of each of the four DD end links that exist between the two feeders to locate the NO tie switch

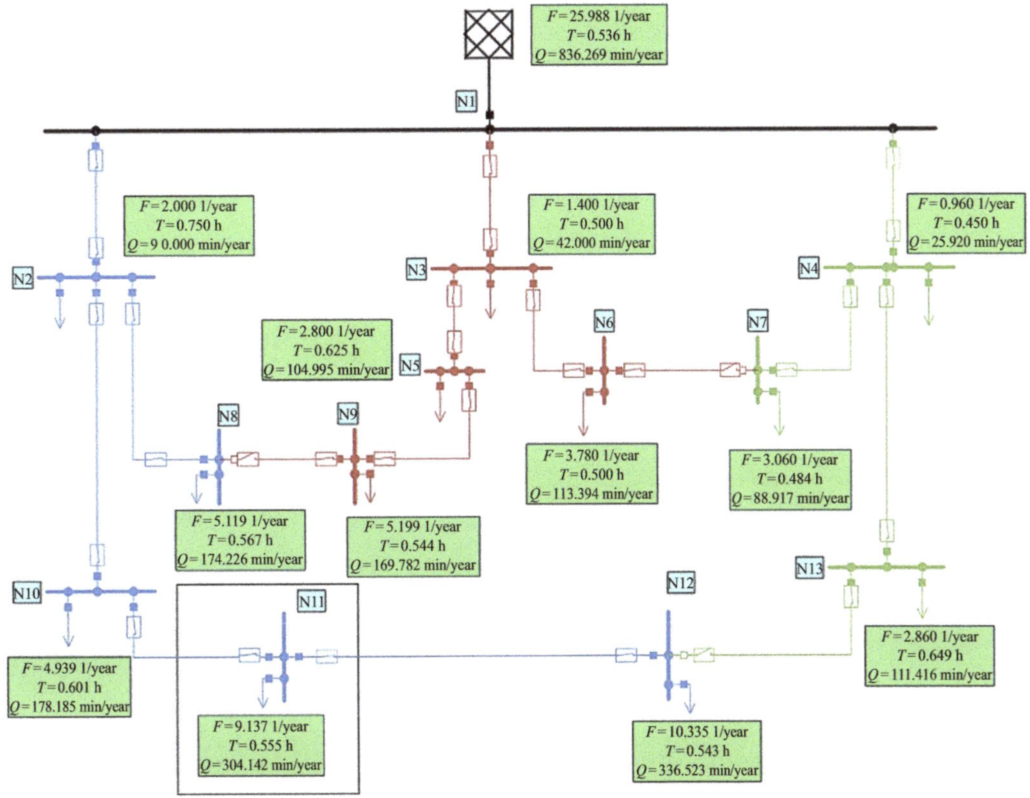

Figure 6.8 Results for scenario (a) of Example 6.2

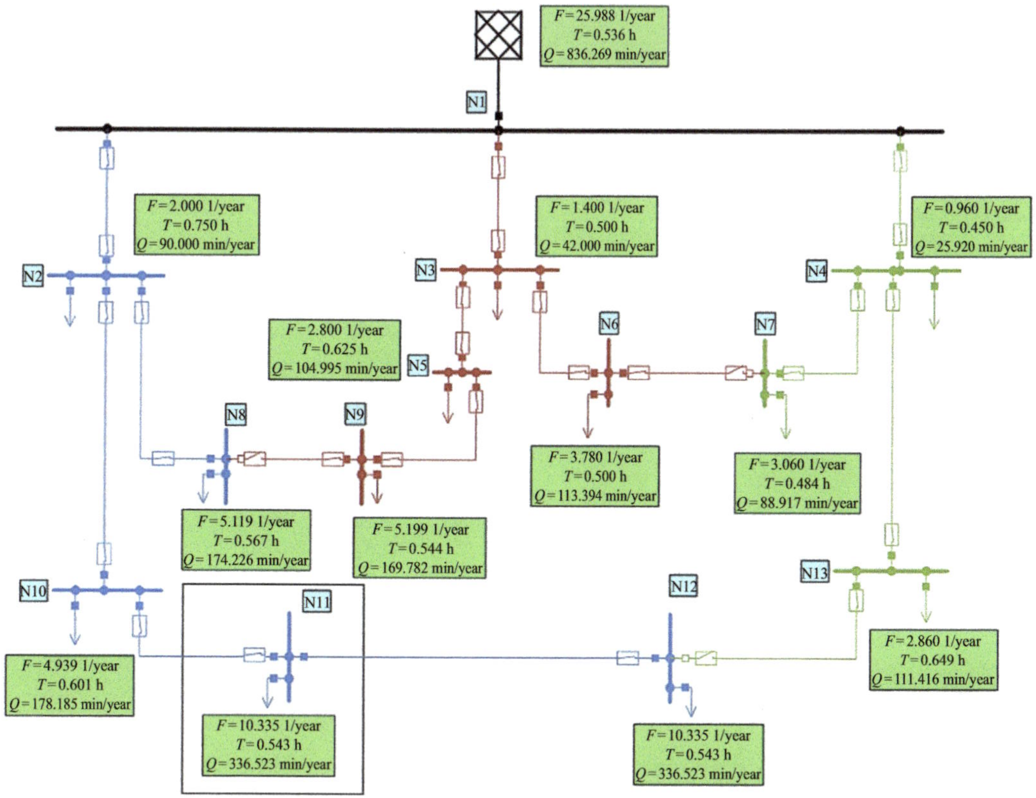

Figure 6.9 Results for scenario (b) of Example 6.2

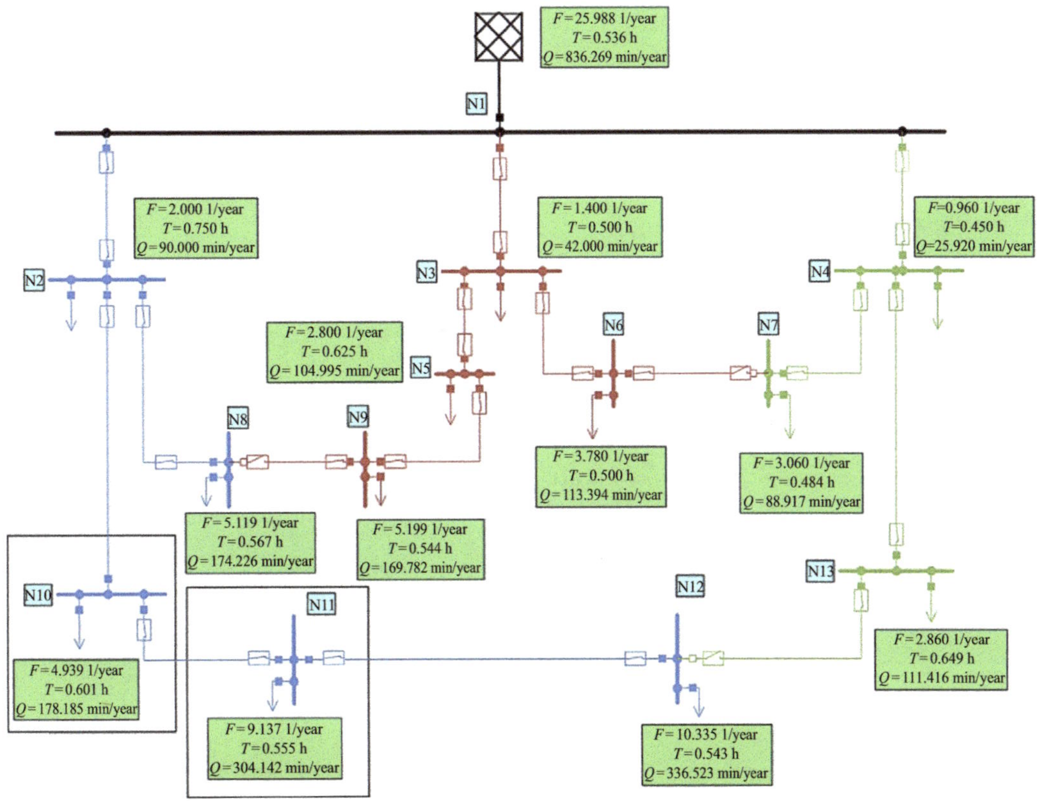

Figure 6.10 Results for scenario (c) of Example 6.2

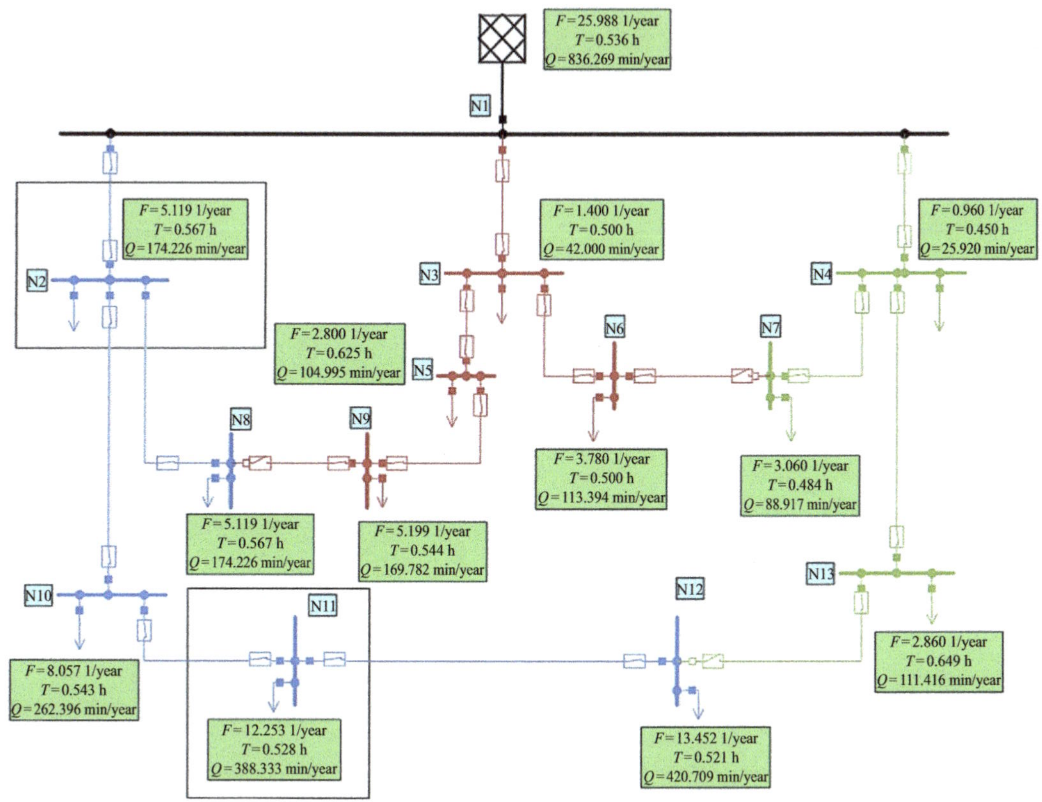

Figure 6.11 Results for scenario (d) of Example 6.2

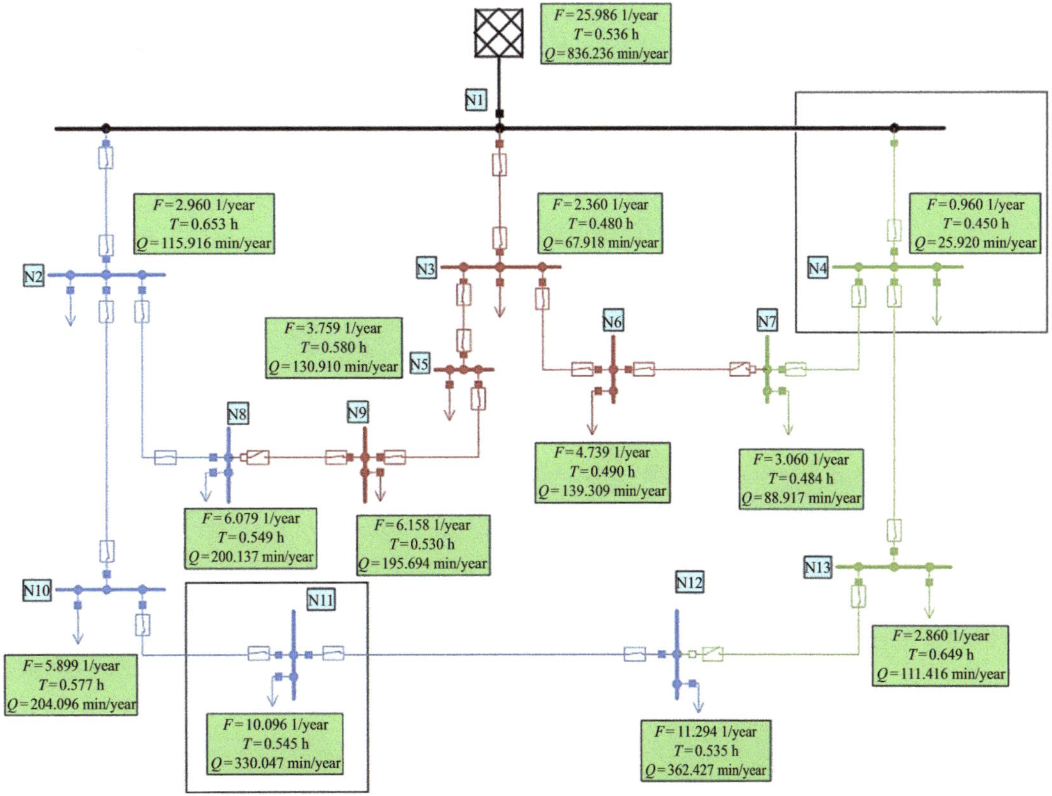

Figure 6.12 Results for scenario (e) of Example 6.2

Table 6.6 Comparison of reliability data for node N11 under different scenarios

	F (1/year)	*T* (h)	*Q* (min/year)	*P* (MW/year)	*W* (MWh/year)
Scenario (a)	9.137	0.555	304.142	45.683	25.345
Scenario (b)	10.335	0.543	336.523	51.676	28.044
Scenario (c)	9.137	0.555	304.142	45.683	25.345
Scenario (d)	12.253	0.528	388.333	61.266	32.361
Scenario (e)	10.096	0.545	330.047	50.478	27.504

Figure 6.13 Location of switches in distribution systems

(or breaker). The purpose is to pick the one that offers the highest flexibility to help restoring the electrical service, for faults in sections 1–3 of the first feeder.

A switch located in the first section of the feeders, such as number 7, would not offer the possibility of any load restoration for faults in any of the three sections of the feeder and of course that location is discarded. The link located in the second section, number 8, would offer the possibility of restoring only the first section, since for faults in section 2 or 3, the NO switch would not be effective at all.

The links located in the third section, identified by numbers 9 and 10, allow the restoration of two-thirds of the feeder load. Therefore, the switches located in this section have the highest flexibility to carry out remotely controlled transfers. It is clear that the link closest to the section switches offers lower current path, and therefore lower losses and better voltage regulation that could be easily validated

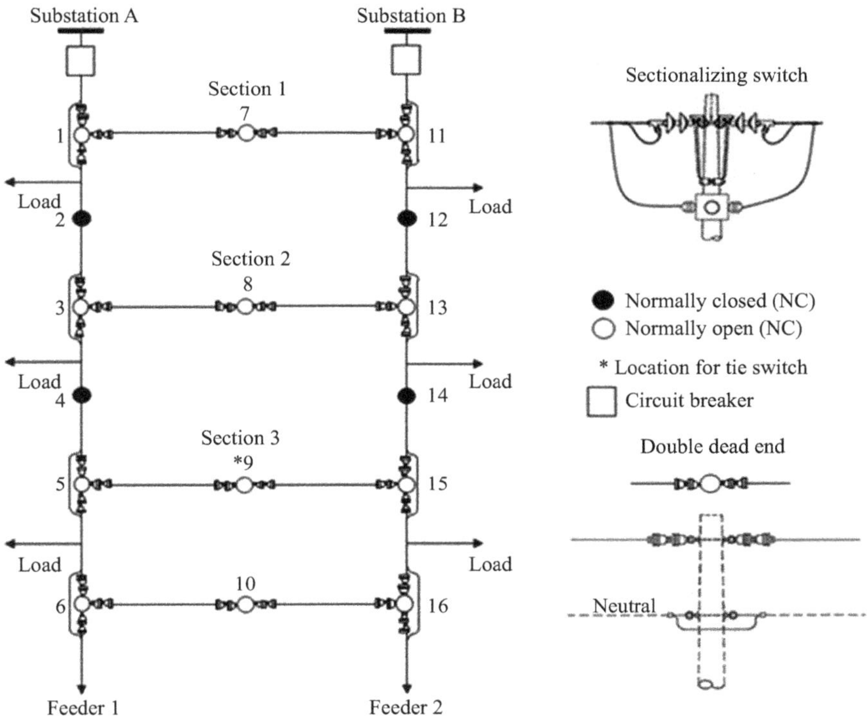

Figure 6.14 Two-feeder system showing location of switches controlled remotely

with load flow runs. That is why location of link 9 is the chosen one to place the NO switch.

The system shown in Figure 6.15 has three substations with one feeder each. The section switches placed after the breaker in the three feeders are all remotely controlled and numbered from 1 to 14. For the reference system (Figure 6.15(a)), the best DD end location to be replaced by a switch is selected (Figure 6.15(b)). In this case, selection is done considering the points in which more sections of each feeder can be transferred in case that an event occurs in the system and is required.

Table 6.7 summarizes the operation of the switches for the different faults indicated in the previous diagram. For some fault locations, especially at the end of the feeders, it is possible that the reconfiguration of tie switches may not be an option. In this case, the operation of the section switches is the only option to implement the upstream restoration. After the system reconfiguration takes place, the time current characteristic settings have to be modified by recalculating the time dial and the pickup values according to the new topology. In order to implement these changes automatically, it is essential to have the availability of multiple setting groups.

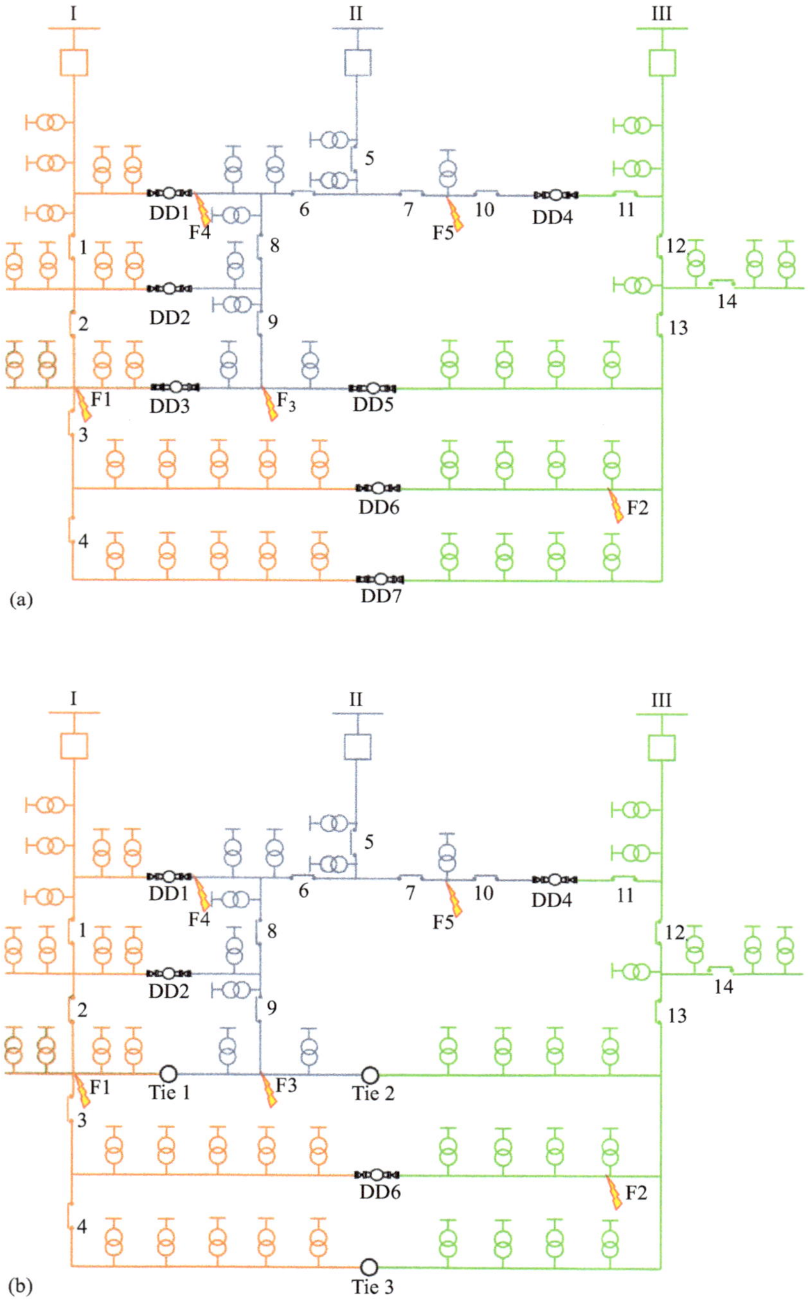

Figure 6.15 Three-feeder system illustrating switch location for service restoration with no tie switches (a) and tie switches (b)

Table 6.7 Reconfiguration options for the previous system

Fault	Open	Close
F1	2 and 3	T3
F2	13	–
F3	9	–
F4	6 and 8	T1 or T2
F5	7	–

6.3 Feeder reconfiguration for improving operating conditions

Feeder reconfiguration consists of the modification of the topology of a network through the closing of a switch that links two feeders and the opening of another switch so as to maintain the radial condition of the feeders. The reconfiguration is carried out for better operation of the network and specifically to reduce the losses due to the Joule effect. Distribution systems should be operated at minimum cost subject to a number of constraints: all loads are served, overcurrent protective devices are coordinated, voltage drops are within limits, radial configuration is maintained, and lines, transformers, and other equipment operate within current capacity.

The exchange process starts by assuming a radial configuration and therefore with the distribution network operating in a radial configuration. One of the tie switches is closed, and then another switch is opened in the loop created, which restores a radial configuration. The switch pairs are chosen through heuristics and approximate formulas to create the desired change in loss reduction. The branch-exchange process is stopped when no more loss reductions are possible.

Figure 6.16 taken from NEPLAN Resources [3] shows a system with four feeders. This figure corresponds to the same network of Example 3.8, Figure 3.33. The reconfiguration brings about savings on significant power losses

6.4 Feeder reconfiguration for service restoration

The other important function in distribution automation is service restoration in the case of a fault in the primary feeder. The main goal of automatic service restoration is the execution of a series of operations of the tie (NO) and section (NC) switches, aimed at restoring the power supply to the maximum number of areas that have been impaired after a fault in a primary feeder or substation. Usually, it would reconfigure the network by transferring loads from the healthy portions of faulty feeders to neighboring feeders that are operating normally.

There are multiple possibilities for reconfiguring the network for service restoration but the group solution is finite; it is defined by the combination of binary states of the section switches and tie switches. This group grows

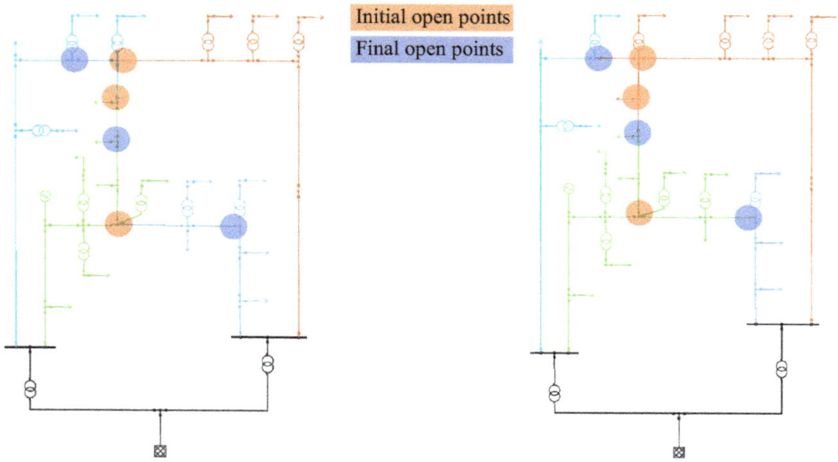

Figure 6.16 Distribution system illustrating loss reduction [3]

exponentially with an increase in equipment. It can become a problem of scale. The solution consists in choosing a combination of states for the equipment in the areas not affected, which satisfies some functional objectives. A system reconfiguration changes the topology, line flows, and short circuit values. Thus, different solutions may be feasible for the same fault.

In these circumstances, equipment, voltage profiles, and equipment loading have to be considered. Relay pickup values and time dials have to be rechecked to avoid nuisance tripping or unwanted pickups. This would require the use of different group settings.

6.4.1 Fault location, isolation, and service restoration

The steps to restore service when a fault happens can be summarized as follows:

1. The corresponding relay operates and trips the breaker. If reclosing units operate and the fault remains, the feeder is open.
2. The fault is located and the associated section switches open to isolate it.
3. The feeder is reenergized up to the location of the first section switch that was open—upstream restoration.
4. The healthy sections are transferred to one or more neighboring feeders by operating NO switches—downstream restoration.
5. The faulty section is repaired by the crew and the system is taken back to the normal configuration.

The first four steps should be completed within a minute to avoid affecting SAIFI and SAIDI indices.

These steps are illustrated by using the system shown in Figure 6.17 that corresponds to the normal configuration of a distribution system with four substations,

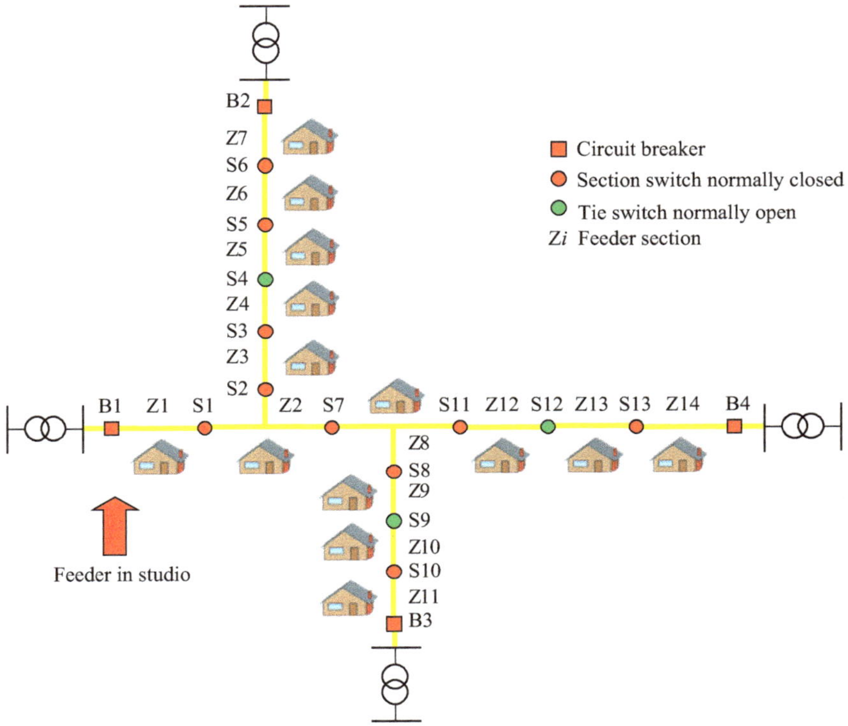

Figure 6.17 Normal configuration

each with one feeder. The system has 14 loading sections labeled Z1–Z14 and 13 switches in total labeled S1–S13. Switches 4, 9, and 12 are NO and the rest are NC. Figures 6.18–6.21 illustrate the first four steps already mentioned, following a fault at section Z2 in order to restore the system as far as possible. The fifth step of the process would be to take the system back to normal.

Example 6.3 The following example shows a distribution system with three feeders coming out from three different substations where switches have been located to increase the flexibility of system operation. The system has in total 13 section switches (NC) and three tie switches (NO) labeled T1, T2, and T3. It can be noted that the tie switches are chosen after the section switches. The five steps for system reconfiguration after fault F1 happens are as follows:

1. Relay clears fault at Feeder 1.
2. Switches 1 and 2 open to isolate the fault.
3. Relay recloses Feeder 1 to reenergize up to the location of Switch 1 (upstream restoration).

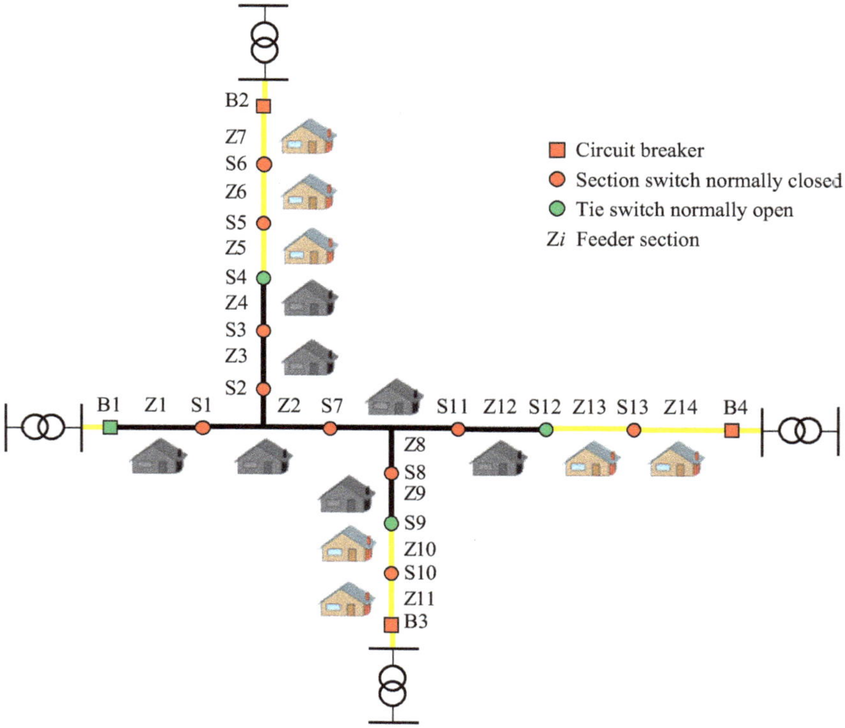

Figure 6.18 Fault clearing by relay at B1—Step 1

4. System reconfigures by choosing the best option from the possibilities indicated in Table 6.8 for fault F1 (downstream restoration).
5. Fault is fixed and system is restored to normal configuration.

The same analysis can be conducted for the other two faults, F2 and F3. Table 6.8 shows the operation of the switches for the three different faults indicated in Figure 6.22. Note that for fault F1, once the section switches 1 and 2 are open, the restoration can be achieved by closing either T1 or T3 or even, a combination involving the opening of switch 3 or 4 and the closing of T1 and T3.

6.4.2 Manual restoration vs. FLISR

When a permanent fault occurs, customers on healthy sections of the feeder may experience a lengthy outage if fault location, isolation, and service restoration (FLISR) is not available. Figure 6.23 is an illustration of the time taken at a typical distribution system to fix a fault and restore service to the users associated with these healthy sections when manual switches have been deployed without any automation. In this case, the maintenance crew has to get the faulty section and

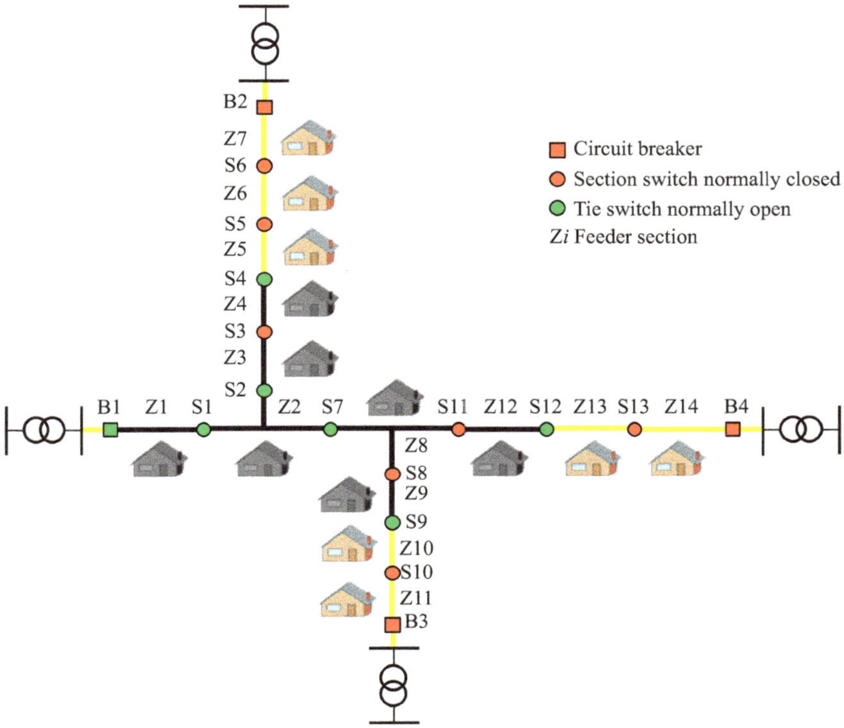

Figure 6.19 Fault isolation by opening corresponding switches—Step 2

perform the required switching to allow both the upstream and downstream restoration as practicable.

6.4.3 Restrictions on restoration

When carrying out the reestablishment, the operations that are executed should allow the system to satisfy the following restrictions:

- The current-carrying capacity of the transformers and lines should be within specified limits.
- The voltage drop should stay inside an established margin.
- The system should continue being radial.
- The number of operations of the equipment stays within limits.
- Important customers have priority.
- System must be balanced as much as possible.
- The coordination of the protection must be maintained.

Example 6.4 To illustrate the nature of service restoration, the same system of Figure 6.17 is used. The total loading of each feeder is shown in Figure 6.24

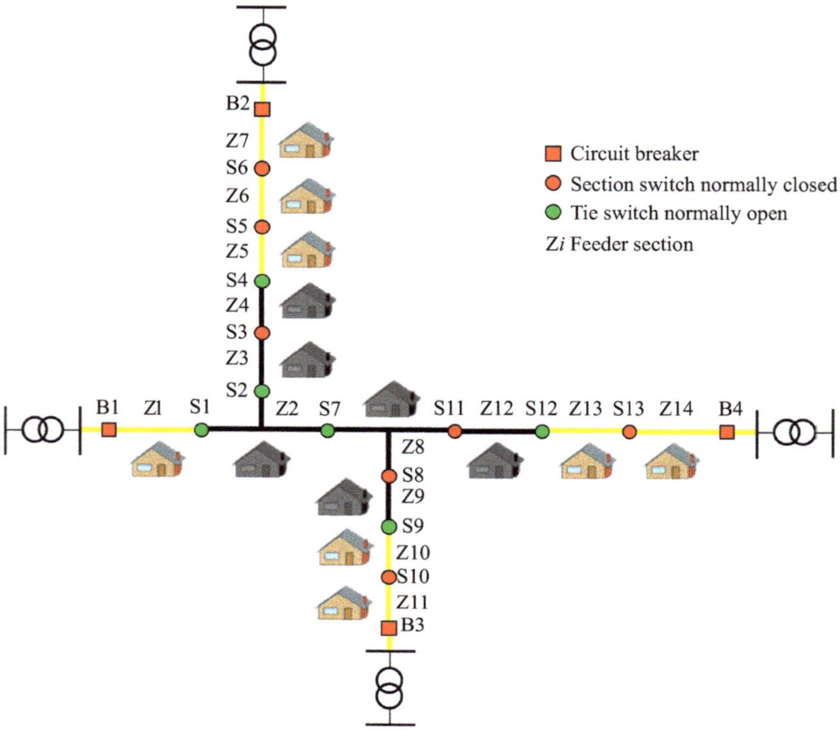

Figure 6.20 Reclosing B1—Step 3 (upstream restoration)

assuming that each section (Z1–Z14) has a loading of 1 pu. Indicate how the operation is affected when the fault happens and the system is reconfigured after Step 4 that was mentioned previously.

In the scenario of a fault in the feeder associated with Breaker 1 (B1), the system should clear and isolate the fault and then energize the healthy sections making the use of neighboring circuits, provided that they have sufficient capacity. The options for solutions grow exponentially with an increase in equipment such as section and tie switches. The restoration algorithm should find a configuration that fulfills the operative restrictions. The reconfigured system is shown in Figure 6.25 where all sections are reenergized except section Z2 that needs to be fixed by the maintenance crew. The reconfigured system of Figure 6.22 clearly has different loadings in Feeders 1–4. Table 6.9 illustrates the section loadings prior to the fault and after the system is reconfigured and also the ratio of currents after the reconfiguration and before the fault.

It is clear that the situation after reconfiguration does not offer any risk to trip Feeder 1. However, it certainly is risky for Feeders 2–4 that experience an important increase in loading that might trip the corresponding breakers if the pickup

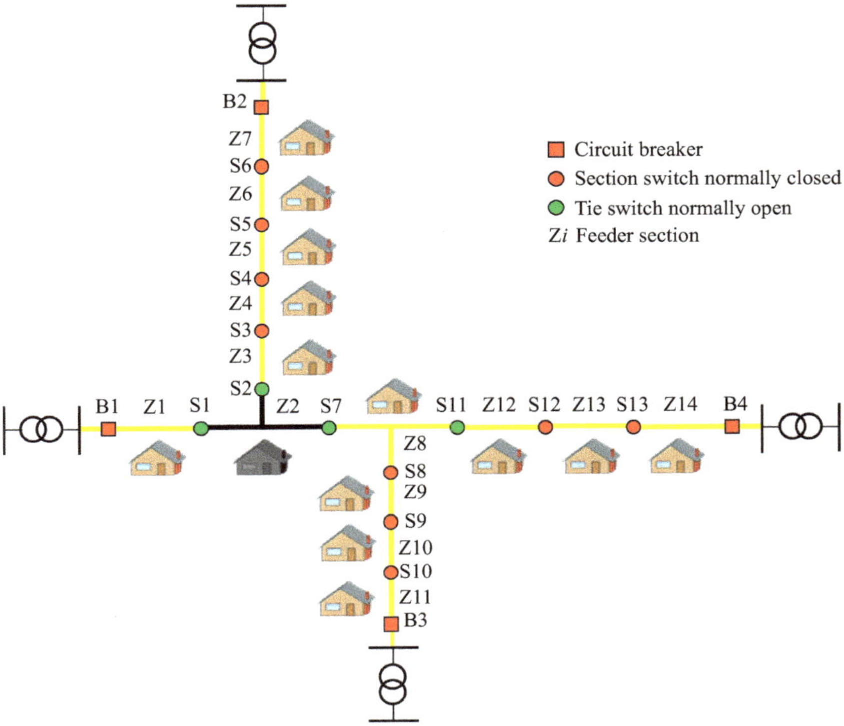

Figure 6.21 Reconfiguration by operating switches NO and NC—Step 4 (downstream restoration)

Table 6.8 Reconfiguration options for system of Example 6.3

Fault	**Open**	**Close**
F1	1 and 2	T1 or T3
F2	8 and 9	T1 or T2
F3	12 and 13	T2 or T3

value of the corresponding relays is exceeded. Measures, including adaptive relaying, have to be taken to avoid any nuisance tripping that could worsen the operation of the system.

6.4.4 FLISR central intelligence

Central intelligence is still the commonest method of automation to control a power system that includes substations. Central intelligence, of course, relies on powerful and fast SCADA systems and therefore requires a very good communication infrastructure. In the case of distribution systems, central intelligence should have

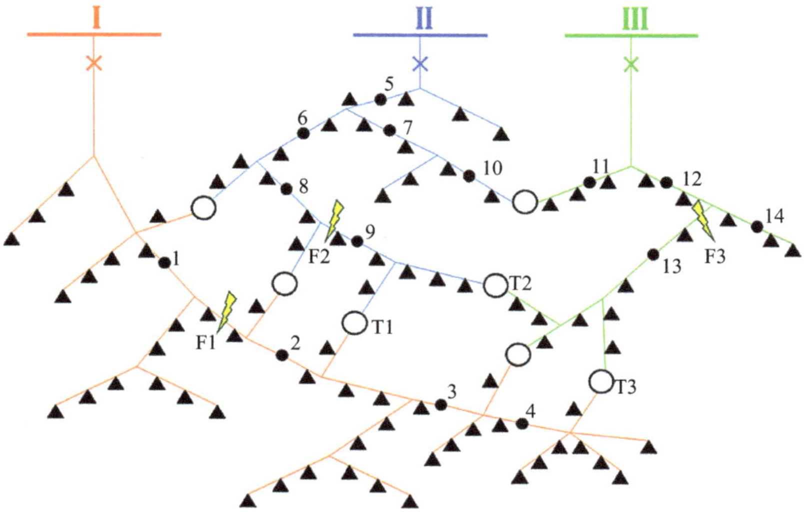

Figure 6.22 Three-feeder system illustrating switch location for service restoration

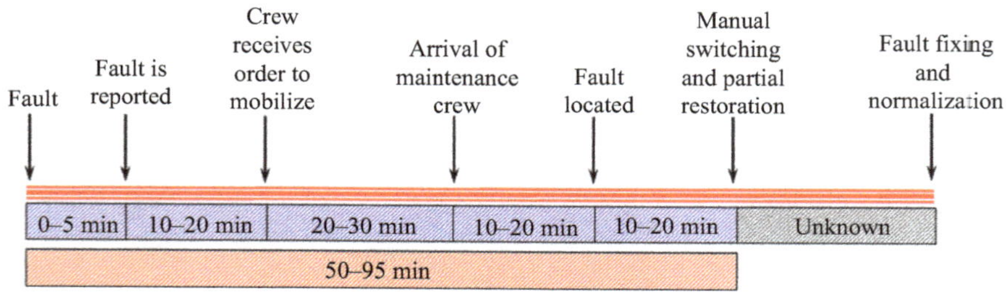

Figure 6.23 Time to fix faults for typical systems

proper control not only of the substation elements but also of all the devices deployed on the feeders. These include breakers, switches, voltage regulators, and capacitors. All of them should be monitored and controlled directly from the control system. This method is very efficient, but it requires a lot of communication.

The central intelligence in FLISR allows one to

- analyze faults based upon input from real-time SCADA (fault detectors and fault currents) and crew reports;
- isolate the fault based upon current state of distribution network;
- generate recommended switching for isolation of a nominated fault and restoration of unaffected customers; and
- propose multiple restoration plans based upon network status, topology, equipment limits, and desired objectives.

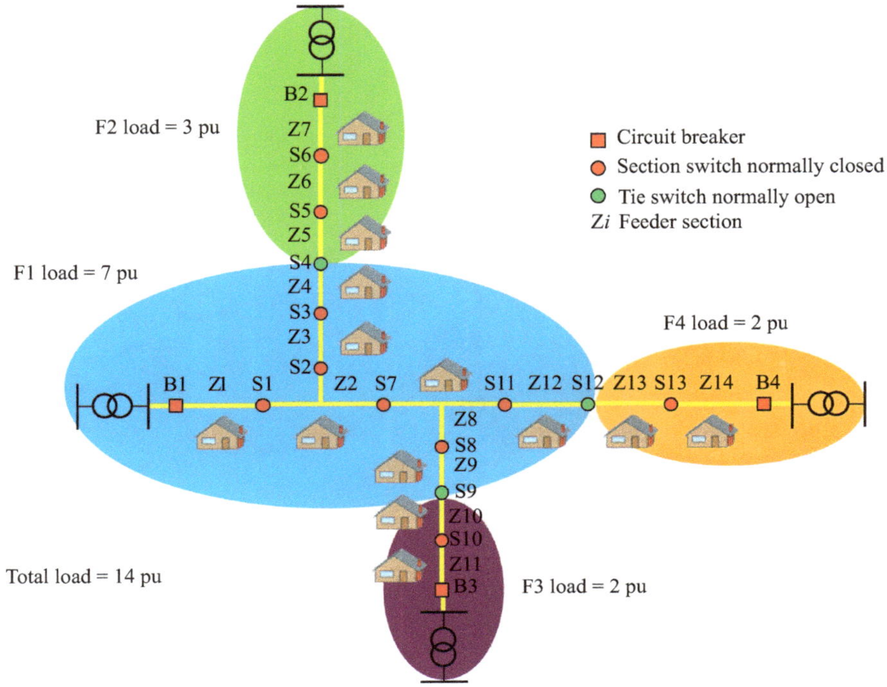

Figure 6.24 Loading for normal condition

An illustration of feeder reconfiguration using central intelligence taken from an ALSTOM case study is given in Figures 6.26–6.31.

6.4.5 FLISR-distributed intelligence

Distributed intelligence automatically isolates faulted distribution segments using a "team" of peer devices. It requires proper communication and protective relay coordination. The objective is to reconfigure adjacent feeders to restore power to customers beyond the fault (to minimize the number of outages, their size, and duration).

Some software applications have been developed to find solutions using FLISR-distributed intelligence. Figure 6.32 shows an application to a faulted segment. Figure 6.32(a) shows the normal configuration of the groups in the distribution system. Figure 6.32(b) presents a fault in Group 6 between the switches F, G, and J. All switches in the affected group experience an overcurrent followed by a loss of voltage. Therefore, switch J opens to isolate the fault. Figure 6.32(c) shows how Group 6 detects the fault and initiates opening of switches F, G, and I based on voltage loss. Finally, Figure 6.32(d) shows how the service is restored to all the unfaulted segments in a relatively/[or just delete the word] short

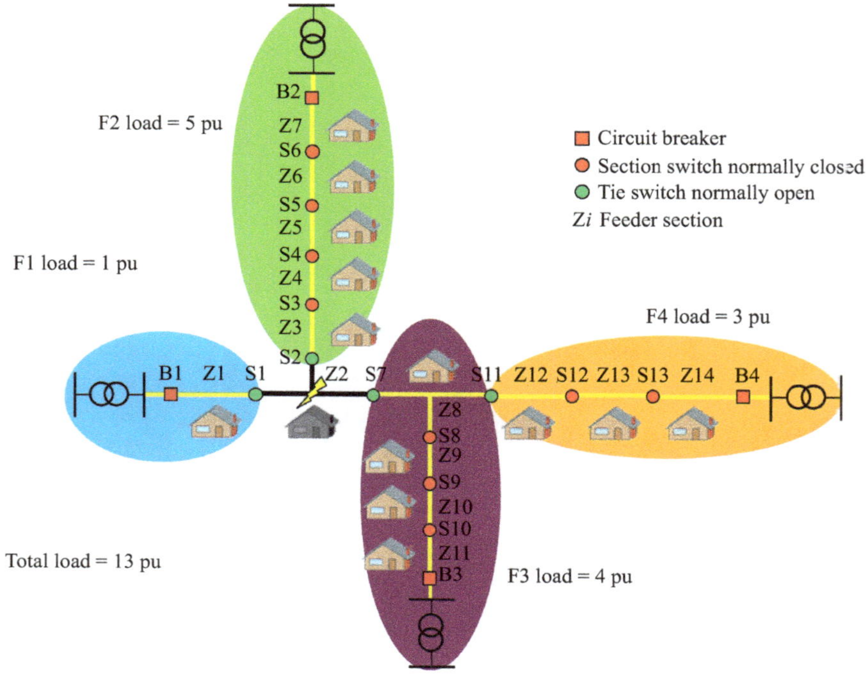

Figure 6.25 Loading after reconfiguration

Table 6.9 Loading values of Example 6.4

Feeder	Sections before	Sections after	I_{after}/I_{before}
B1	7	1	1/7
B2	3	5	5/3
B3	2	4	4/2
B4	2	3	3/2

time: Group 7 closes switch K, Groups 4 and 5 close switch E, and Groups 2 and 3 close switch C.

It is worth noting that other, more advanced algorithms, can locate the fault more accurately. If this were the case, switch I would not need to have been opened at step (c) since the opening of switches F, G, and J isolate the fault.

Distributed intelligence is growing and its application is very popular for systems with long feeders such as those in the United States. Distributed intelligence avoids congesting control centers, since the communication is among elements associated only within groups. However, this method has the disadvantage of not considering the whole system. This means that the optimal topology for the whole system might not be found.

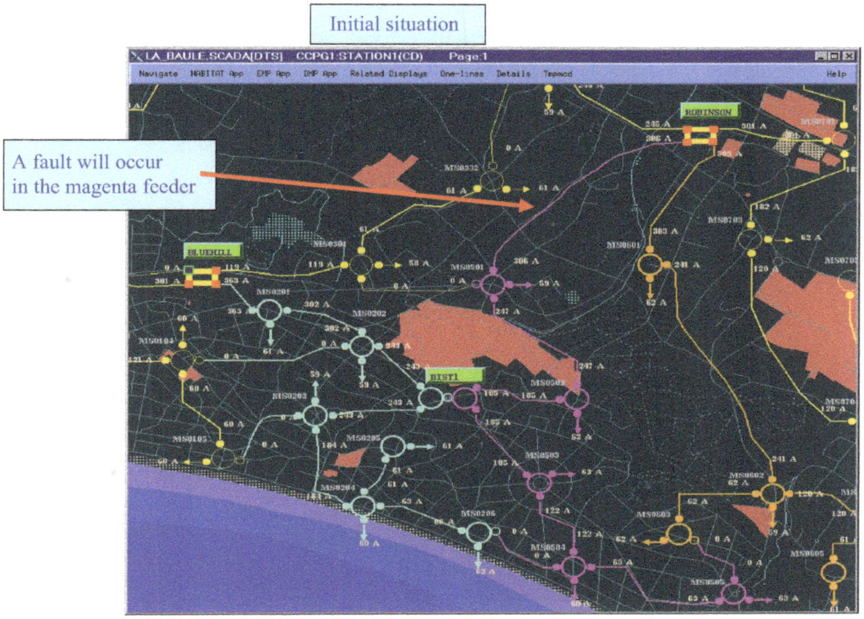

Figure 6.26 *Detection of a fault in an FLISR with central intelligence. Reproduced by permission of ALSTOM*

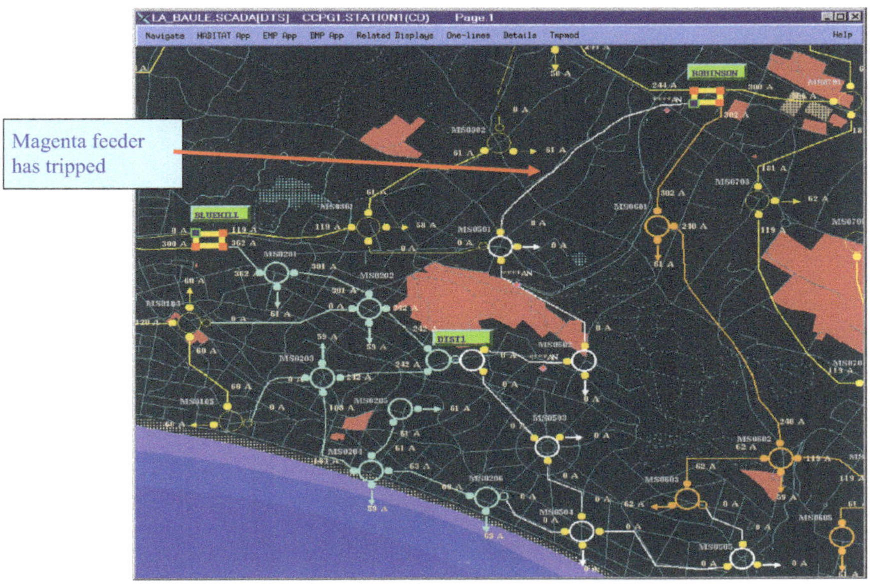

Figure 6.27 *Fault isolation by opening corresponding feeder. Reproduced by permission of ALSTOM*

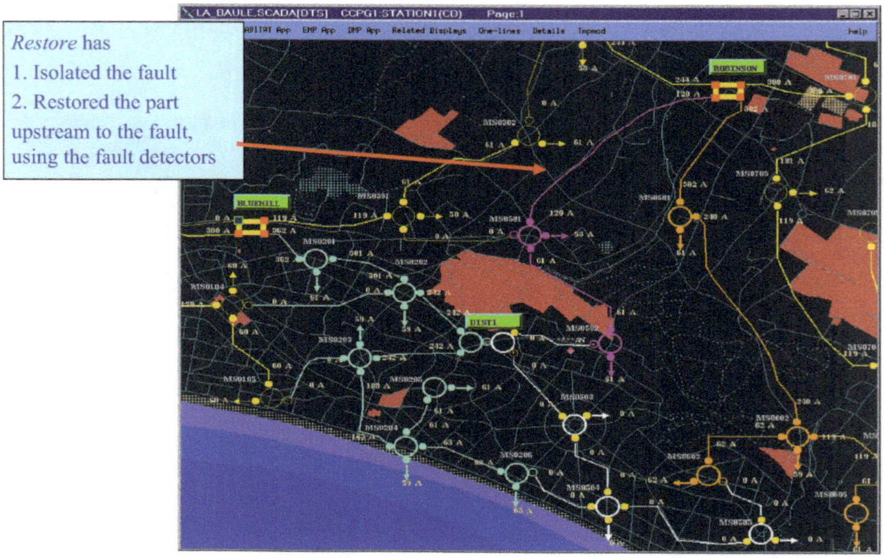

Figure 6.28 Upstream restoration. Reproduced by permission of ALSTOM

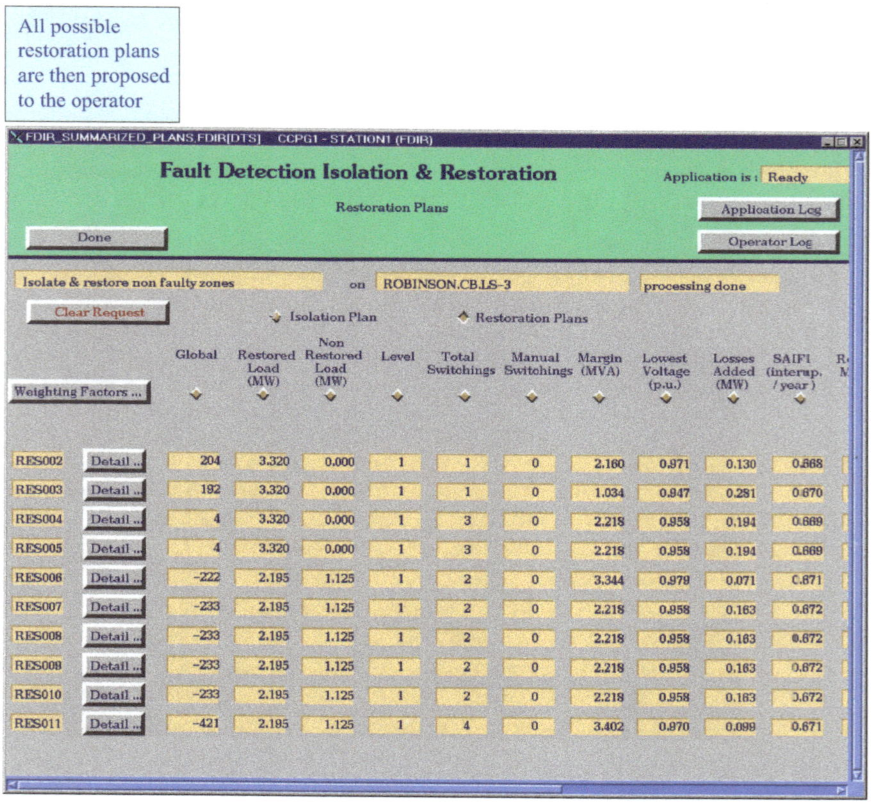

		Global	Restored Load (MW)	Non Restored Load (MW)	Level	Total Switchings	Manual Switchings	Margin (MVA)	Lowest Voltage (p.u.)	Losses Added (MW)	SAIFI (interup. / year)
RES002	Detail ...	204	3.320	0.000	1	1	0	2.160	0.971	0.130	0.668
RES003	Detail ...	192	3.320	0.000	1	1	0	1.034	0.947	0.281	0.670
RES004	Detail ...	4	3.320	0.000	1	3	0	2.218	0.958	0.194	0.669
RES005	Detail ...	4	3.320	0.000	1	3	0	2.218	0.958	0.194	0.669
RES006	Detail ...	−222	2.195	1.125	1	2	0	3.344	0.979	0.071	0.671
RES007	Detail ...	−233	2.195	1.125	1	2	0	2.218	0.958	0.163	0.672
RES008	Detail ...	−233	2.195	1.125	1	2	0	2.218	0.958	0.163	0.672
RES009	Detail ...	−233	2.195	1.125	1	2	0	2.218	0.958	0.163	0.672
RES010	Detail ...	−233	2.195	1.125	1	2	0	2.218	0.958	0.163	0.672
RES011	Detail ...	−421	2.195	1.125	1	4	0	3.402	0.970	0.099	0.671

Figure 6.29 Possible downstream restoration plan with FLISR. Reproduced by permission of ALSTOM

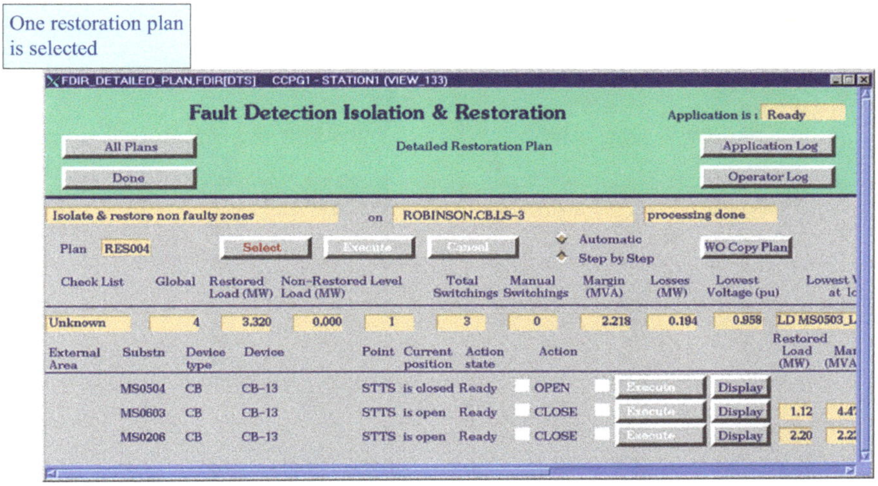

Figure 6.30 *Selection of the best option for restoration. Reproduced by permission of ALSTOM*

Figure 6.31 *Downstream restoration of healthy distributed system section. Reproduced by permission of ALSTOM*

6.4.6 FLISR local intelligence

Local intelligence has been used for many years even before SCADA systems were developed. Local intelligence is extremely simple to use, very reliable and does not require a special communication system, although in some cases, a good communication system helps one to obtain better results. Local intelligence performs actions like opening or closing devices in order to satisfy previously established conditions.

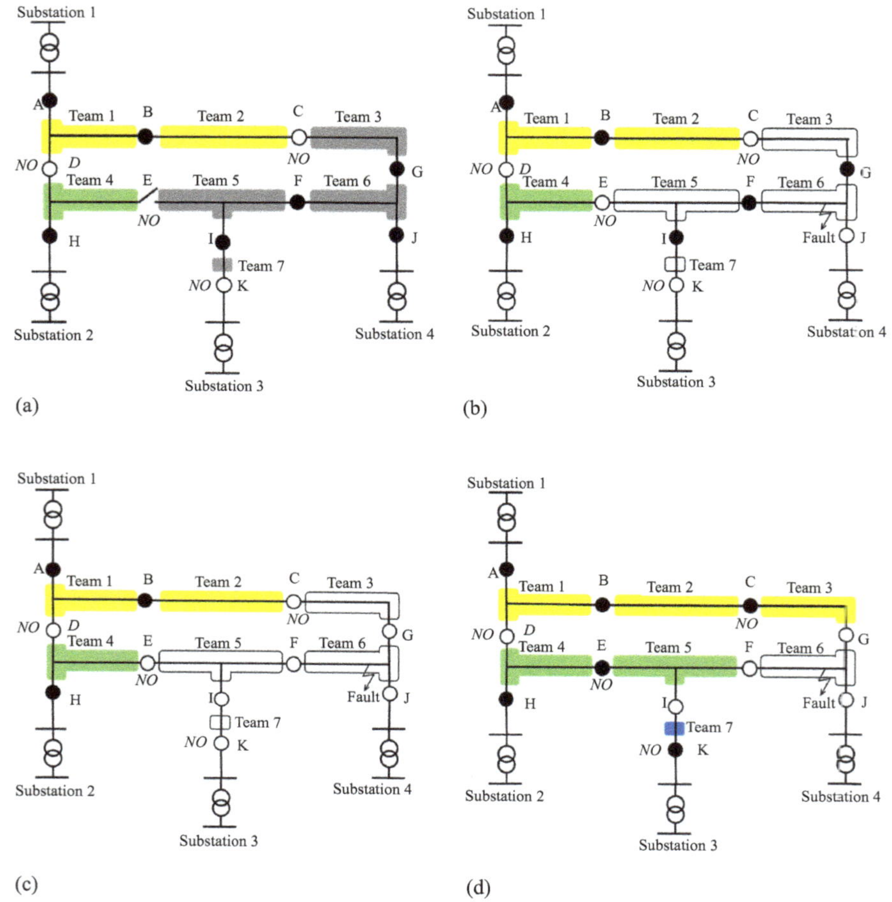

Figure 6.32 Example of an FLISR-distributed intelligence solution sequence: (a) normal configuration; (b) Fault in Team 6; (c) Fault is detected and initiates opening of switches; and (d) Service is restored to all the unfaulted segments

To illustrate the concept of local intelligence, a system with two feeders is presented in Figure 6.33. Each feeder has two switches. The first feeder, shown to the left, is fed from substation A and the second feeder to the right is fed from substation B. Figure 6.34 shows that a fault between switches R03 and R04 has occurred. Switch R04 operates first. Since the fault remains, additional operations are attempted until the switch locks out. Then the other switch operates in connection with the automatic backfeed restoration (ABR) function, attempting to close. The ABR function generates a close action if supply is lost from the load side of a switch that is NO. In this particular case, the fault is still present and the switch eventually locks out. In the end the, fault is cleared. This is shown in Figure 6.35.

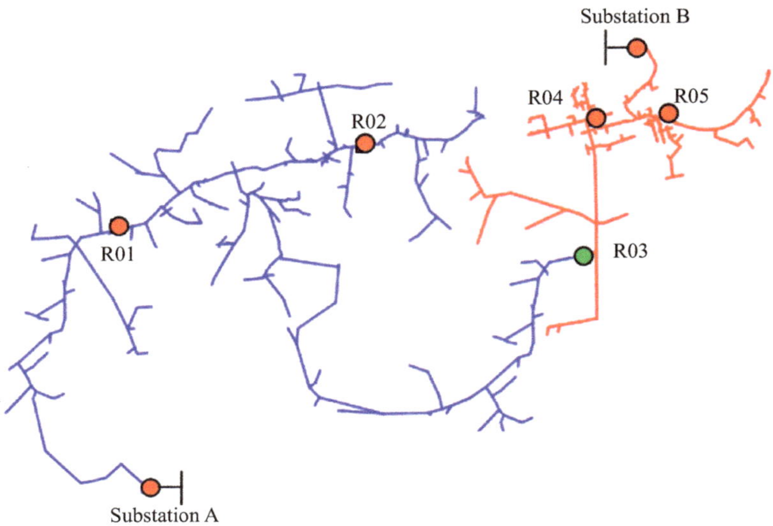

Figure 6.33 Example of a distribution system with FLISR local intelligence

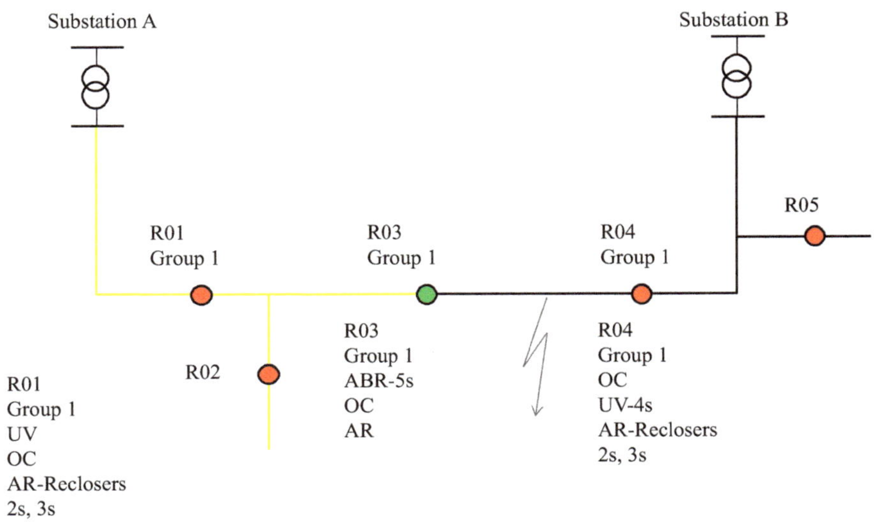

Figure 6.34 Fault between R03 and R04

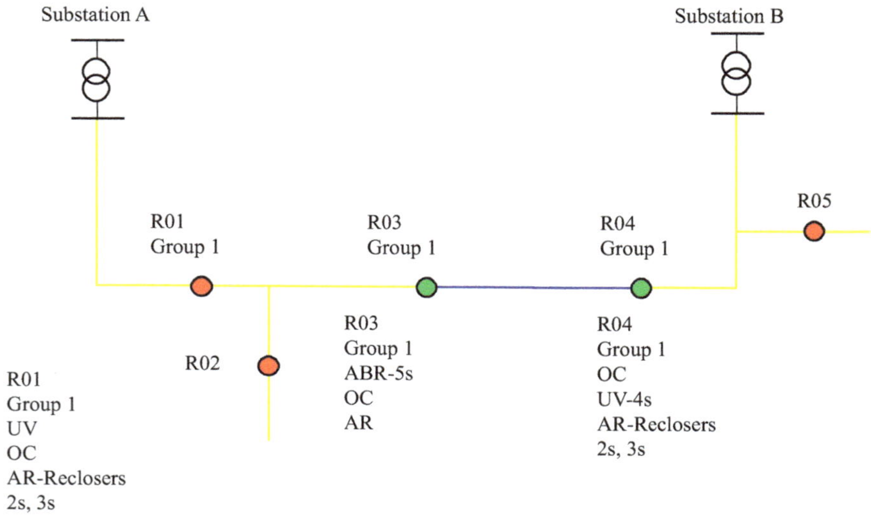

Figure 6.35 Final topology after local intelligence operation

References

[1] Civanlar, S., Grainger, H., Yin, H., and Lee, S. "Distribution feeder reconfiguration for loss reduction," IEEE Transactions on Power Delivery, vol. 3, pp. 1217–1223, 1988.
[2] Shirmohammadi, D., and Hong, H. W. "Reconfiguration of electric distribution networks for resistive line losses reduction," IEEE Transactions on Power Apparatus and Systems, vol. 4, pp. 1492–1498, 1989.
[3] NEPLAN. Switching Optimization Module. 2018.

Further reading

[1] Augugliaro, A., Dusonchet, L., Ippolito, M. G., and Sanseverino, E. R. "Minimum losses reconfiguration of MV distribution networks through local control of tie-switches," IEEE Transactions on Power Delivery, vol. 18, no. 3, pp. 762–771, 2003.
[2] González, A., Echavarren, F. M., Rouco, L., and Gómez, T. "A sensitivities computation method for reconfiguration of radial networks," IEEE Transactions on Power Systems, vol. 27, no. 3, pp. 1294–1301, 2012.
[3] Merlin, A., and Back, H. "Search for a Minimal-Loss Operating Spanning Tree Configuration in an Urban Power Distribution System," in Proc. of PSOC, Kota, Rajasthan, India, 1975.
[4] Billinton, R., and Jonnavithula, S. "Optimal switching device placement in radial distribution systems," IEEE Transactions on Power Delivery, vol. 11, no. 3, pp. 1646–1651, 1996.

[5] Lee, C., Liu, C., Mehrotra, S., and Bie, Z. “Robust distribution network reconfiguration,” IEEE Transactions on Smart Grid, vol. 6, no. 2, pp. 836–842, 2015.
[6] Gutiérrez-Alcaraz, G., and Tovar-Hernández, J. H. “Two-stage heuristic methodology for optimal reconfiguration and Volt/VAr control in the operation of electrical distribution systems,” IET Generation, Transmission & Distribution, vol. 11, no. 16, pp. 3946–3954, 2017.
[7] Viswanadha Raju, G. K., and Bijwe, P. R. “Efficient reconfiguration of balanced and unbalanced distribution systems for loss minimisation,” IET Generation, Transmission & Distribution, vol. 2, no. 1, p. 7, 2008.
[8] Russell, K. J., and Broadwater, R. P. “Model-Based Automated Reconfiguration for Fault Isolation and Restoration,” in 2012 IEEE PES Innovative Smart Grid Technologies (ISGT), Washington, DC, USA, 2012, pp. 1–4.
[9] Lin, C., Chen, C., Ku, T., Tsai, C., and Ho, C. “A multiagent-based distribution automation system for service restoration of fault contingencies,” European Transactions on Electrical Power, vol. 21, pp. 239–253, 2010.
[10] Lo, K. L., and Gers, J. M. “Feeder Reconfiguration for Losses Reduction in Distribution Systems,” in Proc. of the UPEC 94 Conference. University College Galway, Ireland, September 1994.
[11] Munasinghe, M. “Economic Principles and Policy Electricity Loss Reduction,” in Latin American Seminar on Electrical Losses Control, Bogota, DE, 1988.
[12] Pradeep, Y., Seshuraju, P., Khaparde, S. A., and Joshi, R. K. “CIM-based connectivity model for bus-branch topology extraction and exchange,” IEEE Transactions on Smart Grid, vol. 2, no. 2, pp. 24, 2011.
[13] Saini, S., and Mam, M. “Reconfiguration of an unbalanced distribution system for loss reduction by software simulation,” International Journal of Engineering Research & Technology, vol. 2, no. 6, pp. 3314–3318, 2013.
[14] Solanki, J. M., Khushalani, S., and Schulz, N. N. “A multi-agent solution to distribution systems restoration,” IEEE Transactions on Power Systems, vol. 22, no. 3, pp. 1026–1034, 2007.
[15] Uluski, R. W. “Using distribution automation for a self-healing grid,” in Transmission and Distribution Conference and Exposition (T&D), 2012 IEEE PES, Orlando, FL, USA, 7–10 May 2012, pp. 1, 5.
[16] Takenobu, Y., Yasuda, N., Kawano, S., Minato, S. I., and Hayashi, Y. “Evaluation of annual energy loss reduction based on reconfiguration scheduling,” IEEE Transactions on Smart Grid, vol. 9, no. 3, pp. 1–11, 2018.
[17] Wu, Y.-K., Lee, C.-Y., Liu, L.-C., and Tsai, S.-H. “Study of reconfiguration for the distribution system with distributed generators,” IEEE Transactions on Power Delivery, vol. 25, no. 3, pp. 1678–1685, 2010.
[18] Zhou, Q., Shirmohammadi, D., and Edwin-Liu, W. H. “Distribution feeder reconfiguration for operation cost reduction,” IEEE Transactions on Power Systems, vol. 12, no. 2, pp. 730–735, 1997.
[19] Zhou, Q., Shirmohammadi, D., and Edwin-Liu, W. H. “Distribution feeder reconfiguration for service restoration and load balancing,” IEEE Transactions on Power Systems, vol. 12, no. 2, pp. 724–729, 1997.

Chapter 7
Voltage/VAR control

Keeping voltages within predefined ranges is one of the main targets in the operation of power systems. To achieve this goal, reactive flow has to be monitored since the two variables are closely related as it was mentioned in Chapter 3. Therefore, it is common to treat both topics simultaneously.

In recent years and in particular with the development of Smart Grids, the term "VVC" was introduced. The term "VVC" stands for Volt/VAR control and refers to the technique of using voltage-regulating devices and reactive power controls to maintain voltage levels within the accepted ranges at all points of the distribution system under all loading conditions. Modern software techniques and the progress on communication technologies allow nowadays achieving these goals and therefore the service quality has improved remarkably in most utilities throughout the world.

Voltage regulation and reactive power control are performed by switching capacitor banks installed along the lines, substation transformers load tap changers (LTCs), shunt capacitor banks installed at substations, and voltage regulators. The control of the devices is discussed later in this chapter.

The main objectives of implementing VVC can be summarized as follows:

- Maintain acceptable voltages at all points along the feeder under all loading conditions.
- Increase the efficiency of distribution systems without violating any loading and voltage constrains.
- Support the reactive power needs of the bulks power system during system emergencies.
- Keep power factor within the accepted ranges that normally are above 0.9 inductive.

Voltages are the nominal single-phase supply voltages. All voltages are rms and therefore the peak AC voltage is greater by a factor of $\sqrt{2}$. In most countries, if not all, they are in the range of 100 V–240 V.

For the case of the United States, the normal rated voltage is 120 V according to ANSI C84.1-2016 [1]. Figure 7.1 shows a maximum of 126 V and a minimum of 114 V since the voltage regulation is 5%. In the United Kingdom, the statutory limits of the LV supply are 230 V -6% and 230 V $+10\%$. This represents a range from 216.2 V to 253 V [2].

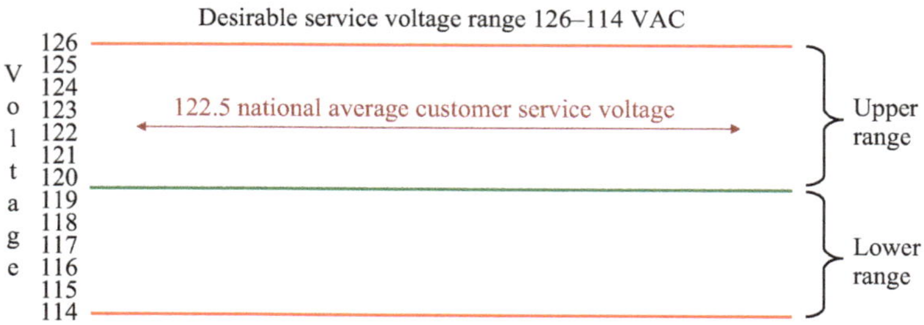

Figure 7.1 Rated voltages specified for the United States

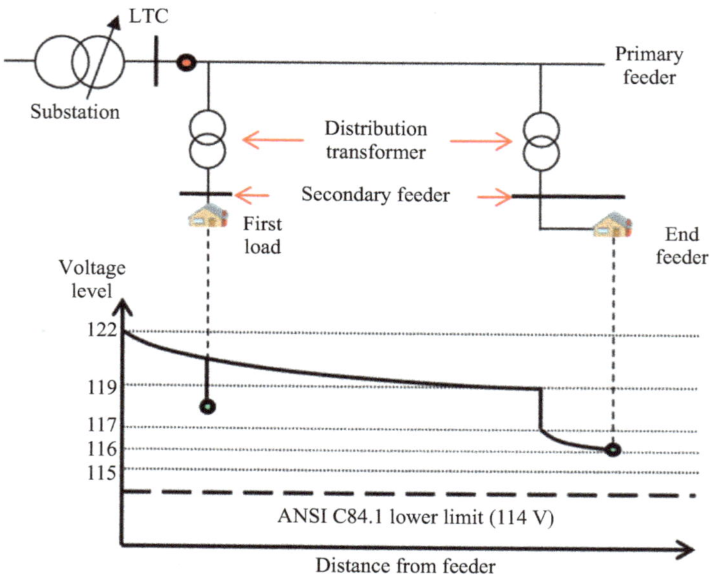

Figure 7.2 Voltage regulation limits in the United States

Figure 7.2 illustrates a portion of a typical system that illustrates the effect of the loading of a feeder on the voltage profile which risks the fulfillment of the voltage ranges specified by the standard.

Different devices are required to achieve the VVC, including on-LTCs, at the main substation transformers, voltage regulators, and capacitor banks. Distributed energy resources (DER) can also help the VVC on those feeders where they are deployed. flexible AC transmission system (FACTS) equipment (like STATCOM) and distributed generation (DG), if available, should be included under the VVC analysis and be part of it.

Some distribution utilities typically have delivery voltage in the upper portion of the range. This is a problem for the concept of conservation voltage reduction or

CVR, which is maintained to deliver voltage to the customer in the lower portion of the acceptable range.

Now the question is, why is it expected to reduce the voltage? CVR suggests that if the voltage is slightly reduced at the customer side end but within the standard limits, energy can be saved and efficiency can also be achieved by a reduction of peak demand and power losses. Load can be decreased by approximately 1.0% for each 1.0% of voltage drop at the substation.

In order to perform voltage reduction at the substation side, the feeder must have a voltage profile sufficiently above the lower voltage level if no compensation is applied. Otherwise, the voltage at the end side will be out of the allowable limits during peak condition.

Therefore, a proper implementation of CVR requires information like load characteristics, load profiles, feeder characteristics, and if applicable, size and location of capacitor banks, voltage regulators, and DER technologies.

A common application of CVR for heavily loaded feeders with a significant voltage drop at the feeder's end is installing capacitor banks, or other voltage-regulation technologies along the distribution feeder. In this way, the voltage profile is supported within acceptable limits, allowing for substation voltage reduction when required.

7.1 Definition of voltage regulation

Voltage regulation can be defined as the percent voltage drop of a line (e.g., a feeder) with respect to the receiving-end voltage.

Therefore

$$\%\,regulation = \frac{|V_s| - |V_r|}{|V_r|} \times 100 \tag{7.1}$$

where $|V_s|$ is the sending voltage, $|V_r|$ is the receiving-end voltage.

7.2 Options to improve voltage regulation

There are numerous ways to improve the distribution system's overall voltage regulation:

- use of generator voltage regulators,
- application of voltage-regulating equipment in the distribution substations,
- application of capacitors in the distribution substation,
- balancing of the loads on the primary feeders,
- increasing feeder conductor size,
- changing feeder sections from single phase to multiphase,
- transferring loads to new feeders,
- installing new substations and primary feeders,
- increasing of primary voltage level,

- application of voltage regulators out on the primary feeders,
- application of shunt capacitors on the primary feeders,
- application of series capacitors on the primary feeders, and
- use of DG or DER.

The commonest way to regulate voltage in power systems is by means of tap changers in transformers. Normally, they are used to control the voltage level at the LV side of transformers associated with transmission or distribution substations. They could control voltage at the HV side in generator plants. However, this is not usual since in this case the voltage is controlled by the excitation of the generators.

Tap changers can be of the no-load operating type or on-load operation type (OLTC). The first one is used mainly on small transformers or nonimportant transformers. OLTC tap changers could be of mechanical type or thyristor assisted. Mechanical tap changers physically make a connection before releasing the previous connection point employing different tap selectors. They avoid the presence of elevated circulating currents through a diverter switch that provides a temporary high impedance value in series with the short-circuited winding section.

Tap changers assisted by thyristors are used to take the load current, while the main contacts switch from one source to the other. This prevents arcs in the main contacts and can result in an extended life cycle in the maintenance activities. They require an additional low voltage supply for the thyristor circuit. This type of tap changers is not very popular due to their complexity and the elevated cost.

Calculations of voltage regulation with tap changes are rather straightforward and were presented in Chapter 3 that has discussed the basics of load flow.

The most economical way of regulating the voltage along the feeders within the required limits is to apply both step voltage regulators and shunt capacitors. Of course, a fixed capacitor is not a voltage regulator and cannot be directly compared to regulators; however, in some cases, automatically switched capacitors can replace conventional step-type voltage regulators for voltage control on distribution feeders. The following sections make specific reference to voltage regulators and capacitors.

7.3 Voltage regulators

A step-type voltage regulator is fundamentally an autotransformer with many taps (or steps) in the series winding that adjusts itself automatically by changing the taps until the desired voltage is obtained. It can be either station-type or distribution-type.

Voltage regulators manufactured by Cooper Power Systems are very popular. They are designed to correct the line voltage from 100% boost to 10% buck (i.e., 10%) in 32 steps, with a 5% or 8% voltage change per step as shown in Figure 7.3. The effect of each step is shown in Figure 7.4.

The voltage regulators can be located at the substation busbar or on the feeders. The settings are calculated as to keep the voltage within certain ranges at a regulating point along the feeders. The procedure to determine the settings has been

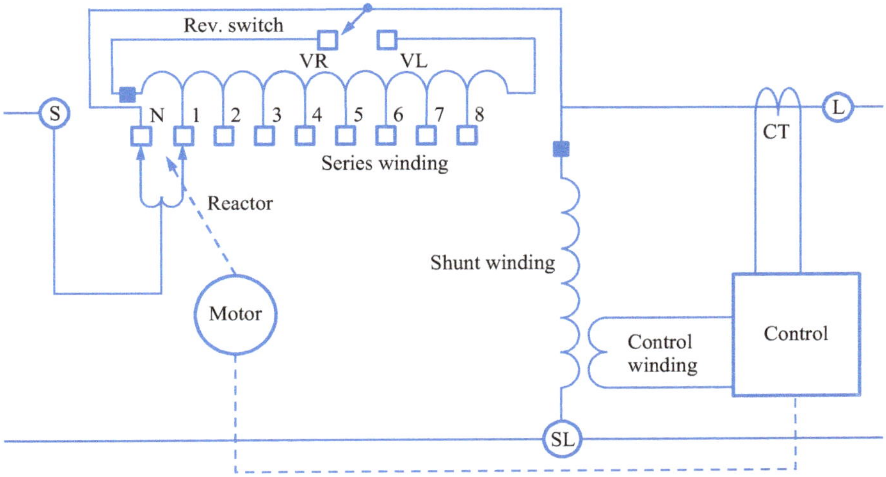

Figure 7.3 Schematic of a single-phase 32-step voltage regulator

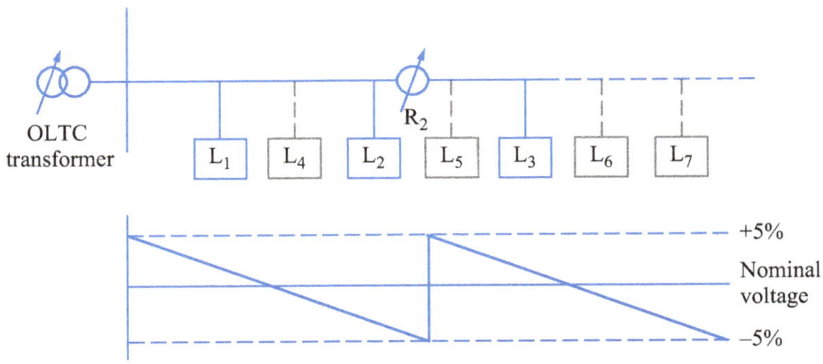

Figure 7.4 Voltage profile with step-type voltage regulators

very well treated in the book *Electric Power Distribution System Engineering* by Gönen [3], from which (7.2)–(7.7) and the corresponding explanations have been taken.

The settings correspond to the dial of the resistance and reactance elements built in the so-called line-drop compensator (LDC) that is in the control panel of the device.

In case, there is no load tapped off between the regulator and the controlled point, then the R dial setting of the LDC is determined from

$$R_{set} = \frac{CT_P}{PT_N} R_{eff} \tag{7.2}$$

where CT_P is the rating of the current transformer's primary, PT_N is the potential transformer's turns ratio (V_{pri}/V_s), R_{eff} is the effective resistance of a feeder conductor from regulator station to regulation point, Ω.

$$R_{eff} = r_a \frac{l - s_1}{2} \tag{7.3}$$

where r_a is the resistance of a feeder conductor from regulator station to regulation point, Ω/mi per conductor; s_1 is the length of three-phase feeder between regulator station and substation, mi (multiply the length by 2 in feeder as in single phase); and l is the primary feeder length, mi.

Also, the *Z*-dial setting of the LDC can be determined from

$$X_{set} = \frac{CT_P}{PT_N} X_{eff} \tag{7.4}$$

where X_{eff} is the effective reactance of a feeder conductor from regulator to regulation point, Ω.

$$X_{eff} = x_L \frac{l - s_1}{2} \tag{7.5}$$

and

$$x_L = x_a + x_d \ (\Omega/\text{mi}) \tag{7.6}$$

where x_a is the inductive reactance of individual phase conductor of feeder at 12-in spacing, x_d is the Inductive-reactance spacing, and x_L is the Inductive reactance of feeder conductor.

The difference between the two voltage values is the total voltage drop between the regulator and the regulation point, which can also be defined as

$$VD = |I_L| \times R_{eff} \times \cos\theta + |I_L| \times X_{eff} \times \sin\theta \tag{7.7}$$

from which R_{eff} and X_{eff} values can be determined easily if the load power factor of the feeder and the average *R*/*X* ratio of the feeder conductors between the regulator and the regulation point are known.

Automatic voltage regulation is provided by bus regulation at the substation, or individual feeder regulation in the substation, or supplementary regulation in the main by regulators mounted on poles or a combination of the previous instances.

Many utilities have experienced that the most economical way of regulating the voltage within the required limits is to apply both step voltage regulators and shunt capacitors.

7.4 Capacitor application in distribution systems

Capacitor location in distribution networks is a very important option to reduce electrical losses and therefore it has been widely used. Capacitors not only help one

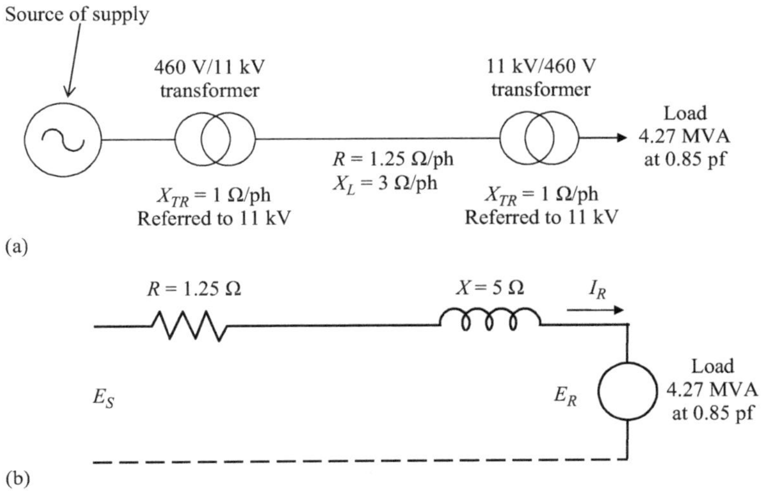

Figure 7.5 Diagrams for Example 7.1: (a) system diagram and (b) equivalent circuit

to save losses but also play an important role in power factor correction and in the improvement of the voltage profile, especially in long feeders. Even with the benefits of capacitor location in distribution systems, their location has to be analyzed carefully not only for the high costs involved but also for the overvoltages that capacitors may produce in networks with harmonic circulation when resonance conditions may occur.

A capacitor is a device consisting essentially of two electrodes separated by a dielectric insulating material that is capable of supplying magnetizing kVAR to the system. The capacitive reactance has the nature of a negative inductive reactance. This property is utilized in electrical circuits to compensate the effects of inductive reactance and the lagging reactive kVA of inductive loads.

Capacitors can be classified as series or parallel according to the type of connection they have. Series capacitors are connected in series with lines to compensate for inductive reactance. Shunt capacitors are connected in parallel with lines to compensate the reactive power or current required by an inductive load.

Example 7.1 Consider a simple 11 kV radial line transmitting power by an overhead system to a lagging power factor load. Figure 7.5 presents the system, the system diagram, and the equivalent circuit.

Analyze the sending and receiving-end conditions for each of the following three cases:

(a) Without capacitors: The line-to-line receiving-end voltage is assumed constant at 11 kV. All calculations are referred to 11 kV base voltage.

(b) With shunt capacitors: Three single-phase capacitor units, each having a reactance of 45.7 Ω, are connected between phase and neutral adjacent to the load. Calculate the current taken by each capacitor, I_C, I_R, and E_s.
(c) With series capacitors: In this case, a capacitor unit is connected in series with each phase, the reactance of each unit being 5 Ω. Calculate I_R and E_s.

For the case (a):

The line-to-line receiving-end voltage is assumed constant at 11 kV. All calculations are referred to 11 kV base voltage.

$$I_R = \frac{4.27 \text{ MVA}}{\sqrt{3} \cdot 11 \text{ kV} \cdot 0.85} \angle(-\cos^{-1}(0.85)) = 264\angle(-31.78°) \text{ A per phase}$$

Taking E_R as reference vector,

$$\begin{aligned} E_S &= E_R + \left[\sqrt{3}(I_R\angle(\varphi_R))(R + jX_L) \cdot 10^{-3}\right] \\ &= 11 + \left[\sqrt{3}(264\angle(-31.78°)) \cdot (1.25 + j5) \cdot 10^{-3}\right] \text{ (in kV)} \\ &= 11 + [2.36\angle(44.18°)] \\ &= 12.79\angle 7.37° \end{aligned}$$

Figures 7.6 and 7.7 show the vector diagram for the current case. That is the base case for the other points.

For the case (b):

Three single-phase capacitor units, each having a reactance of 45.7 Ω, are connected between phase and neutral adjacent to the load.

Then, current taken by each capacitor:

$$I_C = \frac{11{,}000}{\sqrt{3} \cdot 45.7} \angle(90°) \text{ (in A)}$$

where $I_C = 139\angle(90°)$ A leads the applied voltage by 90°.

$$\begin{aligned} \text{The inductive component of the load current} &= 264 \sin(31.78°)\angle(-90°) \\ &= 139\angle(-90°) \end{aligned}$$

lagging the applied voltage by 90°

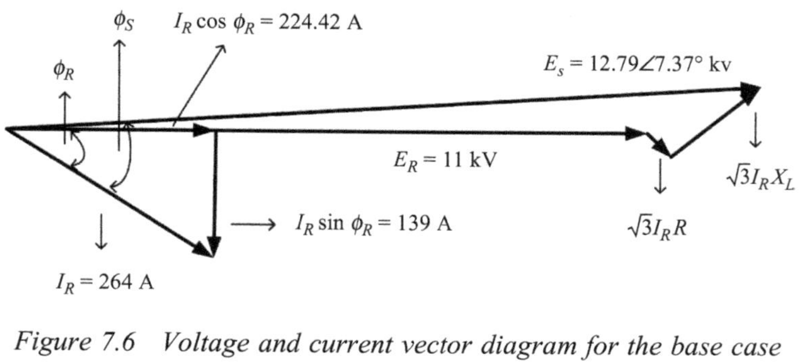

Figure 7.6 Voltage and current vector diagram for the base case

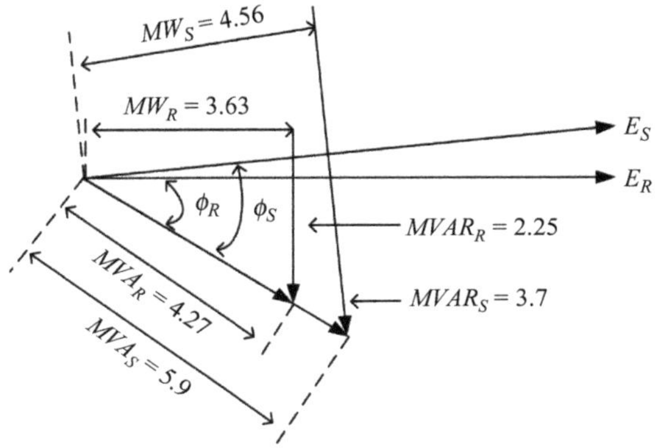

Figure 7.7 Power diagram for the base case

Therefore, the current taken by the capacitor will neutralize the inductive component of the load current, and the actual current taken from the line will be only 224.42 A, in phase with the applied voltage:

$$I_R = 224.42\angle(0^\circ)$$
$$E_S = 11 + \left[\sqrt{3} \cdot 224.42 \cdot (1.25 + j5) \cdot 10^{-3}\right] \text{ kV}$$
$$E_S = 11.65\angle(9.6^\circ) \text{ kV}$$

For the case (c):

In this case, a capacitor unit is connected in series with each phase, the reactance of each unit being 5 Ω.

$$\begin{aligned} I_R &= 264\angle(-31.78^\circ) \\ E_S &= E_R + \left[\sqrt{3}(I_R\angle(\varphi_R))(R + j(X_L - X_C)) \cdot 10^{-3}\right] \\ &= 11 + \left[\sqrt{3}(264\angle(-31.78^\circ)) \cdot (1.25 + j(5 - 5)) \cdot 10^{-3}\right] \text{ (in kV)} \\ &= 11 + [0.57\angle(-31.78^\circ)] \\ &= 11.48\angle(-1.5^\circ) \end{aligned}$$

For the purpose of comparison, the relevant sending and receiving-end conditions for the three cases are shown in Table 7.1.

The following can be seen in the previous table:

1. Both shunt and series capacitors reduce the voltage drop and also the MVAR demand at source of supply.
2. The MVAR rating or the shunt capacitor bank for the three phases
$$= 3 \cdot I_C{}^2 \cdot X_C \cdot 10^{-6}\,\text{MVAR} = 3 \cdot (224.42)^2 \cdot 45.7 \cdot 10^{-6}\,\text{kVAR} = 6.9\text{ MVAR}$$

The MVAR rating of the series capacitor bank $= 3 \cdot (264)^2 \cdot 5 \cdot 10^{-6}$ MVAR
$= 1.05$ VAR

Table 7.1 Results comparison for Example 7.1

Type of system	Line voltage		Voltage drop	Power factor		Sending-end power			Receiving-end power			Active power loss
	Sending end (kV)	Receiving end (kV)		Sending end	Receiving end	Active power (MW)	Reactive power (MVAR)	Apparent power (MVA)	Active power (MW)	Reactive power (MVAR)	Apparent power (MVA)	
Case (a) without capacitors	12.79	11	1.8 kV (14.0%)	0.78 lag	0.85 lag	4.56	3.7 lag	5.9	4.3	2.7 lag	5.1	0.8 MW (13.6%)
Case (b) with shunt capacitors	11.65	11	1.65 kV (5.9%)	0.986 lag	1	4.47	0.76 lag	4.53	4.3	Nil	4.3	0.17 MW (3.9%)
Case (c) with series capacitors	11.48	11	0.48 kV (4.4%)	0.864 lag	0.85 lag	4.53	2.65 lag	5.3	4.3	2.7 lag	5.1	0.23 MW (5.35%)

Thus, for the same improvement in voltage regulation, the series capacitor bank is much smaller than the shunt capacitor bank. At the same time, it should be noted that series capacitors reduce MVAR demand at source far less than shunt capacitors do.

3. Shunt capacitors reduce the line power losses by reducing the receiving-end current.

Table 7.2 shows a comparison between the shunt capacitor and the series capacitor applications.

7.4.1 Feeder model

A general model is considered here, which corresponds to a distributed load along the feeder with the possibility of having a concentrated load at the end, the value of which can also be set to zero, as proposed by Neagle and Samson [4]. If the total feeder current at the substation end is I_1 and the concentrated load at the other end takes a current I_2, the expression for the current along the feeder depends on the distance from the substation according to the following equation:

$$i(x) = I_1 - (I_1 - I_2)x \tag{7.8}$$

If a factor $p = I_2/I_1$ is introduced, the equation becomes

$$i(x) = I_1[(p-1)x + 1] \tag{7.9}$$

The profile corresponding to that expression is shown in Figure 7.8. If the load is uniformly distributed along the feeder and there is not concentrated load at the end, I_2 and therefore p is equal to zero, and (7.9) becomes

$$i(x) = I_1(1 - x) \tag{7.10}$$

Likewise, if the load is concentrated at the end of the feeder, $p = 1$ and therefore the expression for the current is

$$i(x) = I_1 = I_2 \tag{7.11}$$

Table 7.2 Comparison between shunt capacitor and series capacitor

	Shunt capacitor	**Series capacitor**
Size (Ω)	45.7	5
Voltage regulation	Good	Good
Losses reduction	Very good	Nothing
Power factor	Very good	Fair
Stability	Good	Very good
Cost	Reasonable	High

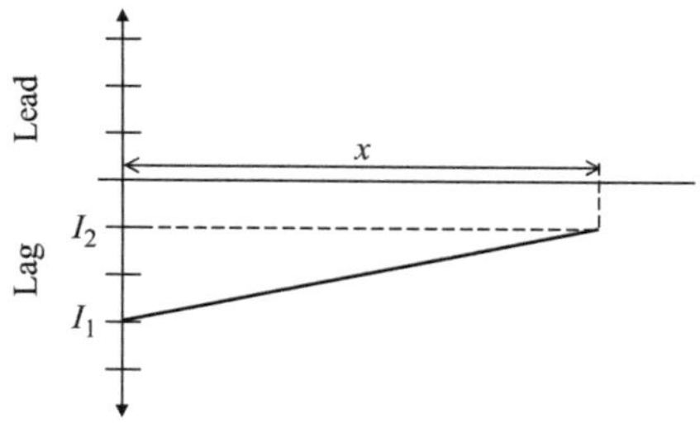

Figure 7.8 Current profiles for feeders with uniformly distributed loads

7.4.2 Capacitor location and sizing

The reduction in losses by using network reconfiguration can be further enhanced by placing capacitors along the feeders. The following sections will develop theories for such applications that have been proposed in several works. The different magnitudes, i.e., power, energy, impedance, and current, will be dealt with in pu values.

The power loss dissipated in a circuit with a total current I, a resistance value R, and a power factor angle ϕ can be expressed as

$$P_1 = I^2R = (I\cos\phi)^2R + (I\sin\phi)^2R \tag{7.12}$$

If a capacitor with a current I_c is installed, the reactive part of the current will be compensated and therefore the new total losses (P_2) can be expressed as

$$P_2 = I^2R = (I\cos\phi)^2R + (I\sin\phi - I_C)^2R \tag{7.13}$$

Therefore, the losses reduction can be found as

$$\Delta P = P_1 - P_2 = 2IRI_C\sin\phi - I_C{}^2R \tag{7.14}$$

Equation (7.14) shows that only the reactive component of the load current is required for power loss reduction studies. Therefore, in the rest of this section that component will be referred to as I_{1R}.

If a feeder has a uniformly distributed loading whose current is that given by (7.9), the total losses due to the *reactive* component without capacitors will again be referred to as P_1 and can be found with the following equation:

$$P_1 = 3\int_0^1 [I_{1R}(1 + (p-1)x)]^2 R\,dx \tag{7.15}$$

where p is the I_{2R}/I_{1R}, I_{1R} is the reactive current at the substation end, I_{2R} is the reactive current at the feeder end, and x is the distance along the feeder ranging from 0 to 1.

The result is given in the following equation:

$$P_1 = RI_{1R}^2 \left(p^2 + p + 1 \right) \tag{7.16}$$

7.4.3 Reduction in power losses with one capacitor bank

Figure 7.9 shows a distribution feeder with one capacitor bank installed at a distance a from the source.

If the capacitor bank takes a current I_c as shown in Figure 7.9, the new total power losses P_2, due to the reactive current, including the capacitor bank, are calculated as

$$P_2 = 3 \int_0^a \left[I_{1R}(1 + (p - 1)x) - I_C \right]^2 R dx + 3 \int_a^1 \left[I_{1R}(1 + (p - 1)x) \right]^2 R dx \tag{7.17}$$

The result of this expression is

$$P_2 = RI_{1R}^2 \left(p^2 + p + 1 \right) - 3RI_{1R}^2 \left[(p - 1) \frac{I_C}{I_{1R}} a^2 + 2 \frac{I_C}{I_{1R}} a - \left(\frac{I_C}{I_{1R}} \right)^2 a \right] \tag{7.18}$$

The power loss reduction due to the capacitor installation is found by subtracting (7.18) from (7.16) as follows:

$$\Delta P = P_1 - P_2 = 3RI_{1R}^2 \left[(p - 1) \frac{I_C}{I_{1R}} a^2 + 2 \frac{I_C}{I_{1R}} a - \left(\frac{I_C}{I_{1R}} \right)^2 a \right] \tag{7.19}$$

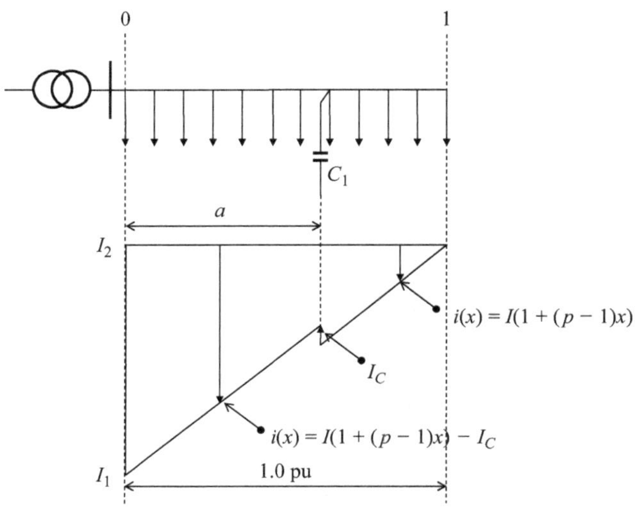

Figure 7.9 Distribution feeder with one capacitor bank

The optimum sizing and location of one capacitor bank to reduce power losses is found by taking partial derivatives of (7.19) with respect to I_c and the distance a, as follows:

$$\frac{\partial \Delta P}{\partial I_C} = 3RI_{1R}{}^2\left[(p-1)\frac{1}{I_{1R}}a^2 + 2\frac{1}{I_{1R}}a - 2\frac{I_C}{I_{1R}{}^2}a\right] \tag{7.20}$$

$$\frac{\partial \Delta P}{\partial a} = 3RI_{1R}{}^2\left[2(p-1)\frac{I_C}{I_{1R}}a + 2\frac{I_C}{I_{1R}} - \left(\frac{I_C}{I_{1R}}\right)^2\right] \tag{7.21}$$

By equating (7.20) and (7.21) to zero and solving simultaneously for I_c and a, the following values are obtained:

$$I_C = \frac{2}{3}I_{1R} \tag{7.22}$$

$$a = \frac{2}{3(1-p)} \tag{7.23}$$

The maximum power loss reduction with one capacitor is obtained by replacing (7.22) and (7.23) into (7.19). The corresponding expression is

$$\Delta P_{\mathrm{MAX}} = RI_{1R}^2\frac{8}{9(1-p)} \tag{7.24}$$

Equations (7.23) and (7.24) are valid only if $p \le 1/3$. For larger values of p, the magnitude of a becomes greater than 1 or even negative, which does not have physical meaning. Therefore, for $p > 1/3$, the capacitor bank should be located at the end of the feeder that corresponds to a value of a that is equal to 1. If this value of a is introduced in (7.19), the value of the capacitor current is

$$I_c = \frac{1}{2}I_{1R}(p+1) \quad \text{for } p > 1/3 \tag{7.25}$$

Likewise, if the value of I_c given in (7.23) and $a = 1$ are introduced in (7.19), the reduction in losses when a capacitor bank is installed and $p > 1/3$ is given by the following equation:

$$\Delta P = \frac{3}{4}RI_{1R}^2(p+1)^2 \quad \text{for } p > 1/3 \tag{7.26}$$

7.4.4 Reduction in power losses with two capacitor banks

Figure 7.10 shows a distribution feeder with two capacitor banks: one installed at a distance a and the other at a distance b from the source.

Initial losses:

$$P_1 = 3\int_0^1 [I(1-x)]^2 R\,dx = I^2R \tag{7.27}$$

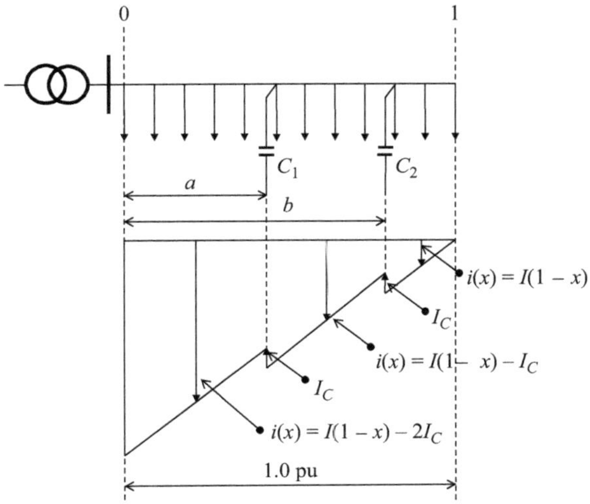

Figure 7.10 Distribution feeder with two capacitor banks

Losses if capacitors are connected:

$$P_2 = 3\int_0^a R[I(1-x) - 2I_C]^2 dx + 3\int_a^b R[I(1-x) - I_C]^2 dx + 3\int_b^1 R[I(1-x)]^2 dx$$
$$= 3R\left(II_C a^2 - 2II_C a + 3I_C{}^2 a - 2II_C b + II_C b^2 + I_C{}^2 b + \frac{I^2}{3}\right) \quad (7.28)$$

Therefore, the loss reductions are expressed as

$$\Delta P = 3R\left(a\left(2II_C - II_C a - 3I_C{}^2\right) + b\left(2II_C - II_C b - I_C{}^2\right)\right) \quad (7.29)$$

It can be demonstrated that the optimal values of a and b *are*

$$a = 1 - \frac{3I_C}{2I} \quad (7.30)$$

$$b = 1 - \frac{I_C}{2I} \quad (7.31)$$

7.4.5 Losses reduction with three capacitor banks

$$\Delta P = 3R\left(a\left(2II_C - II_C a - 5I_C{}^2\right) + b\left(2II_C - II_C b - 3I_C{}^2\right) + c\left(2II_C - II_C c - I_C{}^2\right)\right) \quad (7.32)$$

The optimum localization of capacitors:

$$a = 1 - \frac{5I_C}{2I} \tag{7.33}$$

$$b = 1 - \frac{3I_C}{2I} \tag{7.34}$$

$$c = 1 - \frac{I_C}{2I} \tag{7.35}$$

7.4.6 *Consideration of several capacitor banks*

The expressions presented in (7.19), (7.29), and (7.32) can be generalized for any feeder with n sections to find the power and energy loss reductions when capacitors are located in different places.

For more than one segment on the feeder, expressions (7.19), (7.29), and (7.32) are still valid, although new techniques that consider heuristic strategies have been proposed. Under these circumstances, the expressions can be used by putting $a = 1$, which is then directly applicable to one segment at a time. Therefore, when a capacitor is installed downstream at a segment n, the reduction in the power losses is given as a function of that capacitor current, and of currents at the beginning and at the end of the segment, respectively, as follows:

$$\Delta P_m = P_{1m} - P_{2m} = 3R_m\left[(I_{1Rm} + I_{2Rm})I_{cn} - I_{cn}^2\right] \tag{7.36}$$

For the segment where a capacitor is located, the expressions for power and energy have already been developed. General expressions are obtained by considering the effect of one capacitor at a time for every section toward the source as follows:

$$\Delta P = \Delta P_n + \sum_{m=1}^{n-1} \Delta P_m \tag{7.37}$$

That equation means that for a capacitor installed in a segment n, the power loss reduction is

$$\Delta P = 3R_n I_{1Rn}^2 \left[(p-1)\frac{I_{cn}}{I_{1Rn}}a^2 + 2\frac{I_{cn}}{I_{1Rn}}a - \left(\frac{I_{cn}}{I_{1Rn}}\right)^2 a\right] + \sum_{m=1}^{n-1} 3\left[(I_{1Rm} + I_{2Rm})I_{cn} - I_{cn}^2\right]R_m \tag{7.38}$$

Likewise, if a number of k banks are considered,

$$\Delta P = \sum_{n=1}^{k}\left\{3R_n I_{1Rn}^2\left[(p-1)\frac{I_{cn}}{I_{1Rn}}a^2 + 2\frac{I_{cn}}{I_{1Rn}}a - \left(\frac{I_{cn}}{I_{1Rn}}\right)^2 a\right] + \sum_{m=1}^{n-1} 3\left[(I_{1Rm}+I_{2Rm})I_{cn} - I_{cn}^2\right]R_m\right\}$$

7.4.7 Capacitor sizing and location using software

In order to perform the calculation of the optimal number and size of capacitor banks, the following information is required:

- maximum size allowed (kVAR) for each installation,
- available places (poles),
- maximum number of banks,
- step in kVAR of the bank,
- load factor, and
- existing banks.

Figure 7.11 corresponds to a real distribution system with 12 feeders. The losses of the system are high and the power factor low. Therefore, possibilities of improving the system are considered by installing capacitor banks in the most critical feeders.

One of the feeders is considered to determine the number and size of capacitor banks which is shown in Figure 7.12. Two conditions of load factors are analyzed: 0.6 and 0.9. Two locations are allowed of banks having steps of 50 kVAR. A maximum of 2 MVAR has been established.

Figure 7.11 Distribution system to illustrate application of capacitors

After running the analysis with a software package, two banks are considered for each loading factor as shown in Table 7.3. An economical analysis not shown in this example is recommended to define the option to select from the two presented in the results.

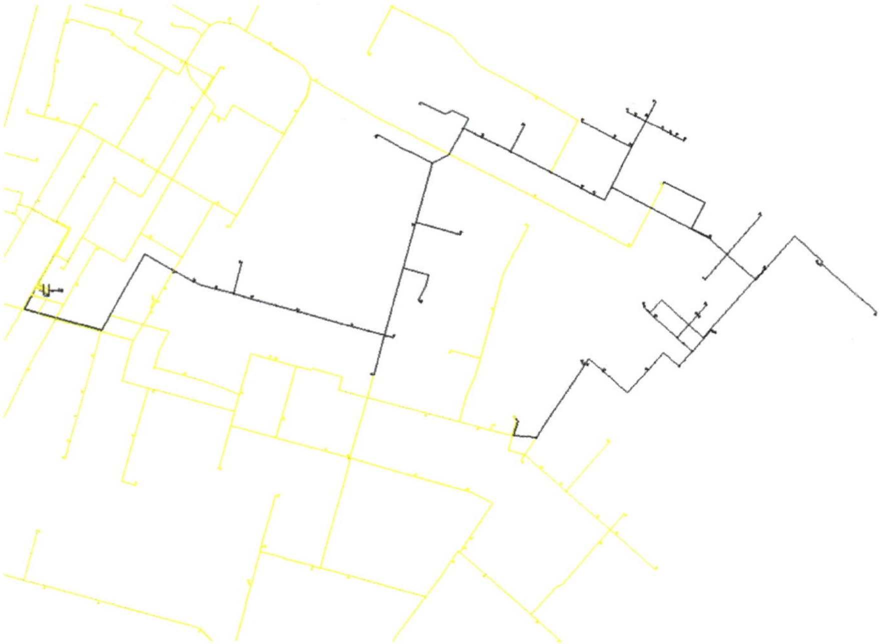

Figure 7.12 Feeder used for the illustration

Table 7.3 Results of software simulation

Load factor	Bus name	Size (kVAR)	Loss (MW)	Additional loss reduction (%)
0.6			**0.2502**	
	306	1,150	0.1972	21.1
	280	850	0.1821	6.03
0.9			**0.6208**	
	306	1,750	0.4839	22
	301	250	0.4709	2.09

Notes: Bold values represent the reference case of losses for the corresponding load factor. Each line shows the losses reduction when the indicated capacitor bank is installed in the system. Therefore, last line shows the losses in the systems after the integration of all the capacitor banks (0.1821 MW for a load factor of 0.6 and 0.4709 MW for a load factor of 0.9).

7.5 Modeling of distribution feeders, including VVC equipment

Modeling of distribution feeders involving the elements previously mentioned can be a comprehensive task due in particular to the nonsymmetry that distribution systems have. All the details of modeling distribution lines are beyond the scope of this book.

In order to illustrate the modeling, a small system taken from IEEE will be considered [5]. The data from the model are introduced in the illustration presented later. The system has the following characteristics:

1. short and relatively highly loaded for a 4.16 kV feeder,
2. one substation voltage regulator consisting of three single-phase units connected in wye,
3. overhead and underground lines with variety of phasing,
4. shunt capacitor banks,
5. in-line transformer, and
6. unbalanced spot and distributed loads.

Figure 7.13 presents a radial feeder taken from IEEE that will be used to illustrate the modeling and the type of analysis that can be performed.

Figure 7.14 shows the modeling of the voltage regulator installed in the system considered, taken from NEPLAN resources [6]. The modeling requires a previous analysis of the device used and the setting of the taps, based on the system requirements.

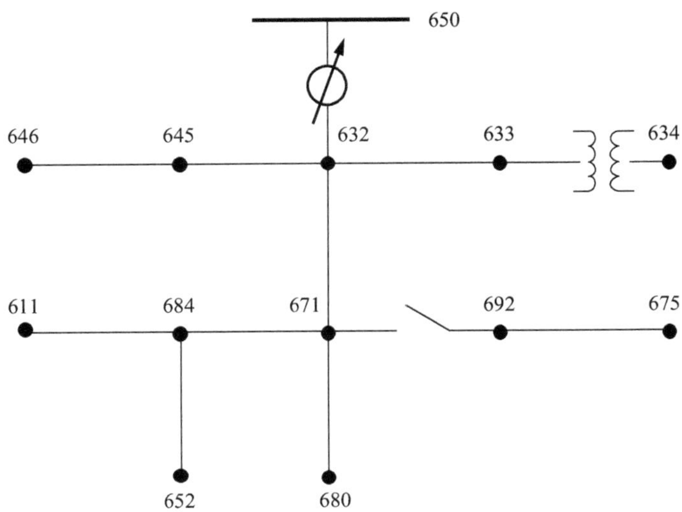

Figure 7.13 Layout of the feeder used in the modeling [5]

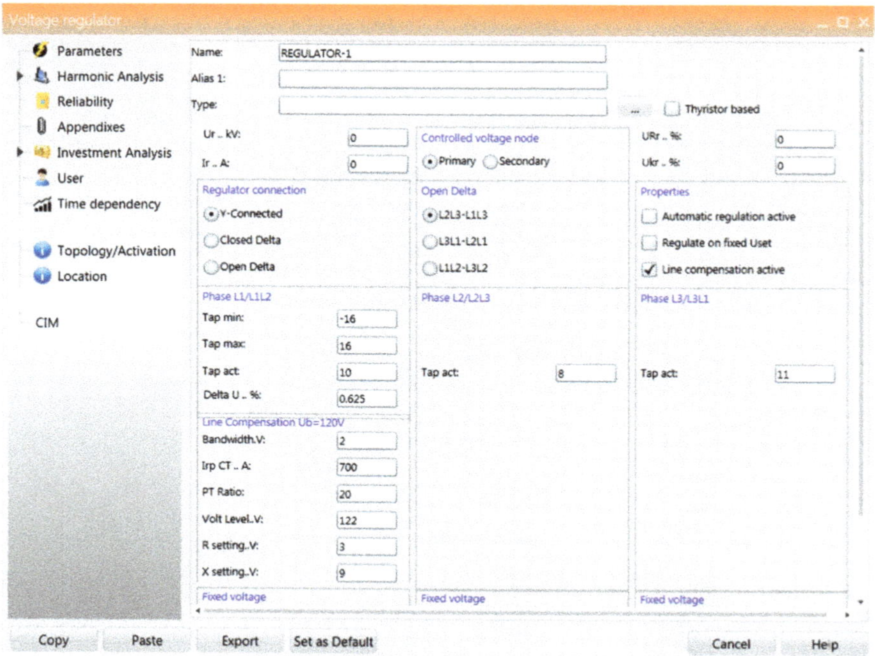

Figure 7.14 Typical voltage regulator modeling input data [6]

Figure 7.15 shows the load flow results for the system modeled, including voltage levels at the different nodes and active and reactive flow through all the elements.

7.6 Voltage/VAR control considering SCADA

Years ago, the information to monitor and command equipment in charge of the control of the voltage and reactive power came from the current and voltage transformers feeding them. These days, much better results are achieved by employing software packages interacting with SCADA systems. Figure 7.16 illustrates various elements of a VVC which are controlled and monitored by a SCADA system.

Distributed resources could present a challenge in the implementation of VVC. The impact on feeders can be remarkable based on their position in the system and the generator size; consequently, corresponding coordination with the VVC devices must be accomplished.

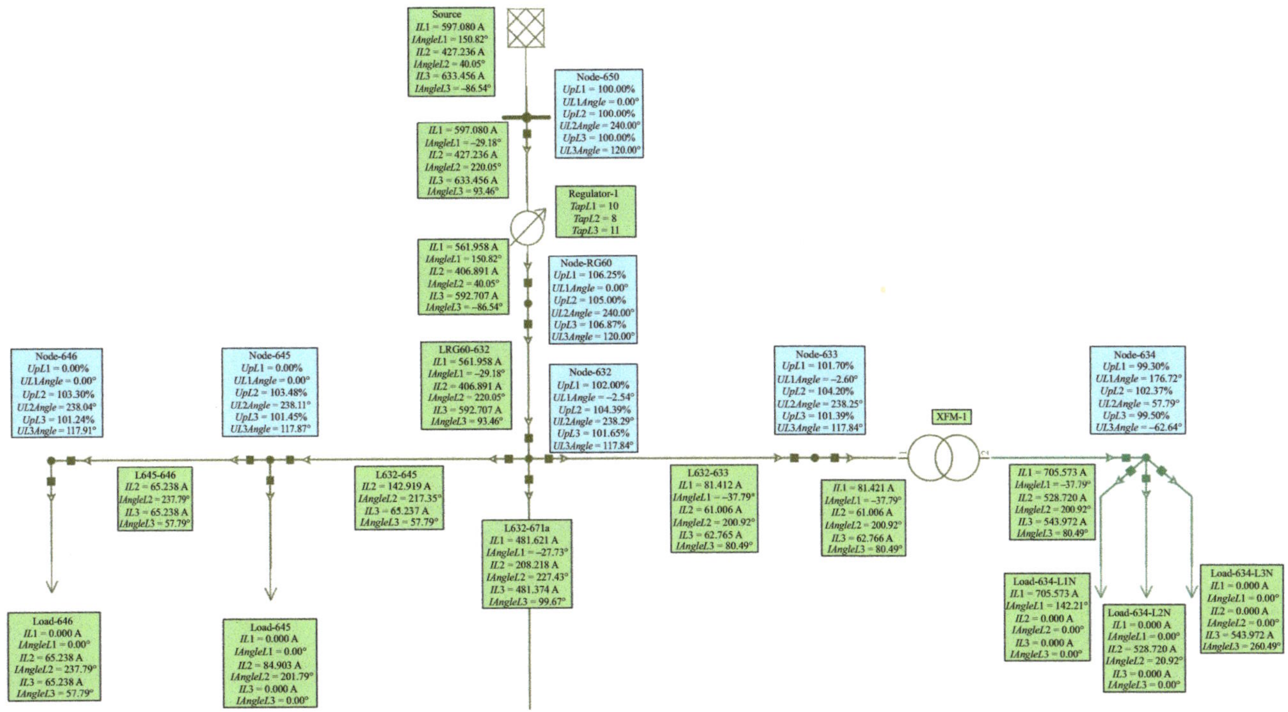

Figure 7.15 Load flow run for the system used with VVC [6]

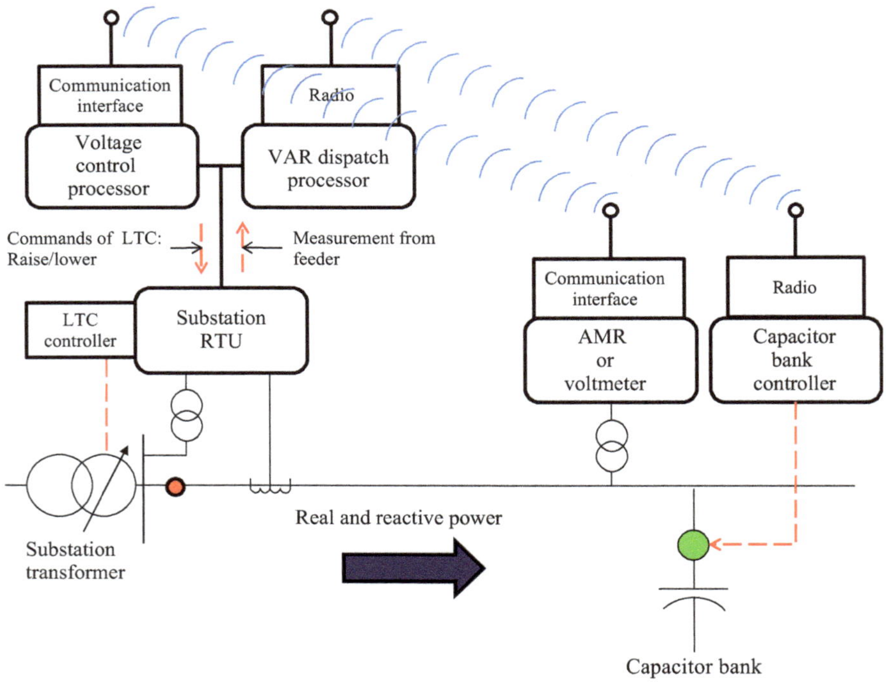

Figure 7.16 Components of a VVC assisted with SCADA

Example 7.2 In order to illustrate how VVC is done, the following system will be used as reference.

The voltage profile across the complete feeder is shown in Figure 7.17. It is shown how the performance and losses can be increased and reduced, respectively, by implementing the voltage-control processor and the VAR dispatch processor. The first one is used to implement the CVR. The second one helps one to improve the reactive power requirements and therefore, the voltage profile becomes more flat. In this case, the following sample rules will be defined:

1. The "candidate" capacitor banks for switching is identified considering that the capacitor bank is currently "off" and the rating value is less than the measured reactive power flow at head end of the feeder.
2. Once all the "candidate" capacitor banks are checked, the one is selected that has the lowest measured local voltage.
3. The switch of the chosen bank is changed to the "ON" position.

The first module control for VAR dispatch is used. It is shown how the voltage profile increases by the end of the feeder and also how the losses and the reactive

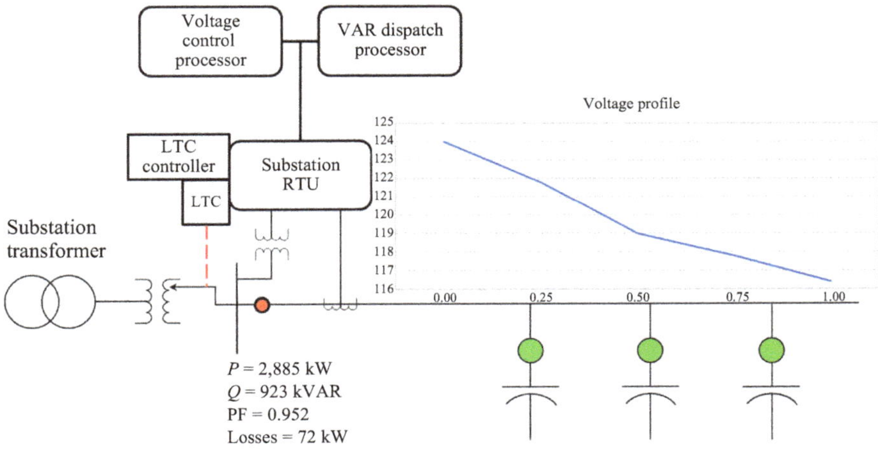

Figure 7.17 System for Example 7.2

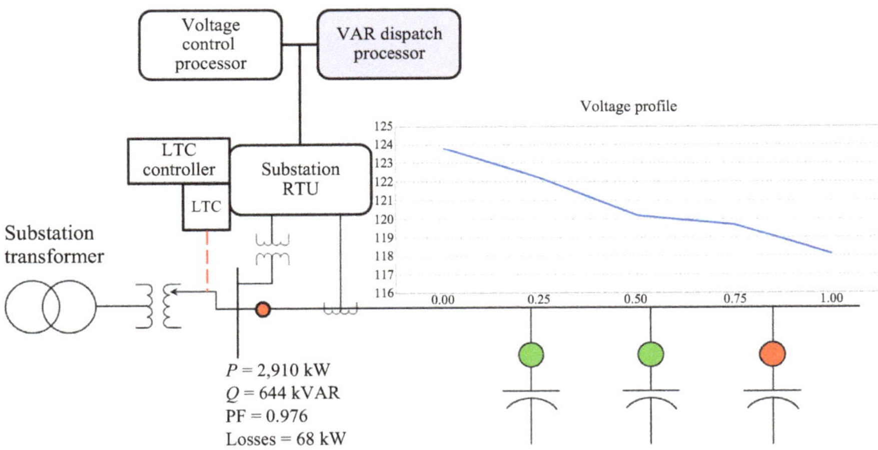

Figure 7.18 VAR dispatch processor control module for Example 7.2

power requirements improve. Also, the power factor improves considerably. This is shown in Figure 7.18.

Then, all the capacitor banks are connected, which is shown in Figure 7.19. Here, a comparison of the system with and without the VAR dispatch processor module is shown. An important improvement for the overall power losses and the reactive power requirements is obtained.

It is shown how a flat voltage profile considerably improves the overall efficiency of the system with the transferred active power increase while the reactive

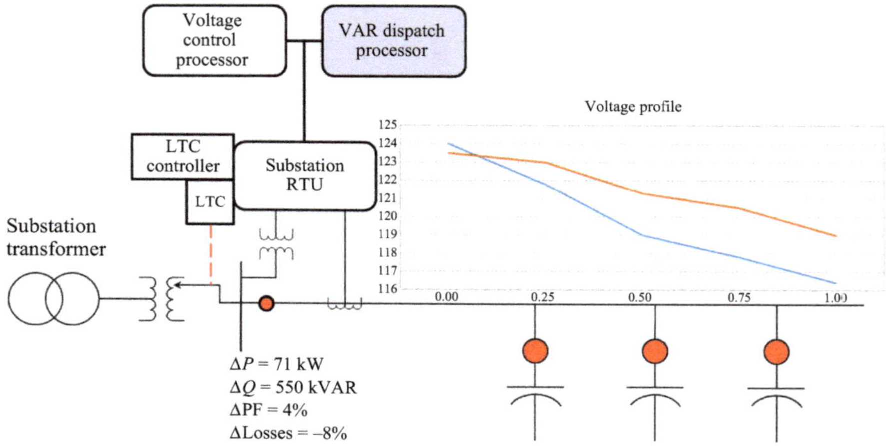

Figure 7.19 VAR dispatch processor control module for all capacitor banks in Example 7.2

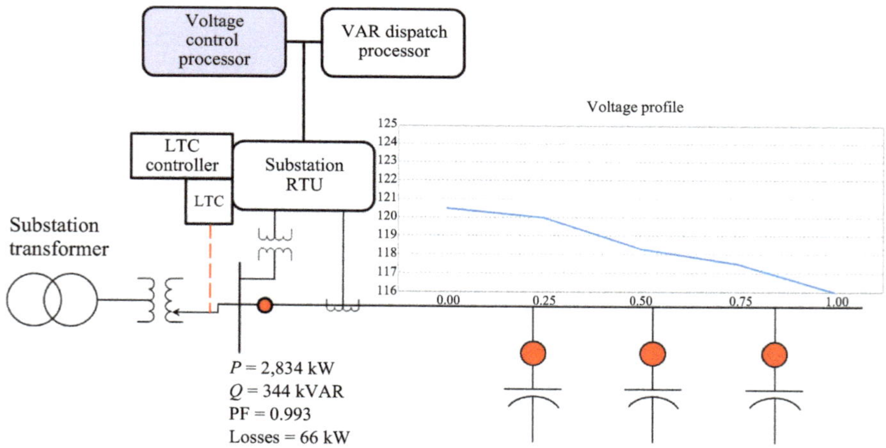

Figure 7.20 Voltage-control processor module for Example 7.2

power reduces, which improves the power factor of the system and reduces the losses.

Now, the voltage control processor module will be used. This is done in order to implement the CVR mentioned previously. For this case, the following sample rule for voltage reduction is implemented: if voltage at head end of the feeder exceeds LTC set point, then it lowers the voltage level. As was stated earlier, this is done in order to improve the efficiency of the transferred power (Figure 7.20).

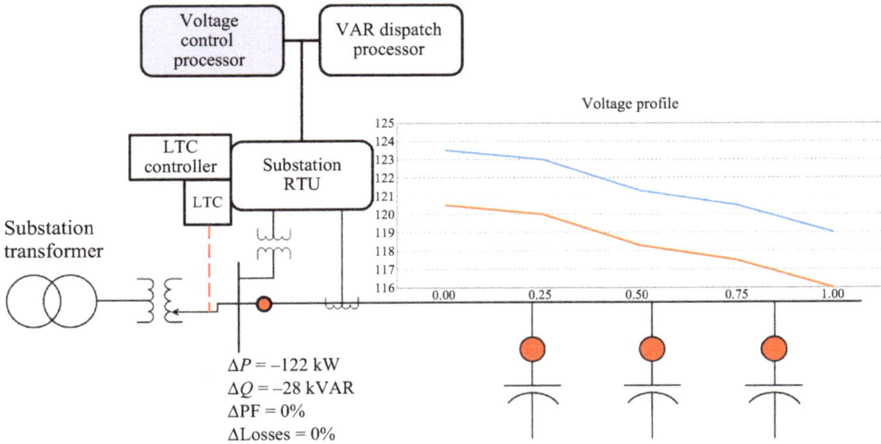

Figure 7.21 Voltage-control processor results comparison for Example 7.2

For the first moment, the LTC changes its position and now the overall profile reduces across the distribution system. The voltage remains in the allowed limits and there is a reduction in the transferred active and reactive power supplied by the system.

The idea is to implement the CVR scheme. Therefore, it is expected to reduce the voltage as much as possible (it means that the lower limit is the voltage at the end of the feeder). Again, there is an important reduction in the transferred active and reactive power supplied by the system.

Comparing this with the previous condition (with the maximum position in the LTC and all the capacitor banks used), it is shown how there is an increase of efficiency for the required power in the overall system. This is illustrated in Figure 7.21.

The effect of both modules as shown here is to considerably improve the voltage profile and the efficiency of the whole distribution system. The SCADA control is based on the corresponding regulation (Figure 7.22).

7.7 Requirements for Volt/VAR control

VCC implementation is getting more popular due to the favorable experiences in service quality and increased efficiency. The system should be able to maintain the range of voltages specified by the user on feeders and branches. Normally, power factors above 0.9 are required, although some utilities request 0.95 and above.

It is important to enable the operator to override in the event of topology changes as a result of an FLISR sequence or reconfiguration to reduce losses under normal conditions. On the other hand, various setting conditions must be developed for the different operational scenarios particular to the system. It helps the operator overview the actions, especially in the event of emergencies where time is crucial.

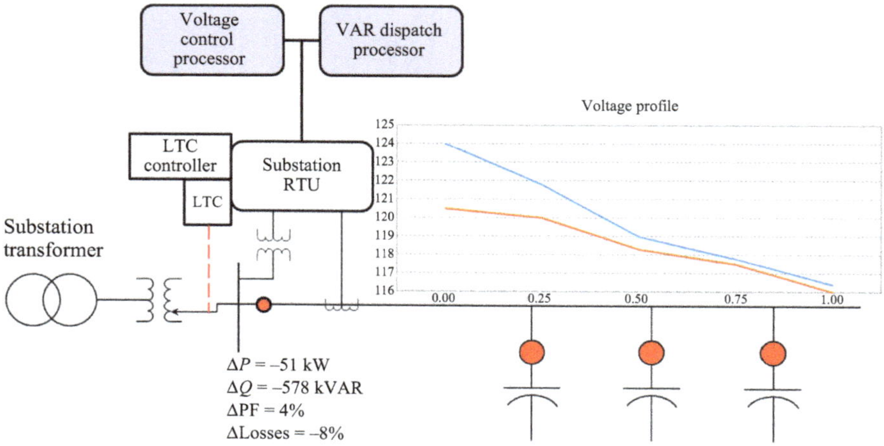

Figure 7.22 Volt/VAR modules applied for Example 7.2

The current number of tap changers, their location in the distribution system, and capacitor bank controls are initial potential information contribution to the database generated in the IEDs control systems. Therefore, devices must operate with various communication protocols that provide a wide range of possibilities of transmitting information.

An important coordination between tap changers connected in series is required in a distribution system. The various commutator control functions that could be considered for the coordination are types and time settings, line drop compensations, $R + X$ and Z type, inverse power configurations, reference points and voltage ranges, and voltage-reduction operation.

A coordination example is the timed-control configuration. Typically, the voltage control manipulates the tap changer when voltage exceeds the voltage ranges for a period longer than the timer setting. Various tap changers allow the voltage to return within the permissible range. Normalized distribution loads causes major voltage changes the further they are from the circuit due to increase of the source/line impedance ratio. As a result, the first-line tap changer upstream of the load (further from the circuit) will operate first and control voltages accordingly. However, assuming a voltage fluctuation at the source, all tap changer controls will experience the same change simultaneously. Aimed to avoid multiple or unnecessary tap changer operations, the time in each one (upstream) should be faster than the following control (upstream), where a time of 10 s–15 s between elements will allow an adequate coordination.

Another control approach to be considered is the inverse power configuration control. If a reverse power situation is presented in any circuit with multiple tap changers connected in series, attention must be focused on the appropriate control actions and coordination times. These operations and the coordination are very different for the radial topology or DG schemes. If the voltage-reduction techniques are applied to a distribution system, it should be applied simultaneously in all tap changer controls in series. Some characteristics have been implemented in the most common commutator controls. These characteristics aid in the tap changers coordination and voltage capacitor controls for more efficient Volt/VAR management. The characteristics are designed to utilize local intelligence, providing a better managed Volt/VAR.

An additional example of commutator coordination control is LDC configuration. The line drop compensation controls R and X characteristics and regulates voltage at the end of the defined line section (corresponding to load center). It is conducted calculating the line voltage drop and voltage-control configuration change to maintain destination voltage at load center. On the other hand, the value of impedance Z is adjusted as a function of load magnitude with the impedance values for the defined section. In the real implementation, the impedance values utilized in the first compensator type are very precise and the load center is not defined accurately. Both the issues cause unexpected voltages during operation. It is not necessary to coordinate the tap changer that controls the compensator configuration. However, the upstream settings could have a greater effect over the location requirements for the next downstream installation.

Finally, a voltage range–control configuration could be set. If the upstream range control is set to reduce the tap changers, these could be changed from the upstream control to the downstream control. Since the upstream tap changer voltage range affects downstream voltages, the tap changer general maintenance could be reduced in one to increase the next one. The lower the range in the upper section of voltage variation, the more stable will be the voltage source downstream, which results in a lower quantity of tap changers.

7.8 Integrated Volt/VAR control

The integrated VCC (IVVC) is considered an exceptional function of distributed automation that helps one to identify the most suitable control procedures for the devices involved in voltage regulation and VAR control to help one to achieve the specified operational goals of utilities avoiding violations to the fundamental operational constraints like high/low voltage limits and load limits.

The established VVC operating goals considers conditions like minimal electrical losses, minimal electrical demand, and reduced energy consumption. The decision criterion of IVVC requires the employment of an online power flow.

The IVVC offers improvements compared to VVC based on the fact that it can accommodate to scenarios resulting from FLISR or feeder reconfiguration. It could include modeling the dynamic effects of DER. Figure 7.23 illustrates the system employing IVVC along with DER.

IVVC control can be achieved similar to feeders, i.e., in a centralized or decentralized controller way.

For the case of centralized Volt/VAR-controlled system, the distribution system SCADA integrates the IVVC that determines how to proceed and provide information for the recording system. Figure 7.24 presents the aforementioned case. The solution

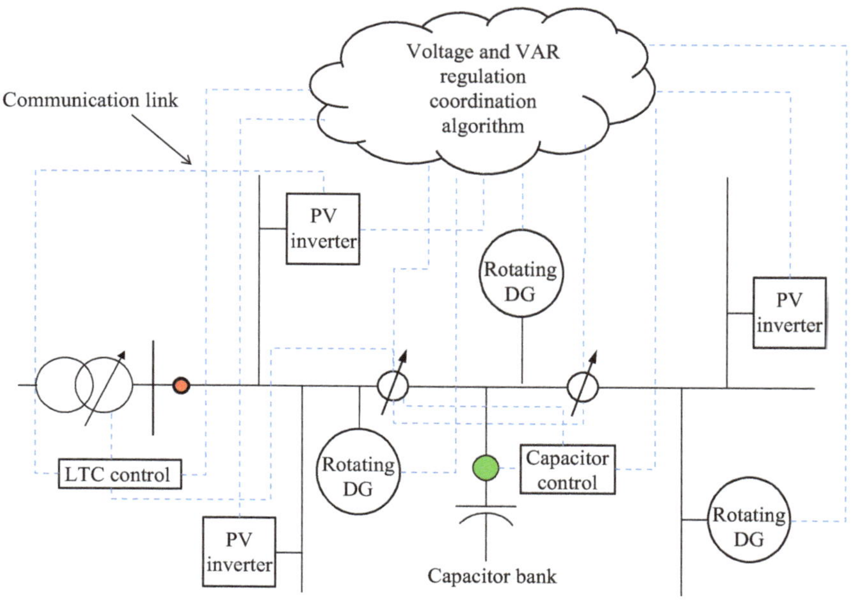

Figure 7.23 Integrated Volt/VAR control, including DG

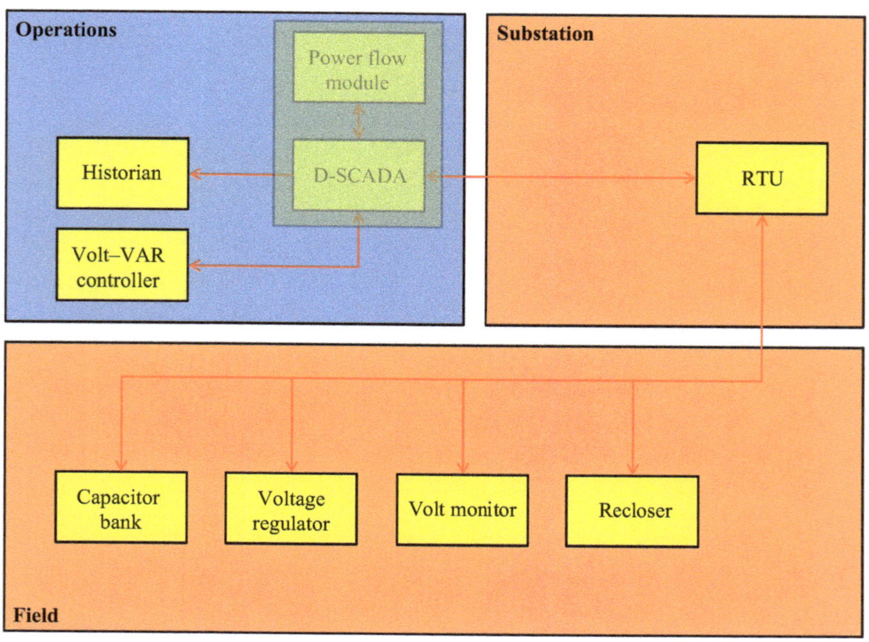

Figure 7.24 Context diagram for centralized integrated Volt/VAR control

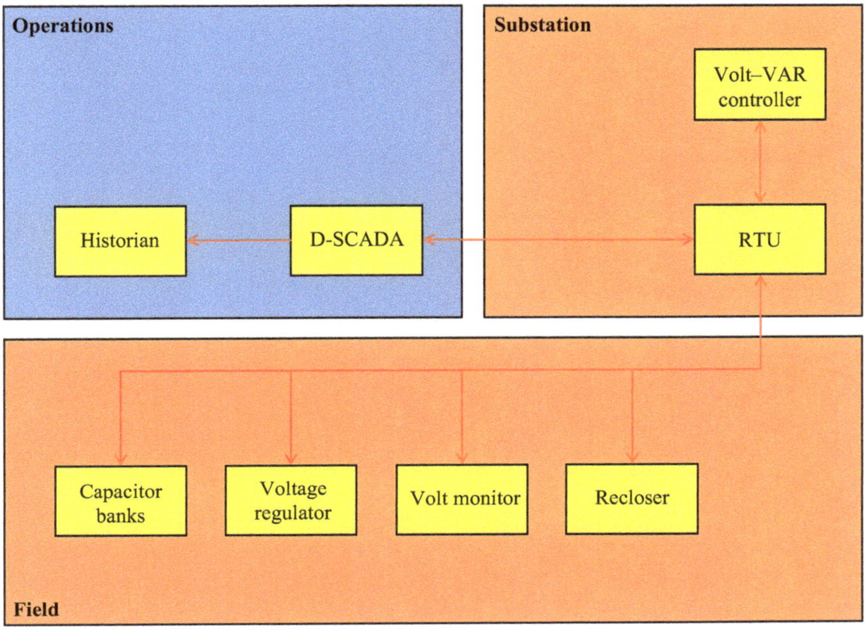

Figure 7.25 Context diagram for decentralized integrated Volt/VAR control

is considered very reliable and offers an overall illustration of the system. However, it is more expensive and could possibly congest the SCADA system.

On the other hand, the decentralized Volt/VAR controller is a stand-alone system that does not involve the overall system SCADA. It relies on local interaction with various devices associated with IVVC as observed in Figure 7.25. The recording takes place once all operations have been completed.

7.9 Proposed exercises

1. A 13.2 kV feeder has a total load of 5.5 MW and 3.2 MVAR. The load is uniformly distributed. The following calculations are performed:
   (i) Location and size of the optimal capacitor bank to achieve the maximum losses reduction for load factors of 0.9 and 0.6, respectively.
   (ii) Consider the same feeder if two capacitor banks were to be installed.
2. Consider a simple 33 kV radial line transmitting power by an overhead system to a lagging power factor load. Figure 7.26 presents the system, the system diagram, and the equivalent circuit.

Calculate shunt and series capacitors (size and impedance) to improve the power factor to 0.85 for each case. The line-to-line receiving-end voltage is assumed constant at 33 kV. All calculations are referred to 33 kV base voltage.

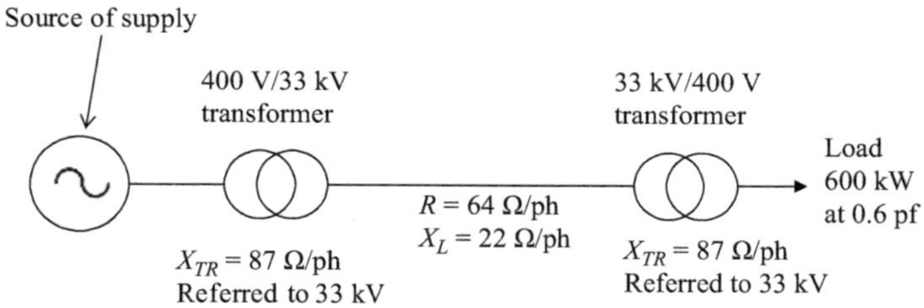

Figure 7.26 Diagrams of list item 2 in Section 7.9

References

[1] ANSI. ANSI C84.1-2016. American National Standard for Electric Power Systems and Equipment – Voltage Ratings (60 Hz), 2016.
[2] Statutory Instruments. The Electricity Safety, Quality and Continuity Regulations 2002 No. 2665. 2002.
[3] Gönen, T. Electric Power Distribution System Engineering, 2nd ed. CRC Press, Boca Raton, FL, USA, 2008.
[4] Neagle, N. M., and Samson, D. R. “Loss reduction from capacitor installed on primary feeders,” Power Apparatus and Systems, Part III. Transactions of the American Institute of Electrical Engineers, vol. 75, no. 3, pp. 950–959, 1956.
[5] IEEE. IEEE 13 Node Test Feeder. Distribution System Analysis Subcommittee, 1992.
[6] NEPLAN. IEEE 13 Node Test Feeder. 2018.

Further reading

[1] Baghzouz, Y. “Effects of nonlinear loads on optimal capacitor placement in radial feeders,” IEEE Transactions on Power Delivery, vol. 6, no. 1, pp. 245–251, 1991.
[2] Ayoubi, M., Hooshmand, R.-A., and Esfahani, M. T. “Optimal capacitor placement in distorted distribution systems considering resonance constraint using multi-swarm particle swarm optimisation algorithm,” Generation Transmission & Distribution IET, vol. 11, no. 13, pp. 3210–3221, 2017.
[3] Baran, M. E., and Wu, F. F. “Optimal capacitor placement on radial distribution systems,” IEEE Transactions on Power Delivery, vol. 4, no. 1, pp. 725–734, 1989.
[4] Eajal, A. A., El-Saadany, E. F., and AlHajri, M. F. “Distributed Generation Planning in Smart Distribution Grids via a Meta-Heuristic Approach,” in Electrical and Computer Engineering (CCECE) 2014 IEEE 27th Canadian Conference on, 2014, pp. 1–6.
[5] Cook, R. F. “Calculating loss reduction afforded by shunt capacitor application,” IEEE Transactions on Power Apparatus and Systems, vol. 83, no. 12, pp. 1227–1230, 1964.

[6] Askarzadeh, A. "Capacitor placement in distribution systems for power loss reduction and voltage improvement: a new methodology," Generation Transmission & Distribution IET, vol. 10, no. 14, pp. 3631–3638, 2016.
[7] Cook, R. F. "Optimizing the application of shunt capacitors for reactive-volt-ampere control and loss reduction," Power Apparatus and Systems, Part III. Transactions of the American Institute of Electrical Engineers, vol. 80, no. 3, pp. 430–441, 1961.
[8] Du, L., He, L., and Harley, R. G. "A Survey of Methods for Placing Shunt Capacitor Banks in Power Network With Harmonic Distortion," in IECON 2012 – 38th Annual Conference on IEEE Industrial Electronics Society, 2012, pp. 1198–1203.
[9] Ceylan, O., and Paudyal, S. "Optimal Capacitor Placement and Sizing Considering Load Profile Variations Using Moth-Flame Optimization Algorithm," in Modern Power Systems (MPS) 2017 International Conference on, 2017, pp. 1–6.
[10] Grainger, J. J., and Civanlar, S. "Volt-VAR control on distribution systems with lateral branches using shunt capacitors and voltage regulators, Part I: The overall problem," IEEE Transactions on Power Apparatus and Systems, vol. PAS-104, no. 11, pp. 3278–3283, 1985.
[11] Grainger, J. J., and Civanlar, S. "Volt-VAR control on distribution systems with lateral branches using shunt capacitors and voltage regulators, Part II: The solution method," IEEE Transactions on Power Apparatus and Systems, vol. PAS-104, no. 11, pp. 3284–3290, 1985.
[12] Grainger, J. J., and Civanlar, S. "Volt-VAR control on distribution systems with lateral branches using shunt capacitors and voltage regulators, Part III: The numerical results," IEEE Transactions on Power Apparatus and Systems, vol. PAS-104, no. 11, pp. 3291–3297, 1985.
[13] IEEE. IEEE Std 18-2012 – IEEE Standard for Shunt Power Capacitors, 2012.
[14] Maxwell, M. "The economic application of capacitors to distribution feeders," Power Apparatus and Systems, Part III. Transactions of the American Institute of Electrical Engineers, vol. 79, no. 3, pp. 353–358, 1960.
[15] Gutiérrez-Alcaraz, G., and Tovar-Hernández, J. H. "Two-stage heuristic methodology for optimal reconfiguration and Volt/VAR control in the operation of electrical distribution systems," Generation Transmission & Distribution IET, vol. 11, no. 16, pp. 3946–3954, 2017.
[16] Schmill, J. V. "Optimum size and location of shunt capacitors on distribution feeders," IEEE Transactions on Power Apparatus and Systems, vol. 84, no. 9, pp. 825–832, 1965.
[17] Aman, M. M., Jasmon, G. B., Bakar, A. H. A., Mokhlis, H., and Karimi, M. "Optimum shunt capacitor placement in distribution system—a review and comparative study," Renewable and Sustainable Energy Reviews, vol. 30, pp. 429, 2014.
[18] Uluski, R. W. "VVC in the Smart Grid Era," in Power and Energy Society General Meeting, 2010 IEEE, Minneapolis, MN, USA, 25–29 July 2010, pp. 1, 7.

[19] Krok, M. J., and Genc, S. "A Coordinated Optimization Approach to Volt/VAR Control for Large Power Distribution Networks," in American Control Conference (ACC) 2011, San Francisco, CA, USA, 2011, pp. 1145–1150.

[20] Švenda, V. "Practical Load Reduction in Distribution Network," in Power and Energy Society General Meeting (PESGM) 2016, Chicago, IL, USA, 2016, pp. 1–6.

[21] Wang, Z., and Wang, J. "Review on implementation and assessment of conservation voltage reduction," IEEE Transactions on Power Systems, vol. 29, no. 3, pp. 1306–1315, 2014.

[22] Sen, P. K., and Lee, K. H. "Conservation voltage reduction technique: an application guideline for smarter grid," IEEE Transactions on Industry Applications, vol. 52, no. 3, pp. 2122–2128, 2016.

[23] Anilkumar, R., Devriese, G., and Srivastava, A. K. "Intelligent Volt/VAR Control Algorithm for Active Power Distribution System to Maximize the Energy Savings," in 2015 IEEE Industry Applications Society Annual Meeting, Addison, TX, 2015, pp. 1–8.

[24] Zavoda, F., Perreault, C., and Lemire, A. "The Impact of a Volt & VAR Control System (VVC) on PQ and Customer's Equipment," in IEEE PES T&D 2010, New Orleans, LA, 2010, pp. 1–6.

[25] Anilkumar, R., Devriese, G., and Srivastava, A. K. "Voltage and reactive power control to maximize the energy savings in power distribution system with wind energy," IEEE Transactions on Industry Applications, vol. 54, no. 1, pp. 656–664, 2018.

[26] Azzouz, M. A., and El-Saadany, E. F. "Optimal Coordinated Volt/VAR Control in Active Distribution Networks," in 2014 IEEE PES General Meeting – Conference and Exposition, National Harbor, MD, 2014, pp. 1–5.

Chapter 8
Harmonic analysis

Harmonics have become one of the most critical issues in power systems due to the high penetration of them and the magnitudes. This is particularly critical in distribution systems. Harmonics create problems that affect considerably the power quality of a system and therefore are treated in that field.

Power quality can be defined as the goodness of the electric power quality supply in terms of its voltage waveshape, its current waveshape, its frequency, its voltage regulation, as well as level of impulses and noise, and the absence of momentary outages.

Other definition defines power quality as the measure, analysis, and improvement of bus voltage, usually a load bus voltage, to maintain that voltage to be a pure sinusoid and at rated levels of voltage and frequency. A proper quality level is ultimately a compromise of the utility, the end user, and the equipment manufacturer.

Power quality involves several categories of phenomena, which are better presented in Table 8.1.

The categories and characteristics are presented on Table 8.2 taken from the IEEE Std. 1159-2009. Category 5 deals with waveform distortion that includes the harmonic handling, which is the topic developed in this chapter.

8.1 General considerations about harmonics

Harmonics are voltage or current sinusoidal signals with frequencies that are multiples of the power frequency, which can produce harmful conditions in power systems.

The harmonic signals in power and distribution systems are produced by different equipment like arc furnaces, nonlinear loads, AC/DC rectifiers, adjustable speed drives, and static uninterruptible power source (UPS) systems among others.

Nonlinear loads cause harmonic currents to change from a sinusoidal current to a nonsinusoidal current by drawing short bursts of current each cycle or interrupting the current during a cycle.

Harmonic distortion is a growing concern for many customers and for the overall power system due to increasing application of power electronics equipment.

In low voltage, there are other harmonic sources like electric welders, devices with saturated cores, electronic ballasts for fluorescent lighting, and UPS for

Table 8.1 Summary of power quality variation categories. Reproduced by permission of Dranetz

Example waveshape or rms variation	Power quality variation and category	Method of characterizing	Typical causes	Example power conditions solutions
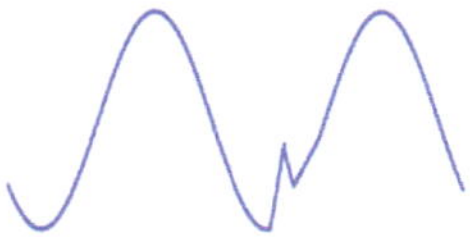	Impulsive Transients Transients disturbance	• Peak magnitude • Rise time • Duration	• Lightning • Electrostatic discharge • Load switching • Capacitor switching	• Surge arresters • Filters • Isolation transformers
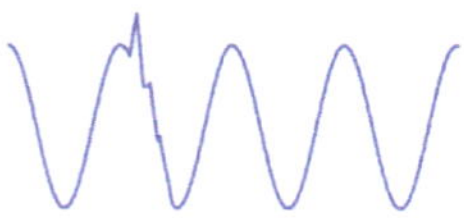	Oscillatory Transients Transients disturbance	• Waveforms • Peak magnitude • Frequency components	• Line/Cable switching • Capacitor switching • Load switching	• Surge arresters • Filters • Isolation transformers
	Sags/swells rms disturbance	• rms versus time • Magnitude • Duration	• Remote system • Faults	• Ferroresonant transformers • Energy storage technologies • UPS
	Interruptions rms disturbance	• Duration	• System protection • Breakers • Fuses • Maintenance	• Energy/storage technologies • UPS • Backup generators
	Undervoltages/ overvoltages Steady-state variation	• rms versus time • Statistics	• Motor starting • Load variations • Load dropping	• Voltage regulators • Ferroresonant transformers
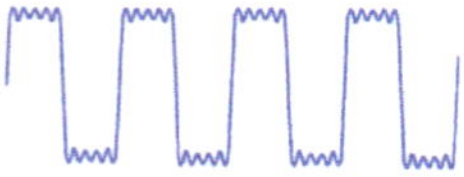	Harmonic distortion Steady-state variation	• Harmonic spectrum • Total harmonic distortion • Statistics	• Nonlinear loads • System resonance	• Active or passive filters • Transformers with cancelation or zero sequence components
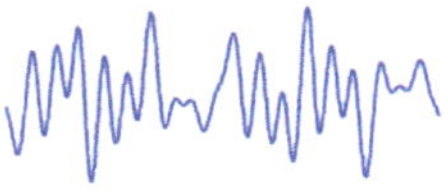	Voltage flicker Steady-state variation	• Variation magnitude • Frequency of occurrence • Modulation frequency	• Intermittent loads • Motor starting • Arc furnaces	• Static VAR systems

Table 8.2 Categories and typical characteristics of power system electromagnetic phenomena. Taken from IEEE Std. 1159-2009

Categories	Typical spectral content	Typical duration	Typical voltage magnitude
Transients			
Impulsive			
Nanosecond	5 ns rise	<50 ns	
Microsecond	1 μs rise	50 ns–1 ms	
Millisecond	0.1 ms rise	>1 ms	
Oscillatory			
Low frequency	<5 kHz	0.3 ms–50 ms	0 pu–4 pu[a]
Medium frequency	5 kHz–500 kHz	20 μs	0 pu–8 pu
High frequency	0.5 MHz–5 MHz	5 μs	0 pu–4 pu
Short-duration root-mean-square (rms) variations			
Instantaneous			
Sag		0.5–30 cycles	0.1 pu–0.9 pu
Swell		0.5–30 cycles	1.1 pu–1.8 pu
Momentary			
Interruption		0.5 cycles—3 s	<0.1 pu
Sag		30 cycles—3 s	0.1 pu–0.9 pu
Swell		30 cycles—3 s	1.1 pu–1.4 pu
Temporary			
Interruption		>3 s–1 min	<0.1 pu
Sag		>3 s–1 min	0.1 pu–0.9 pu
Swell		>3 s–1 min	1.1 pu–1.2 pu
Long-duration rms variations			
Interruption, sustained		>1 min	0.0 pu
Undervoltages		>1 min	0.8 pu–0.9 pu
Overvoltages		>1 min	1.1 pu–1.2 pu
Current overload		>1 min	
Imbalance			
Voltage		Steady state	0.5%–2%
Current		Steady state	1.0%–30%
Waveform distortion			
DC offset		Steady state	0%–0.1%
Harmonics	0 kHz–9 kHz	Steady state	0 kHz–9 kHz
Interharmonics	0 kHz–9 kHz	Steady state	0%–2%
Notching		Steady state	
Noise	Broadband	Steady state	0%–1%
Voltage fluctuations	<25 Hz	Intermittent	0.1%–7% 0.2 Pst–2 Pst[b]
Power frequency variations		<10 s	±0.10 Hz

[a]The quantity pu refers to per unit. The nominal condition is often considered to be 1.0 pu. In this table, the nominal peak value is used as the base for transients and the nominal rms value is used as the base for rms variations.
[b]Flicker severity index Pst as defined in IEC 61000-4-15:2010 and IEEE Std 1453™.

computers. The latter although have lower values, their high number of units has increased the effect on distribution systems.

The harmonics produce, in general, negative effects such as overheating of cables, transformers, and rotate machines; mal-operation of protection, control and measurement devices especially if they are electronic made; and overvoltages especially when resonance conditions appear due to capacitors.

8.2 Mathematical background

Fourier series allows to represent any continuous periodic function by a sinusoidal component plus a series of sinusoidal harmonics with frequencies multiple of the fundamental one.

Any periodical signal can be expanded by Fourier series if it satisfies Dirichlet's conditions:

- should have a definite number of discontinuities in a period,
- should have a finite number of maximums and minimum in a period, and
- the integral of the function in a period should have a finite value.

Under these conditions, a function $f(t)$ with a period 2π is represented by Fourier series as follows:

$$f(t) = \frac{A_0}{2} + \sum_{n=1}^{\infty} (A_n \cos(n\omega t) + B_n \sin(n\omega t)) \tag{8.1}$$

where

$$A_0 = \frac{2}{T}\int_{-T/2}^{T/2} f(t)dt$$

$$A_n = \frac{2}{T}\int_{-T/2}^{T/2} f(t)\cos(n\omega t)dt$$

$$B_n = \frac{2}{T}\int_{-T/2}^{T/2} f(t)\sin(n\omega t)dt$$

$$n = 1, 2, 3, \ldots$$

$$\omega = \frac{2\,\pi}{T} = \text{angular frequency}$$

This equation can be written as follows:

$$f(t) = \frac{A_0}{2} + \sum_{n=1}^{\infty} C_n \cos(n\omega t) \tag{8.2}$$

where

$$C_n \sqrt{(A_n^2 + B_n^2)}$$

$$\alpha_n = \text{Arctg}\left(\frac{B_n}{A_n}\right)$$

8.3 Verification of harmonic values

With the results obtained from the load flow at power frequency and the harmonic load flow, the rms voltages of the nodes and currents at the branches and capacitors are calculated, as well as the total harmonic distortion (THD) or total demand distortion (TDD) as defined in the Standard IEEE 519-2014.

The following expressions are applied:

$$V_{\text{rms}} = \sqrt{V_1^2 + V_2^2 + \cdots + V_n^2} \tag{8.3}$$

$$I_{\text{rms}} = \sqrt{I_1^2 + I_2^2 + \cdots + I_n^2} \tag{8.4}$$

$$THD_V = \frac{\sqrt{V_2^2 + V_3^2 + \cdots + V_n^2}}{V_1} \times 100\% \tag{8.5}$$

$$THD_I = \frac{\sqrt{I_2^2 + I_3^2 + \cdots + I_n^2}}{I_1} \times 100\% \tag{8.6}$$

$$TDD = \frac{\sqrt{I_2^2 + I_3^2 + \cdots + I_n^2}}{I_D} \times 100\% \tag{8.7}$$

where V_{rms} is the rms voltage of a node, I_{rms} is the rms current through a branch or element of the system, THD_V is the total harmonic voltage distortion in a node, THD_I is the total harmonic current distortion of the current through a branch or element of the system, TDD is the total harmonic current distortion of the current through a branch or element of the system based on the maximum demand current, V_1 is the fundamental component of the voltage in a node, I_1 is the fundamental component of the current through a branch or element of the system, V_n is the harmonic component of order n of the voltage in a node, I_n is the harmonic component of order n of the current through a branch or element of the system, and I_D is the calculated maximum demand current through a branch or element of the system.

8.4 Parallel resonance

A parallel resonance can occur when a capacitor is installed at the same busbar that has the harmonic source. Parallel resonance produces very high impedance to the harmonic current of the resonant frequency. Since most loads that generate harmonics can be treated as current sources, very high voltages and currents occur in the elements connected in parallel. This can be explained with the help of Figure 8.1.

If the equivalent impedance behind a busbar is X_{th} and a capacitor of impedance X_c is installed, the equivalent impedance is

$$Z_{eq} = \frac{X_{th}X_c}{X_{th} + X_c} \tag{8.8}$$

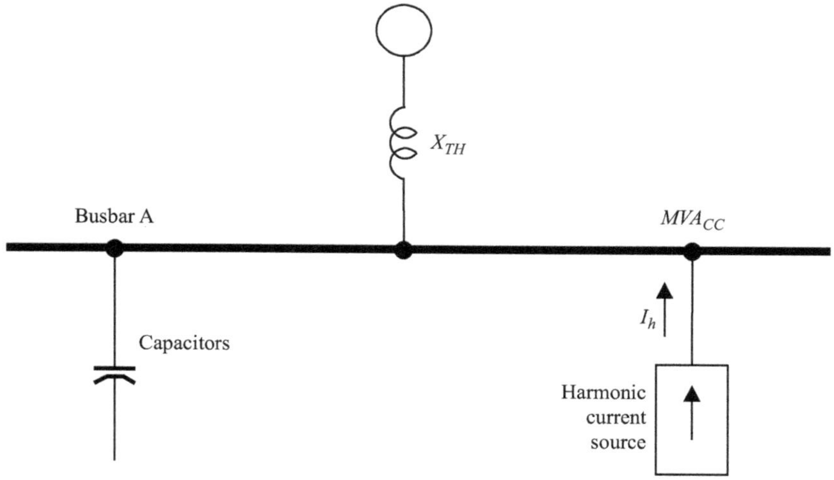

Figure 8.1 Connection of equipment producing parallel resonance

The parallel resonant condition occurs when the denominator of the previous expression becomes zero:

$$X_{th} + X_c = 0$$
$$X_{th} = -X_c$$

The reactances calculated to the angular frequency of resonance (ω_n) are calculated as follows:

$$X_{th} = \omega_n L \text{ and } X_c = \frac{1}{\omega_n C}$$

where L is the inductance of the system equivalent and C is the capacitance of the capacitor bank.

Besides, with the impedance corresponding to the fundamental frequency (ω), the following expressions apply:

$$\omega L = \frac{V^2}{\text{MV A}_{sc}}$$
$$\frac{1}{\omega C} = \frac{V^2}{\text{MV A}_{cap}}$$

where V is the nominal line voltage in kV, MV A_{SC} is the three-phase short circuit level at busbar in MV A, and MV A_{cap} is the total reactive power of the capacitor bank at fundamental frequency in MV A.

Solving for the values of L and C and replacing for the resonant condition, the following is obtained:

$$\frac{\omega_n V^2}{\omega \text{MV A}_{SC}} = \frac{\omega V^2}{\omega_n \text{MV A}_{cap}}$$

Solving for the parallel resonance $f_p\left(\omega_n = 2\pi f_p\right)$, the following is obtained:

$$f_p = f\sqrt{\frac{\text{MV A}_{sc}}{\text{MV A}_{cap}}} \tag{8.9}$$

where f_p is the parallel resonant frequency (Hz) and f is the fundamental frequency (Hz).

8.5 Series resonance

A series resonance can occur when a capacitor is installed in series with an inductive element and the impedances are the same for a specific harmonic. In this case, the equivalent impedance is zero. Under series resonance conditions, the system offers very low impedance to harmonic voltages at resonance frequency. Harmonic filters then are designed to introduce a series resonance in the shunt element connected to the busbar of the system. They are made up normally by a reactance and a capacitor with equal reactance values but opposite in sign at the resonance frequency. This allows ground harmonic currents injected by harmonic sources. Figure 8.2 shows the connection of elements when series resonance occurs, while Figure 8.3 illustrates the curve of impedance versus frequency, which shows the resonance frequency value for a general case.

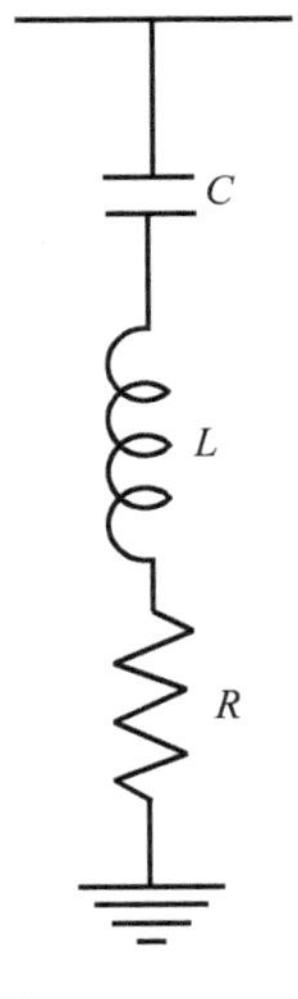

Figure 8.2 Connection of equipment producing series resonance

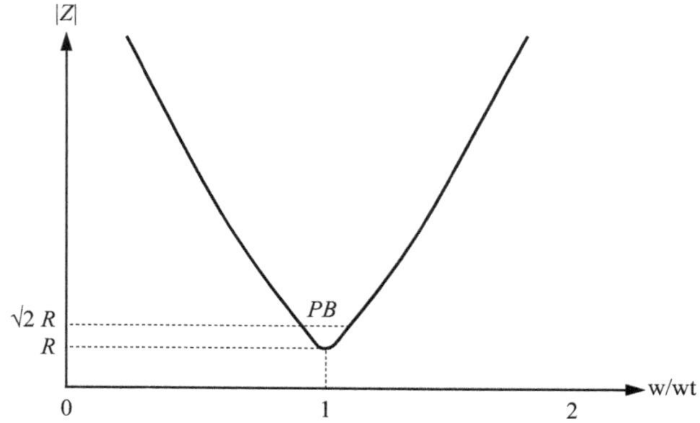

Figure 8.3 Z versus W in series resonance filters

8.6 Validation of harmonic values

8.6.1 Harmonic limits

The most important concepts to be illustrated involve the evaluation of harmonic current limits at individual customers and harmonic voltage limits on the overall system. These limits are typically evaluated at the point of common coupling (PCC) between the supplier and the customer.

8.6.2 Voltage distortion limits

The utility is responsible for maintaining the quality of voltage on the overall system. Table 8.3 summarizes the voltage distortion guidelines for different system voltage levels given by the IEEE Std. 519-2014.

$$THD_{V_n} = \frac{\sqrt{\sum_{h=2}^{\infty} V_h^2}}{V_n} \times 100\% \tag{8.10}$$

where V_h is the magnitude of individual harmonic components (rms V), n is the harmonic order, and V_n is the nominal system rms voltage (rms V).

8.6.3 Current distortion limits

The harmonic currents from an individual customer are evaluated at the PCC where the utility can supply other customers. Table 8.4 summarizes the current distortion limits given by the IEEE Std. 519-2014. The limits are dependent on the customer load in relation to the system short circuit capacity at the PCC. Note that all current limits are expressed as a percentage of the customer's average maximum demand load current. Table 8.5 shows the current distortion limits for different voltage levels as established by IEEE Std. 519-2014.

Table 8.3 Harmonic voltage distortion limits in % of nominal fundamental frequency voltage. Taken from IEEE Std. 519-2014

Bus voltage at PCC (V_n)	**Individual harmonic voltage distortion (%)**	**Total voltage harmonic distortion—THD_{Vn} (%)**
$V \leq 1.0$ kV	5.0	8.0
1 kV $< V \leq$ 69 kV	3.0	5.0
69 kV $< V \leq$ 161 kV	1.5	2.5
161 kV $< V$	1.0	1.5

Table 8.4 Current distortion limits. Taken from IEEE Std. 519-2014

	I_{sc}/I_L	$3 \leq h < 11$	$11 \leq h < 17$	$17 \leq h < 23$	$23 \leq h < 35$	$35 \leq h \leq 50$	TDD
$V_n \leq 69$ kV	<20	4.0	2.0	1.5	0.6	0.3	5.0
	20 < 50	7.0	3.5	2.5	1.0	0.5	8.0
	50 < 100	10.0	4.5	4.0	1.5	0.7	12.0
	100 < 1,000	12.0	5.5	5.0	2.0	1.0	15.0
	>1,000	15.0	7.0	6.0	2.5	1.4	20.0
69 kV $< V_n \leq$ 161 kV	<20*	2.0	1.0	0.75	0.3	0.15	2.5
	20 < 50	3.5	1.75	1.25	0.5	0.25	4.0
	50 < 100	5.0	2.25	2.0	0.75	0.35	6.0
	100 < 1,000	6.0	2.75	2.5	1.0	0.5	7.5
	>1,000	7.5	3.5	3.0	1.25	0.7	10.0
$V_n > 161$ kV	<25	1.0	0.5	0.38	0.15	0.1	1.5
	25 < 50	2.0	1.0	0.75	0.3	0.15	2.5
	≥ 50	3.0	1.5	1.15	0.45	0.22	3.75

*All power generation equipment are limited to these values of current distortion, regardless of actual I_{sc}/I_L.

8.7 Verification of harmonic values

For capacitor application, Standard IEEE 18-2012 should be applied from which the following conditions are taken:

- The maximum permanent voltage, including harmonics of any capacitor bank, should not be higher than 110% of its nominal voltage.
- The crest voltage should not exceed $1.2 \times \sqrt{2}$ of rated rms voltage, including harmonics but excluding transients.
- The maximum permanent current through any capacitor bank should not be higher than 135% of its nominal current based on rated kVAR and rated voltage.
- The maximum permanent power, including harmonics of any capacitor bank, should not be higher than 135% of its nominal power.

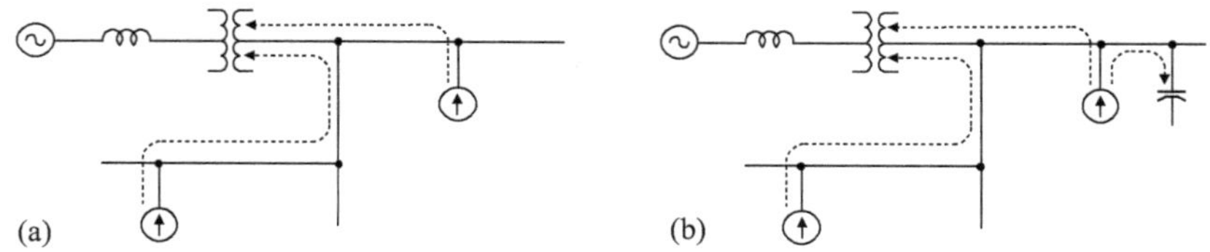

Figure 8.4 General flow of harmonic currents in radial power system: (a) without power capacitor and (b) with power capacitor

If any of the previous limits is overcome, changes in the capacitor size and location should be considered to avoid exposing the system to serious difficulties. Figure 8.4 shows the effect on the harmonic currents when a capacitor is placed in the power system.

8.8 Resizing and relocation of capacitor banks

If the feeders that are going to give up load in a reconfiguration process have harmonic sources, especial care has to be considered for resonance with the capacitors installed in the feeders that are going to receive the loads.

Whenever a resonance situation occurs in a distribution system due to a capacitor bank, the solutions to be considered first are the relocation and/or resizing of the banks in order to vary the resonance frequency to a value that does not correspond to any generated by the harmonic sources. On the other hand, the short circuit level varies too, which also changes the resonance frequency.

A novel algorithm is proposed here for the analysis of distribution systems which can be applied in conjunction with the reconfiguration software or as a stand-alone analysis tool. The sources and loads are connected to earth for every harmonic under consideration, except for the fundamental frequency. The algorithm is applied for every harmonic source existing in the system and can be summarized in the following steps:

First, a harmonic value is selected from one of the existing sources in the system and the feeder is organized from the node that has the harmonic source up to last node that in this case will be the power source.

Once the feeder is ordered, the load impedances of each node are calculated with the active and reactive powers of the loads and the same for the capacitors connected to the same node. For this, it is important to consider the impact of the frequencies of the respective harmonics into the impedance calculations.

Afterward, a value of 1.0 V is given to every node for the harmonic frequency under analysis. With this voltage value, the load currents are calculated in all elements, starting from the furthest away node to the harmonic source and then the current values are accumulated, taking into account the connectivity of the nodes. This routine could be called the “upstream iteration.”

With the currents so calculated and starting from the node with the harmonic source, the voltages in every node are updated by calculating the voltage drop in all branches for the harmonic frequency under analysis. This routine could be called the "downstream iteration."

The new harmonic voltages, so obtained in each node, are used to recalculate the load currents. The cycle is repeated until convergence is obtained, i.e., when the variation in the voltage nodes is less than a predetermined value, 0.0001 V for this case.

With the accumulated current of last iteration, the current supplied by the harmonic source is calculated. The actual value of the current source is divided by the calculated value and a factor is so obtained. The solution of the load flow is then obtained by multiplying the voltages and currents of the last iteration by this factor. The whole process is repeated for each harmonic frequency of every harmonic source.

The previous process is illustrated in Figure 8.5(a) that shows a feeder with 17 nodes with current harmonic sources in three of them. For the sake of the illustration, it is supposed that the nodes with harmonic sources have also a capacitor and a load made up by active and reactive elements.

According to the process mentioned, one of the harmonic sources is chosen first and then the feeder is reordered from it. Then the impedances are calculated as shown in the two nodes simplified diagram of Figure 8.5(b) for the harmonic under consideration.

Afterward, a voltage of 1.0 V is given to both nodes as shown again in Figure 8.5(b). With this voltage, the current at node B is first calculated, since it is the furthest away node to the harmonic source. This current is the same to that circulating through branch A–B. The current of the capacitor and the reactive and resistive loads of node A can also be calculated knowing the voltage at A. With the branch current, a voltage drop can be calculated and therefore a new voltage for node B. This allows the calculation of a new current for node B which is the same for branch A–B. Then a new voltage drop through the branch is calculated and the voltage at node B updated.

The previous process is repeated until no further variation of voltage at node B is found. With the final current of the branch and the currents of the elements of node A, the total current of the harmonic source is calculated. The real value of the current of the source is compared with the calculated value and the factor so obtained is used to calculate the real current values of the branch A–B and the currents of the elements connected to nodes A and B. The voltages of nodes A and B are also calculated with the same factor.

8.9 Models

Harmonics are analyzed by means of harmonic load flows that require a proper model for the harmonic sources, for the system equivalent, for the loads, and for the branches, which considers the frequency of the respective harmonic.

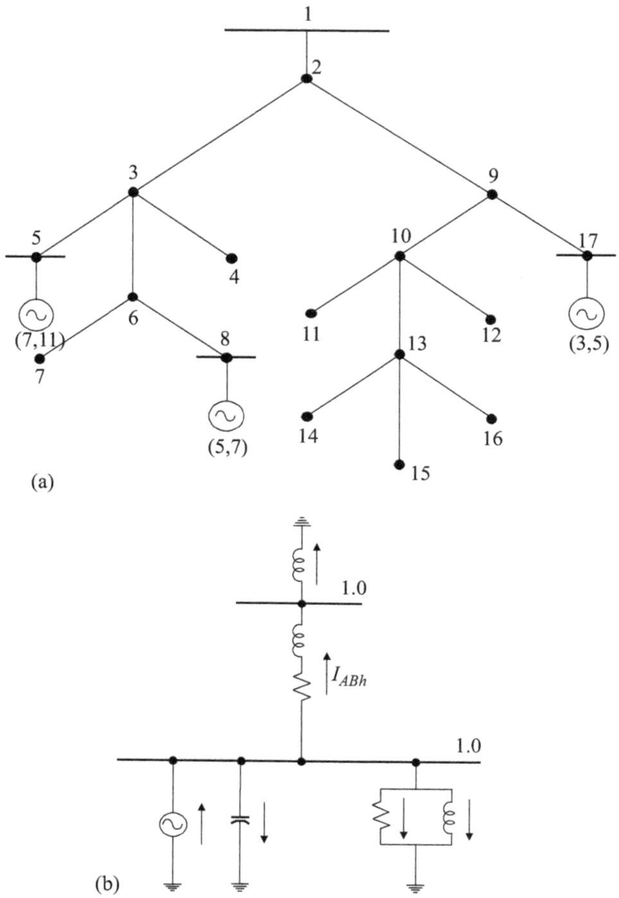

Figure 8.5 System example for illustrate harmonic evaluation procedure: (a) Feeder with harmonic sources and (b) Two-node system equivalent

The harmonic sources considered in the harmonic load flow are represented as independent current sources, each with a determined number of harmonic frequency. For each harmonic frequency, the source has a definite magnitude in Amps.

In order to obtain a result that combines all the harmonic sources, the superposition theorem is applied.

Depending on the source location, a branch can have different currents and, therefore, it is essential to consider the phasor angles of voltages and currents.

The harmonic load flow requires a proper model for the harmonic sources, for the system equivalent, for the loads, and for the branches, which considers the frequency of the respective harmonic. The modeling for each case is explained in the following sections.

8.9.1 Harmonic sources

The harmonic sources considered in the harmonic load flow are represented as independent current sources, each with a determined number of harmonic frequency. For each harmonic frequency, the source has a definite magnitude in Amps.

In order to obtain a result that combines all the harmonic sources, the superposition theorem is applied. Depending on the source location, a branch can have different currents, and, therefore, it is essential to consider the phasor angles of voltages and currents.

8.9.2 System model

The system behind the substation busbar where the feeder is connected is represented by a Thévenin equivalent reactance represented as follows:

$$X_{Th} = \frac{hV^2}{S_{SC}} \tag{8.11}$$

where $X_{Th}(f)$ is the Thévenin equivalent reactance of the system as a frequency function, h is the harmonic order, V is the nominal system voltage, and S_{SC} is the three-phase short circuit power at the substation busbar.

8.9.3 Load model

The load model as proposed in reference considers a parallel of a resistance and an inductive reactance. For each one of the components of the load model, the following expressions are used:

$$R = \frac{V^2}{P_{60}} \tag{8.12}$$

$$X(f) = \frac{hV^2}{Q_{60}} \tag{8.13}$$

where R is the resistive load component, $X(f)$ is the reactive load component as a frequency function, V is the nominal load voltage, P_{60} is the nominal active power load at the power frequency, Q_{60} is the nominal reactive power load at the power frequency, and h is the harmonic order.

8.9.4 Branch model

The branches connecting the nodes in distribution systems are represented by an impedance made up by a resistance and an inductive reactance connected in series. The resistance is considered as independent from the frequency but not the reactance whose expression is as follows:

$$X_L(f) = 2\pi fL = hX_{L60} \tag{8.14}$$

where $X_L(f)$ is the branch reactive component as a frequency function, f is the frequency, L is the branch inductance, h is the harmonic order, and X_{L60} is the branch inductive reactance at power frequency.

It can be noted that the parallel admittance of the branch that has the capacitive effect has been neglected, which a valid consideration in distribution systems, especially those is made up by overhead lines, rather than underground cables.

Example 8.1 A distribution feeder at 13.2 kV, shown schematically in Figure 8.6, will be considered in which a bank of 1,300 kVAR will be installed. The three-phase short circuit level in the respective node is 7 kA, and the load current has 20% of the 5th harmonic, 14.3% of the 7th harmonic, 9.1% of 11th harmonic, and 7.7% of the 13th harmonic. The nominal current in the node is 151.6 A at a $pf = 0.7$, which is assumed to be the same for all harmonics.

Perform the following actions:

- Calculate the possible resonance frequency.
- For the fundamental and harmonics 5th, 7th, 11th, and 13th, calculate the Thévenin and capacitance impedances, harmonic currents, and voltages across the capacitances.
- Calculate the THD for voltages and currents.
- Analyze the size of the capacitor bank with the results obtained.

For the calculations of the currents, bear in mind that the Thévenin equivalent is used only with the fundamental frequency. For the harmonic equivalents, there is no source and the load becomes a current source. The equivalent circuits then are as in Figure 8.7.

The possible resonance frequency for this node is

$$f_r = 60\text{ Hz} \times \sqrt{\frac{\sqrt{3} \times 13.2\text{ kV} \times 7\text{ kA}}{1.3\text{ MVAR}}} = 665.73\text{ Hz}$$

This value corresponds to the harmonic 11.1 that is very close to one of the harmonics injected by the load at that node. The values calculated for this example are presented in the table at the end of the solution.

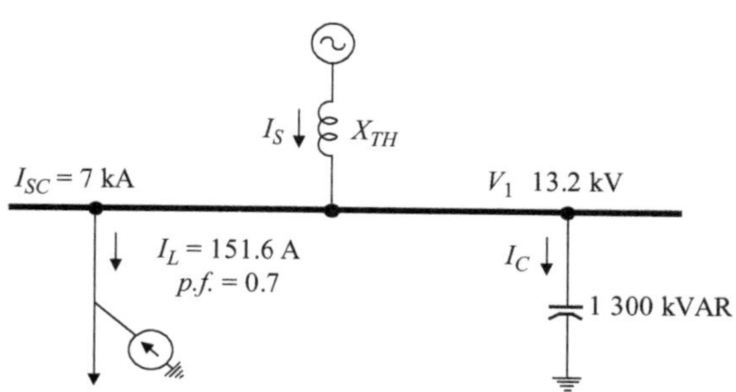

h	Magnitude (%)
5	20.0
7	14.3
11	9.1
13	7.7

Figure 8.6 Simplified diagram of the system

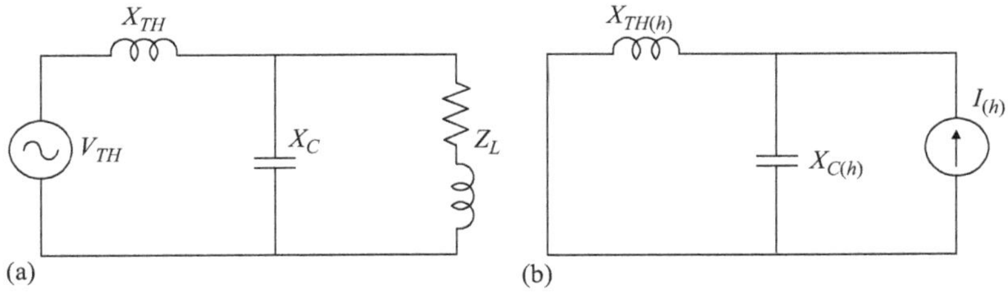

Figure 8.7 (a) Model for fundamental frequency and (b) model for harmonic h

$$
\begin{aligned}
I_L &= 151.6\ \text{A} \\
I_L(5^\circ) &= 20\%I_L = 0.2 \cdot (151.6\ \text{A}) = 30.32\ \text{A} \\
I_L(7^\circ) &= 14.3\%I_L = 0.143 \cdot (151.6\ \text{A}) = 21.67\ \text{A} \\
I_L(11^\circ) &= 9.1\%I_L = 0.091 \cdot (151.6\ \text{A}) = 13.79\ \text{A} \\
I_L(13^\circ) &= 7.7\%I_L = 0.077 \cdot (151.6\ \text{A}) = 11.67\ \text{A} \\
F.P. &= 0.7 \rightarrow \cos^{-1}(0.7) = 45.57 lagging
\end{aligned}
$$

Calculation of reactance impedances:

$$
\begin{aligned}
X_{TH(60)} &= \frac{V_{LN}}{I} = \frac{13.2\ \text{kV}}{\sqrt{3} \cdot 7\ \text{kA}} = j1.088\Omega \\
X_{TH(5\cdot 60)} &= 5 \cdot X_{TH(60)} = 5 \cdot (j1.088) = j5.44\Omega \\
X_{TH(7\cdot 60)} &= 7 \cdot X_{TH(60)} = 7 \cdot (j1.088) = j7.62\Omega \\
X_{TH(11\cdot 60)} &= 11 \cdot X_{TH(60)} = 11 \cdot (j1.088) = j11.97\Omega \\
X_{TH(13\cdot 60)} &= 13 \cdot X_{TH(60)} = 13 \cdot (j1.088) = j14.15\Omega
\end{aligned}
$$

Calculation of capacitance impedances:

$$
\begin{aligned}
X_{C(60)} &= \frac{V^2}{Q} = \frac{(13.2\ \text{kV})^2}{1.3\ \text{MVAR}} = -j134.03\Omega \\
X_{C(60/5)} &= \frac{X_{C(60)}}{5} = \frac{-j134.03}{5} = -j26.80\Omega \\
X_{C(60/7)} &= \frac{X_{C(60)}}{7} = \frac{-j134.03}{7} = -j19.15\Omega \\
X_{C(60/11)} &= \frac{X_{C(60)}}{11} = \frac{-j134.03}{11} = -j12.18\Omega \\
X_{C(60/13)} &= \frac{X_{C(60)}}{13} = \frac{-j134.03}{13} = -j10.31\Omega \\
I_S &= I_C + I_L \\
I_{C(60)} &= \frac{V}{X_C} = \frac{13.2\ \text{kV}}{\sqrt{3} \cdot (-j134.03\ \Omega)} = j56.86\ \text{A} \\
I_{S(60)} &= I_{C(60)} + I_{L(60)} = j56.86 + 151.6\angle - 45.57^\circ \\
&= 117.92\angle - 25.84^\circ\ \text{A}
\end{aligned}
$$

Figure 8.8 shows the equivalent circuit for fundamental frequency.

For the fifth harmonic (300 Hz), the equivalent circuit is as shown in Figure 8.9:

$$Z_{eq} = \frac{X_{TH(s)} \cdot X_{C(s)}}{X_{TH(s)} + X_{C(s)}} = \frac{j5.44 \cdot (-j26.80)}{j5.44 - j26.80} = \frac{145.79}{-j21.36} = j6.82\ \Omega$$

$$V_C = I_L \cdot Z_{eq} = 30.32\angle -45.57^\circ \cdot (j6.82) = 206.78\angle 44.43^\circ \text{ V}$$

$$I_{C(5)} = \frac{V_{LN}}{X_{C(5)}} = \frac{206.78\angle 44.43^\circ}{-j26.80} = 7.71\angle 134.43^\circ \text{ A}$$

$$I_{S(5)} = I_{C(5)} - I_{L(5)} = 7.71\angle 134.43^\circ - 30.32\angle -45.57^\circ = 38.03\angle 134.43^\circ \text{ A}$$

For the seventh harmonic (420 Hz)

$$Z_{eq} = \frac{X_{TH(7)} \cdot X_{C(7)}}{X_{TH(7)} + X_{C(7)}} = \frac{j7.62 \cdot (-j19.15)}{j7.62 - j19.15} = \frac{145.92}{-j11.53} = j12.66\ \Omega$$

$$V_C = I_{L(7)} \cdot Z_{eq} = 21.67\angle -45.57^\circ \cdot (j12.66) = 274.25\angle 44.43^\circ \text{ V}$$

$$I_{C(7)} = \frac{V_{LN}}{X_{C(7)}} = \frac{274.25\angle 44.43^\circ}{-j19.15} = 14.32\angle 134.43^\circ \text{ A}$$

$$I_{S(7)} = I_{C(7)} - I_{L(7)} = 14.32\angle 134.43^\circ - 21.67\angle -45.57^\circ = 36\angle 134.43^\circ \text{ A}$$

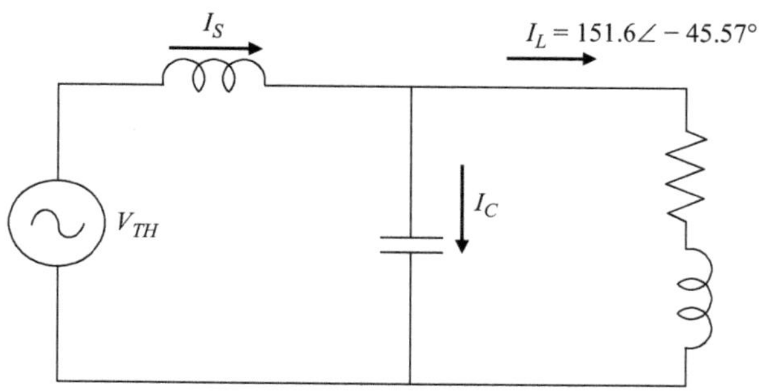

Figure 8.8 Circuit for fundamental frequency

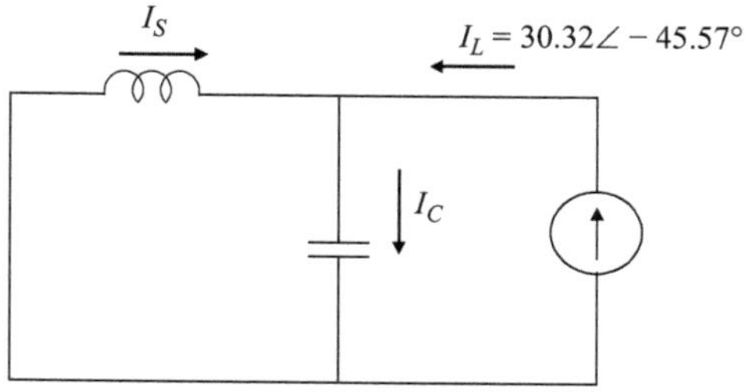

Figure 8.9 Circuit for fifth harmonic

The 11th harmonic (660 Hz) is very critical, since it is close to the resonance frequency.

This is illustrated with the following equations:

$$f_r = f\sqrt{\frac{P_a}{P_{cap}}}$$

$$P_{cc} = \sqrt{3} \cdot V_{nom} \cdot I_{cc} = \sqrt{3} \cdot (13.2\,\text{kV}) \cdot (7\,\text{kA}) = 160.04\ \text{MVA}$$

$$P_{cap} = 1.3\ \text{MVA}$$

$$f_r = 60\sqrt{\frac{160.04\ \text{MVA}}{1.3\ \text{MVA}}} = 665.73\ \text{Hz}$$

This value is very close to the resonance frequency which is at the 11th harmonic (660 Hz).

$$\frac{X_{TH(11)} \cdot X_{C(11)}}{X_{TH(11)} + X_{C(11)}} = \frac{j11.97 \cdot (-j12.18)}{j11.97 - j12.18} = \frac{145.79}{-j0.21} = j694.26\ \Omega$$

$$V_C = I_{L(11)} \cdot Z_{eq} = 13.79\angle -45.57^\circ \cdot (j694.26) = 9,573.84\angle 44.43^\circ\ \text{V}$$

$$I_{C(11)} = \frac{V_{LN}}{X_{C(11)}} = \frac{9,573.84\angle 44.43^\circ}{-j12.18} = 786.03\angle 134.43^\circ\ \text{A}$$

$$I_{S(11)} = I_{C(11)} - I_{L(11)}$$

$$= 786.03\angle 134.43^\circ - 13.79\angle -45.57^\circ = 799.82\angle 134.43^\circ\ \text{A}$$

For the 13th harmonic (780 Hz)

$$Z_{eq} = \frac{X_{TH(13)} \cdot X_{C(13)}}{X_{TH(13)} + X_{C(13)}} = \frac{j14.15 \cdot (-j10.31)}{j14.15 - j10.31} = \frac{145.88}{j3.84} = -j38.0\ \Omega$$

$$V_C = I_{L(13)} \cdot Z_{eq} = 11.67\angle -45.57^\circ \cdot (-j38.0) = 443.46\angle -135.57^\circ\ \text{V}$$

$$I_{C(13)} = \frac{V_{LN}}{X_{C(13)}} = \frac{443.46\angle -135.57^\circ}{-j10.31} = 43.01\angle -45.57^\circ\ \text{A}$$

$$I_{S(13)} = I_{C(13)} - I_{L(13)} = 43.01\angle -45.57^\circ - 11.67\angle -45.57^\circ = 31.34\angle -45.57^\circ\ \text{A}$$

Table 8.5 presents the values of currents, voltages, and impedances calculated for the fundamental and harmonics 5th, 7th, 11th, and 13th.

Table 8.5 Summary of results Example 8.1

h	1	5	7	11	13
I_L (A)	151.6	30.32	21.67	13.79	11.67
I_S (A)	117.92∠−25.84°	38.03∠134.43°	36∠134.43°	799.82∠134.43°	31.34∠45.57°
X_C (Ω)	−*j*134.03	−*j*26.80	−*j*19.15	−*j*12.18	−*j*10.31
X_{TH} (Ω)	*j*1.088	*j*5.44	*j*7.62	*j*11.97	*j*14.15
I_C (A)	56.86∠90°	7.71∠134.43°	14.32∠134.43°	786.03∠134.43°	43.01∠−45.57°
V_C (V)	7,621.02∠0°	206.78∠44.43°	274.25∠44.43°	9,573.84∠44.43°	443.46∠−135.57°

THD values:

$$THD = \frac{\sqrt{A_2^2 + A_3^2 + \cdots}}{A_1} = \frac{\sqrt{A_5^2 + A_7^2 + A_{11}^2 + A_{13}^2}}{A_1}$$

For current:

$$THD = \frac{\sqrt{30.32^2 + 21.67^2 + 13.79^2 + 11.67^2}}{151.6}$$

$$= 0.2732 \cong 27.32\%$$

$$\frac{I_{SC}}{I_L} = \frac{7,000\ \text{A}}{151.6\ \text{A}} = 46.17$$

From the Std. IEEE 519-2014:

For $20 < \frac{I_{SC}}{I_L} < 50 \rightarrow THD = 8\%$

The obtained value of 27.32%→it does not fulfill the standard.

For voltages

$$THD = \frac{\sqrt{0.206^2 + 0.274^2 + 9.573^2 + 0.443^2}}{7.62} = 1.258 \cong 126\%$$

The standard allows 5%→it does not fulfill the standard.

$$V_{\text{rms}_{TOTAL}} = \sqrt{V_1{}^2 + V_5{}^2 + V_7{}^2 + V_{11}{}^2 + V_{13}{}^2}$$

$$V_{\text{rms}_{TOTAL}} = \sqrt{7.62^2 + 0.206^2 + 0.274^2 + 9.573^2 + 0.443^2} = 12.25\ \text{kV}$$

This exceeds 10% regarding the nominal $\cong 7.62 \cdot (1.1) = 8.382$

$$\begin{aligned}
I_{\mathrm{rms}_{TOTAL}} &= \sqrt{I_1{}^2 + I_5{}^2 + I_7{}^2 + I_{11}{}^2 + I_{13}{}^2} \\
I_{\mathrm{rms}_{TOTAL}} &= \sqrt{56.86^2 + 7.71^2 + 14.32^2 + 786.03^2 + 43.01^2} = 789.42\ \mathrm{A} \\
\frac{789.42\ \mathrm{A}}{56.86\ \mathrm{A}} &\cong 1{,}388\% \\
I_{L(\mathrm{rms})} &= 157.15\ \mathrm{A} \\
V_{C(\mathrm{rms})} &= 12.25\ \mathrm{kV} \\
I_{C(\mathrm{rms})} &= 789.42\ \mathrm{A} \\
I_{S(\mathrm{rms})} &= 810.77\ \mathrm{A} \\
THD_{I_L} &= 27.32\% \\
THD_{V_C} &= 126\% \\
THD_{I_C} &= 1{,}385\% \\
THD_{I_S} &= 680\%
\end{aligned}$$

The former results show that the capacitor bank has to be resized, relocated, or both, since there is a clear violation of the standards referred before. In fact, all the *THD* values were exceeded and overvoltages and overcurrents registered at the capacitors.

The best approach in this case would be an iterative process until the standards be satisfied.

8.10 Derating transformers

Transformers serving nonlinear loads register an important rise in eddy current losses associated with the harmonic currents generated by those loads. This increases the temperature of the transformers, which obliges to reduce or derate their nominal rating. In order to determine the derating caused by this phenomenon, the harmonic loss factor for winding eddy currents factor, known as F_{HL}, has been introduced.

In pu, the F_{HL} factor is defined in IEEE Standard C57.110-2008 as follows:

$$F_{HL} = \frac{\sum_{h=1}^{h=h_{\max}} I_h^2 \cdot h^2}{\sum_{h=1}^{h=h_{\max}} I_h^2} \tag{8.15}$$

where I_h is the rms current at harmonic h, in pu, or rated rms load current.

Underwriters Laboratories has introduced another term called the *K*-factor that defines a rating optionally applied to a transformer indicating its suitability for use with loads that have nonsinusoidal currents. The *K*-factor is defined with the following expression:

$$K = \sum_{h=1}^{h=h_{\max}} I_h^2 \cdot h^2 \tag{8.16}$$

From (8.15) and (8.16), it is clear that F_{HL} in pu is equal to the K-factor divided by the summation of the squared of the fundamental and harmonic currents in pu.

Manufactures build special K-factor transformers. Standard K-factor ratings are 4, 9, 13, 20, 30, 40, and 50. For linear loads, the K-factor is always 1.

For nonlinear loads, if harmonic currents are known, the K-factor is calculated and compared against the transformer's nameplate K-factor. As long as the load K-factor is equal to or less than 1, the transformer K-factor, the transformer does not need to be derated.

For transformers, ANSI/IEEE Standard C57.110 provides a method to derate the transformer capacity when supplying nonlinear loads.

$$\textit{Transformer derating} = \sqrt{\frac{1+P_{ec-r}}{1+\left(\sum_{h=1}^{\infty} I_h^2 h^2 / \sum_{h=1}^{\infty} I_h^2\right) \times P_{ec-r}}} \tag{8.17}$$

or

$$\textit{Transformer derating} = \sqrt{\frac{1+P_{ec-r}}{1+K \times P_{ec-r}}} \tag{8.18}$$

where P_{ec-r} is the maximum transformer pu eddy current loss factor (typically, between 0.05 and 0.10 pu for dry-type transformers), I_h is the harmonic current, normalized by dividing it by the fundamental current, and h is the harmonic order.

Example 8.2 Assume that the pu harmonic currents are 1.000, 0.016, 0.261, 0.050, 0.003, 0.089, 0.031, 0.002, 0.048, 0.026, 0.001, 0.033, and 0.021 pu A for harmonic order of 1, 3, 5, 7, 9, 11, 13, 15, 17, 19, 21, 23, and 25, respectively. Also assume that the eddy current loss factor is 8%.

Based on ANSI/IEEE Standard C57.110, determine the following:

1. The F_{HL} and K-factor on the transformer.
2. The transformer derating.

Solution:

1. The K and F_{HL} factors are obtained from Table 8.6.
   Table 8.6 results corresponding to Example 8.2

$$K = \sum_{h=1}^{h=25} I_h^2 \cdot h^2 = 5.712$$

$$F_{HL} = \frac{\sum_{h=1}^{h=25} I_h^2 \cdot h^2}{\sum_{h=1}^{h=25} I_h^2} = \frac{5.712}{1.084} \cong 5.3$$

Table 8.6 Currents for different harmonic orders

h	i (pu)	h^2	i^2	$i^2 \cdot h^2$
1	1	1	1	1
3	0.016	9	0.000256	0.002304
5	0.261	25	0.068121	1.703025
7	0.05	49	0.0025	0.1225
9	0.003	81	0.000009	0.000729
11	0.089	121	0.007921	0.958441
13	0.031	169	0.000961	0.162409
15	0.002	225	0.000004	0.0009
17	0.048	289	0.002304	0.665856
19	0.026	361	0.000676	0.244036
21	0.001	441	0.000001	0.000441
23	0.033	529	0.001089	0.576081
25	0.021	625	0.000441	0.275625
Sum			1.084283	5.712347

2. The transformer derating is obtained as follows:

$$Transformer\ derating = \sqrt{\frac{1 + P_{ec-r}}{1 + F_{HL} \cdot P_{ec-r}}}$$
$$= \sqrt{\frac{1 + 0.08}{1 + 5.3 \times 0.08}} \cong 0.87 \text{ pu or } 87\%$$

Further reading

[1] Arrillaga, J., Bradley, D. A., and Bodger, P. S. Power System Harmonics. John Wiley & Sons, New Delhi, 1985.

[2] IEEE. IEEE Std. 1159-2019 - IEEE Recommended Practice for Monitoring Electric Power Quality, 2019.

[3] IEEE. IEEE Std. 18-2012 – IEEE Standard for Shunt Power Capacitors, 2013.

[4] IEEE. IEEE Std. 519-2014 – IEEE Recommended Practices and Requirements for Harmonic Control in Electrical Power Systems, 2014.

[5] McGranaghan, M. F., Dugan, R. C., King, J. A., and Jewell, W. T. "Distribution feeder harmonic study methodology," IEEE Transaction on Power Apparatus and Systems, vol. PAS-103, no. 12, pp. 3663–3671, 1984.

[6] Ortmeyer, T. H., Hammam, M. S. A. A., Hiyama, T., and Webb, D. B. "Measurement of the harmonic characteristics of radial distribution systems," Power Engineering Journal, vol. 2, no. 3, pp. 163–172, 1988.

[7] Rizy, D. T., Gunther, E. W., and McGranaghan, M. F. "Transient and harmonic voltages associated with automated capacitor switching on distribution systems," IEEE Transactions on Power Systems, vol. PWRS-2, no. 3, pp. 713–723, 1987.

[8] Sabin, D., McGranaghan, M., and Sundaram, A. "A systems approach to power quality monitoring for performance assessment," in Proceedings of the 7th Annual Canadian Electricity Forum on Power Quality and Power Harmonics, Toronto, Canada, May 1997.

Chapter 9
Modern protection of distribution systems

Overcurrent relays are the most common form of protection used to operate only under fault conditions. They should not be installed purely as a means of protecting systems against overloads. The relay settings that are selected are often a compromise in order to cope with both overload and overcurrent conditions.

Overcurrent relays can be classified as definite current, definite time, and inverse time as shown in parts (a)–(c) of Figure 9.1. The time-delay units can work in conjunction with the instantaneous units as shown in Figure 9.1(d).

9.1 Fundamentals of overcurrent protection

9.1.1 Protection coordination principles

Relay coordination is the process of selecting settings that will assure that the relays will operate in a reliable and selective way. In overcurrent relays the coordination is based on the relay time–current characteristics (TCCs) of instantaneous and/or time-delay units.

1. Instantaneous units should be set so they do not trip for fault levels equal or lower to those at busbars or elements protected by downstream instantaneous relays.

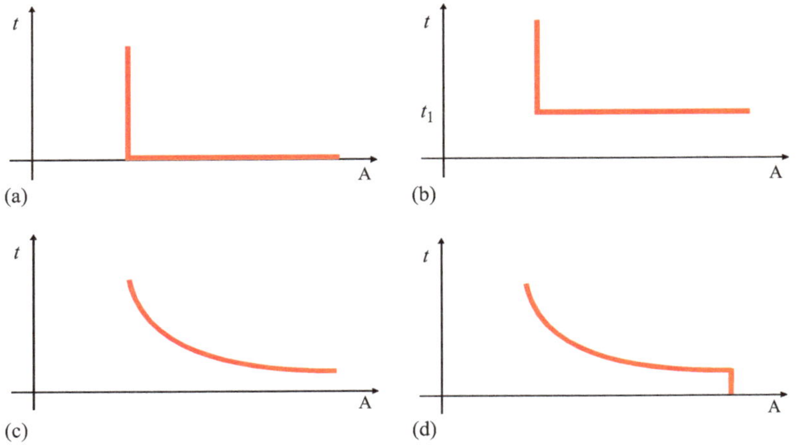

Figure 9.1 Time–current operating characteristics of overcurrent relays: (a) Definite current; (b) Definite time; (c) Inverse time; and (d) Inverse time with instantaneous unit

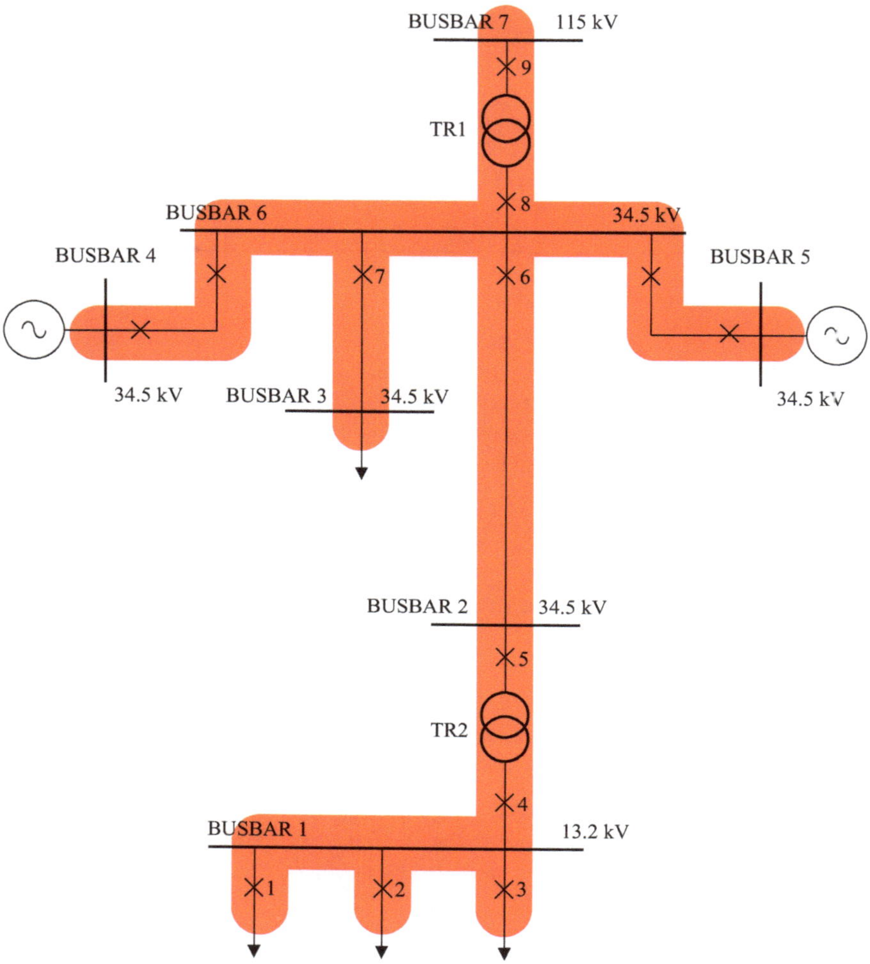

Figure 9.2 Overcurrent relay coordination procedure in a distribution system

2. Time-delay units should be set to clear faults in a selective and reliable way, assuring the proper coverage of the thermal limits of the elements protected.

Figure 9.2 illustrates for a typical distribution substation, the process to carry out the coordination, starting from the relay associated with breakers 1–3 all the way up to breaker 9.

9.1.2 Criteria for setting instantaneous units

Instantaneous units are set by adjusting the pickup level current at which the relays operate. Most numerical relays now have the possibility of setting an operating time, allowing the relay to behave as a definite time unit.

1. Distribution lines
   (i) between six and ten times the maximum circuit rating
   (ii) about 50% of the maximum short circuit at the point of connection of the relay

2. Lines between substations
   (i) about 125% to 150% of the short circuit current existing on the next substation

3. Transformer units
   (i) about 125% to 150% of the short circuit current existing on the LV side
   (ii) the units at the LV side are overridden unless there is communication with the relays protecting the feeders.

9.1.3 Setting time-delay relays

Time-delay units are set by selecting the time-curve characteristic that is defined by two parameters:

1. Tap or pickup value: A value that defines the pickup current of the relay. Current values are expressed as multiples of this value in the TCC curves.
2. Dial: Defines the time curve at which the relay operates for any tap value. Higher dial values represent higher operating times.

Figure 9.3 shows an overcurrent inverse time-relay curves associated with the two breakers on the same feeder.

A margin between two successive devices in the order of 0.2 s–0.4 s should be used to avoid losing selectivity due to one or more of the following reasons:

1. the breaker opening time,
2. the overrun time after the fault has been cleared,
3. variations in fault levels, deviations from the characteristic curves of the relays, and errors in the current transformers (CTs).

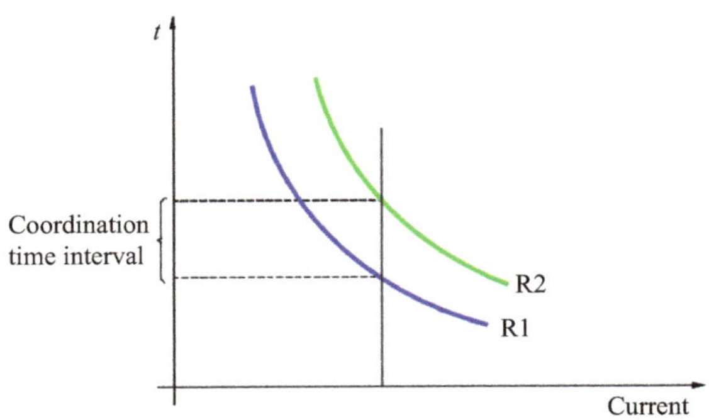

Figure 9.3 Overcurrent inverse-time relay curves associated with the two breakers on the same feeder

For phase relays, the tap or pickup value is determined by

$$TAP = \frac{(OLF \times I_{nom})}{CTR}$$

For ground fault relays, the tap value is determined, with the maximum unbalance, typically around 20%:

$$TAP = \frac{((0.2) \times I_{nom})}{CTR}$$

The overload factor recommended is as follows:

- Motors = 1.05
- HV lines, transformers, and generators = 1.25–1.5
- Distribution feeders = 2.0

For phase relays, three-phase faults and maximum short time overload should be considered. For ground relays, line-to-ground faults and max $3I_o$ should be considered.

The procedure to determine the time dial settings is based on operating time targets corresponding to the multiples of pickup or tap values at the instantaneous values.

The process starts at the furthest downstream relay and finishes with the furthest up relay.

For the furthest downstream relays, the lowest time dial is chosen or that considering cold load pickup conditions.

Normally, the settings are first carried out for phase relays and then for ground (neutral) relays. For the latter, the lowest time dial is selected whenever an open ground circuit is established, like that through Dy transformers.

The process to determine the time dial setting is rather elaborate and is summarized in the following steps:

1. Calculate the multiple of pickup value for the secondary short circuit current corresponding to the instantaneous setting of the relay where the process starts. If the instantaneous unit is overridden, the calculation is carried out with the total secondary short circuit current at the relay location.
2. With the value calculated above, determine the operating time t_1 of the relay for the given time dial.
3. Determine the operating time t_{2a} of the upstream relay with the expression $t_{2a} = t_1 + t_{margin}$.
4. Calculate the multiple of pickup value of the upstream relay using the same short circuit current used in the first relay (Step 1).
5. Knowing t_{2a}, and having calculated the multiple of pickup value of the upstream relay, select the above nearest time dial for that relay.

The process follows the same steps for the next upstream relay and is repeated until the settings of the furthest up relay are calculated.

Operating time defined by IEC 60255 and IEEE C37.112 are as follows:

$$t = \frac{k \cdot \beta}{(I/I_s)^{\alpha} - 1} + L$$

where t is the relay operating time in seconds; k is the time dial, or time multiplier setting; I is the fault current level in s A; I_S is the tap or pickup current selected; L is the constant; α is the slope constant; β is the slope constant.

Table 9.1 shows the constant values of the parameters for curves defined by IEEE C 37.112 and IEC 60255, and Figure 9.4 illustrates these curves.

Table 9.1 IEEE and IEC constants for standards, overcurrent relays

IDMT curve description	Standard	α	β	*L*
Moderately inverse	IEEE	0.02	0.0515	0.114
Very inverse	IEEE	2	19.61	0.491
Extremely inverse	IEEE	2	28.2	0.1217
Inverse	US-CO8	2	5.95	0.18
Short-time inverse	US-CO2	0.02	0.02394	0.01694
Standard inverse	IEC	0.02	0.14	
Very inverse	IEC	1	13.5	
Extremely inverse	IEC	2	80.0	
Long-time inverse	IEC	1	120	

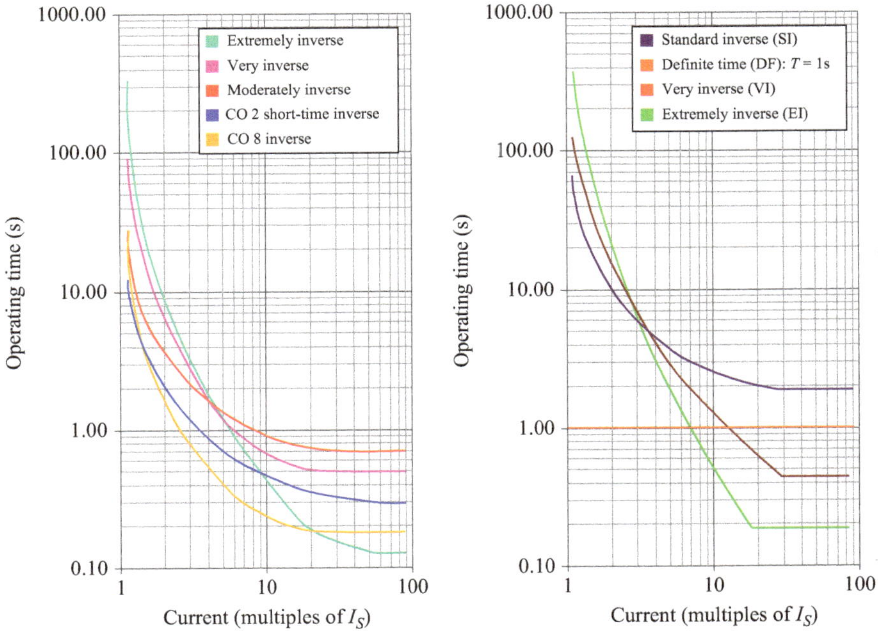

Figure 9.4 Typical ANSI/IEEE and IEC overcurrent relay curves

9.1.4 Setting overcurrent relays using software techniques

1. Locate the fault and obtain the current for setting the relays.
2. Identify the pairs of relays to be set, first determining which one is farther away from the source.
3. Verify that the requirement's thermal capabilities are protected and devices operate for minimum short circuit levels.

9.2 Coordination across Dy transformers

In the case of overcurrent relay coordination for Dy transformers, the distribution of currents in these transformers should be checked for three-phase, phase-to-phase, and single-phase faults on the secondary winding.

The results of the three cases are summarized in Table 9.2. By analyzing the results, it can be seen that the critical case for the coordination of overcurrent relays is the phase-to-phase fault. In this case, the relays installed in the secondary carry a current less than the equivalent current flowing through the primary relays that could lead to a situation where the selectivity between the two relays is at risk. For this reason, the discrimination margin between the relays is based on the operating time of the secondary relays at a current equal to $\sqrt{3}I_f/2$, and on the operating time for the primary relays for the full fault current value I_f, as shown in Figure 9.5.

Example 9.1 The power system in Figure 9.6 calculates the following:

1. The three-phase short circuit levels on busbars 1 and 2.
2. The transformation ratios of the CTs associated with breakers 1–8, given that the number of primary turns is a multiple of 100. The CT for breaker 9 is 250/5.
3. The settings of the instantaneous elements, and the *TAP* and *dial* settings of the relays to guarantee a coordinated protection arrangement, allowing a discrimination margin of 0.4 s.
4. The percentage of the 34.5 kV line protected by the instantaneous element of the overcurrent relay associated with breaker 6.

The pu impedances are calculated on the following bases:

$$V = 34.5\ \text{kV}, \quad P = 100\ \text{MV A}$$

Table 9.2 Summary of fault conditions

Fault	$I_{primary}$	$I_{secondary}$
Three phase	I	I
Phase to phase	I	$\sqrt{3}I/2$
Phase to earth	I	$\sqrt{3}I$

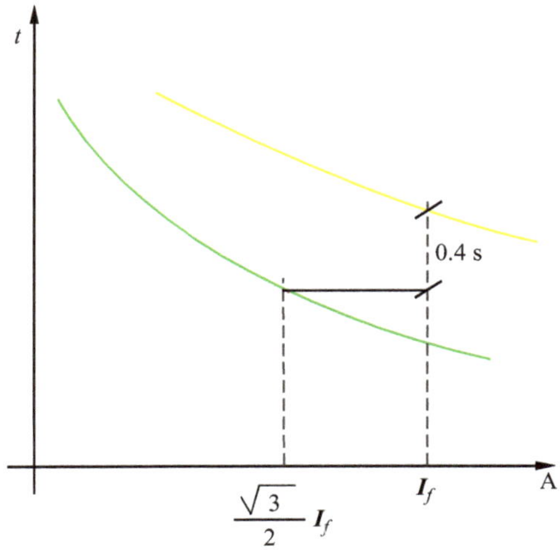

Figure 9.5 Coordination of overcurrent relays for a Dy transformer

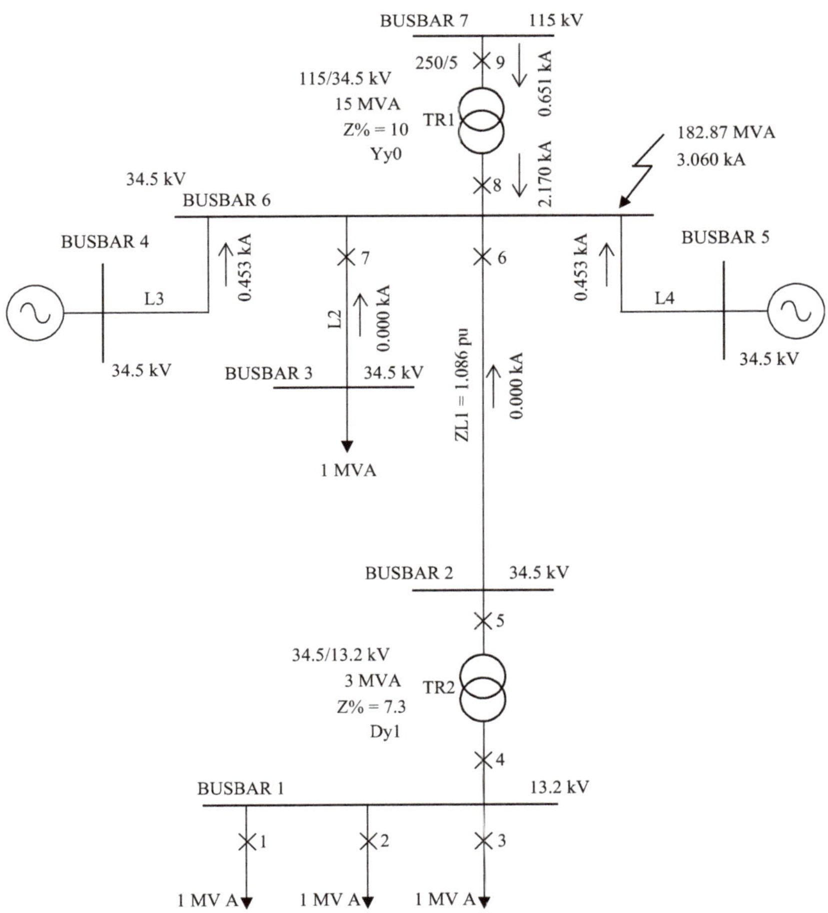

Figure 9.6 Power system of Example 9.1

Consider all the relays to be set are numerical type Beckwith M-7651 with the characteristics as is shown in Figure 9.7. The relays have an extremely inverse TCC with the following constants:

$$\alpha = 2.0, \quad \beta = 80, L = 0$$

The TCC is defined by $t = (Time\ dial \times 80)/\left((MULT)^2 - 1\right)$, where MULT (setting multiplier) is the fault current (in secondary amps)/tap. The following considerations have to be taken into account:

1. For setting of the instantaneous element, a value of ten times the maximum load current is used.
2. The margin time for this relay can be 0.2 s since it is of numerical type.
3. Relay 7 is the same W CO 11 with $TAP = 4$ A, $dial = 5$, and $I_{Inst} = 1{,}100$ A.

The margin time for this relay can be 0.2 s since it is of numerical type. All the parameters of the power system elements are shown in Table 9.3.

Defining the pickup for relays:

$$
\begin{aligned}
I_{load1,2,3} &= 43.74\ \text{A}; Pickup_{1,2,3} = (1.5)(43.74)(5/100) = 3.28\ \text{A}\\
I_{load4} &= 131.22\ \text{A}; Pickup_4 = (1.5)(131.22)(5/200) = 4.92\ \text{A}\\
I_{load5} &= 50.20\ \text{A}; Pickup_5 = (1.5)(50.20)(5/100) = 3.76\ \text{A}\\
I_{load6} &= 50.20\ \text{A}; Pickup_6 = (1.5)(50.20)(5/200) = 1.88\ \text{A}\\
I_{load8} &= 251.02\ \text{A}; Pickup_8 = (1.5)(251.02)(5/300) = 6.28\ \text{A}\\
I_{load9} &= 75.31\ \text{A}; Pickup_9 = (1.5)(75.31)(5/250) = 2.26\ \text{A}
\end{aligned}
$$

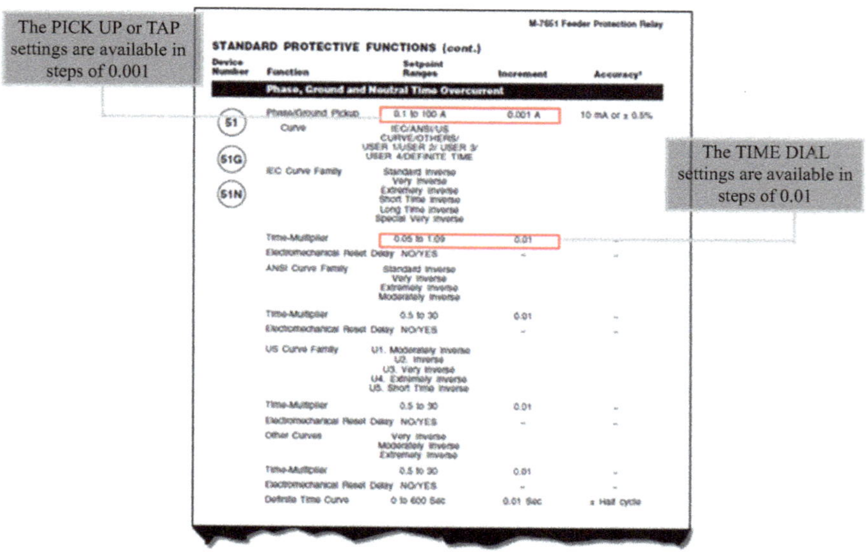

Figure 9.7 Specifications of relay Beckwith M-7651. Reproduced by permission of Beckwith Electric

Table 9.3 Short circuit calculation for power system of Example 9.1

Fault location		Distance from fault	Element name	Type	U_n (kV)	UL-E (RST) (kV)	AU L-E (RST) (°)	$I_{k''}$ (RST) (kA)
From node	To node							
BUSBAR6	Faulted	0			34.5	19.919	180	3.06
BUSBAR6	BUSBAR2		L1	Line				0
BUSBAR6	BUSBAR3		L2	Line				0
BUSBAR6	BUSBAR5		L4	Line				0.453
BUSBAR6	BUSBAR4		L3	Line				0.453
BUSBAR6	BUSBAR7		TR1	2W transformer				2.17
BUSBAR2		1			34.5	19.919	180	
BUSBAR2	BUSBAR6		L1	Line				0
BUSBAR2	BUSBAR1		TR2	2W transformer				0
BUSBAR3		1			34.5	19.919	180	
BUSBAR3	BUSBAR6		L2	Line				0
BUSBAR5		1			34.5	8.281	179.94	
BUSBAR5	BUSBAR6		L4	Line				0.453
BUSBAR5	BUSBAR5		G8	Synchronous machine				0.453
BUSBAR4		1			34.5	8.281	179.94	
BUSBAR4	BUSBAR6		L3	Line				0.453
BUSBAR4	BUSBAR4		G15	Synchronous machine				0.453
BUSBAR7		1			115	8.988	180	
BUSBAR7	BUSBAR6		TR1	2W transformer				0.651
BUSBAR7	BUSBAR7		Equivalent	Network feeder				0.651
BUSBAR1		2			13.2	7.621	180	
BUSBAR1	BUSBAR2		TR2	2W transformer				0

For relays 1–3:

$$I_{inst.trip} = 10 \times I_{nom} \times (1/CTR)$$
$$= 10 \times 43.74 \times (5/100) = 21.87 \text{ A}$$
$$I_{prim.trip} = 21.87(100/5) = 437.4 \text{ A}$$
$$MULT = 21.87/3.28 = 6.668 \text{ times, with } dial = 0.05,$$
$$t = \frac{0.05 \cdot 80}{(6.668)^2 - 1} = 0.092 \text{ s}$$

To coordinate with relays 1–3 at 437.4 A,

$$t_{4a} = 0.092 + 0.2 = 0.292 \text{ s.}$$

For relay 4:

$$MULT_{4a} = (437.4)(5/200)(1/4.92) = 2.223 \text{ times.}$$

At 2.223 times and $t_{4a} = 0.292$ s,

$$dial = \frac{0.292}{80}\left((2.223)^2 - 1\right) = 0.01$$

However, the dial 0.05 is the minimum that the relay has.

The operating time for a line-to-line fault is determined by taking 86% of the three-phase fault current.

$$MULT_{4b} = (0.86)(1,076.06)(5/200)(1/4.92) = 4.702 \text{ times.}$$
$$t_{4b} = \frac{0.05 \cdot 80}{(4.627)^2 - 1} = 0.190 \text{ s}$$

This relay has no setting for the instantaneous.

The procedure is the same for all the relays of the power. Table 9.4 shows the corresponding settings. The curves obtained are shown in Figure 9.8.

Table 9.4 Summary of relay settings for Example 9.1

Relay no.	**CT ratio**	**Pickup**	**Dial**	**Instantaneous (A)**
1,2,3	100/5	3.28	0.05	21.87
4	200/5	4.92	0.05	–
5	100/5	3.76	0.14	25.73
6	200/5	1.88	0.25	32.05
7	200/5	4	5	27.5
8	300/5	6.28	0.05	–
9	250/5	2.26	0.13	16.27

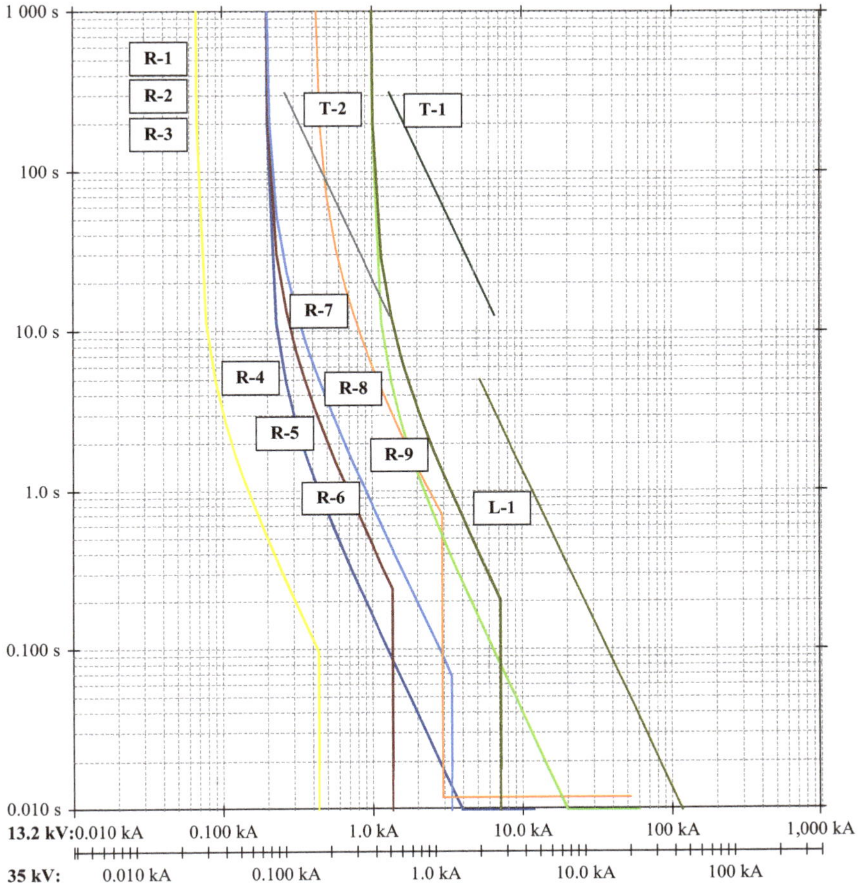

Figure 9.8 Relay coordination curves for Example 9.1

9.3 Protection equipment installed along the feeders

The devices most used for distribution system protection are overcurrent relays, reclosers, sectionalizers, and fuses.

The coordination of overcurrent relays was dealt with in detail in the previous section, and this part will cover the other three devices referred to above. The three last devices can be mounted as stand-alone or as part of other switchgear like those known as VFI. This type of switchgear can be arranged in different configurations where the type of equipment can be different (breakers or fuses) and also the number of inputs and outputs can vary. Figure 9.9 illustrates a VFI with seven different configurations from S&C. Figure 9.10 is a picture of a typical Pad-Mounted Gear also from S&C.

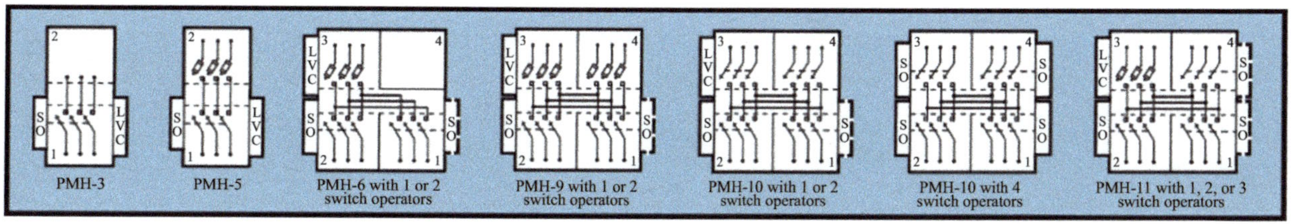

Figure 9.9 Remote supervisory PMH models. Reproduced by permission of S&C

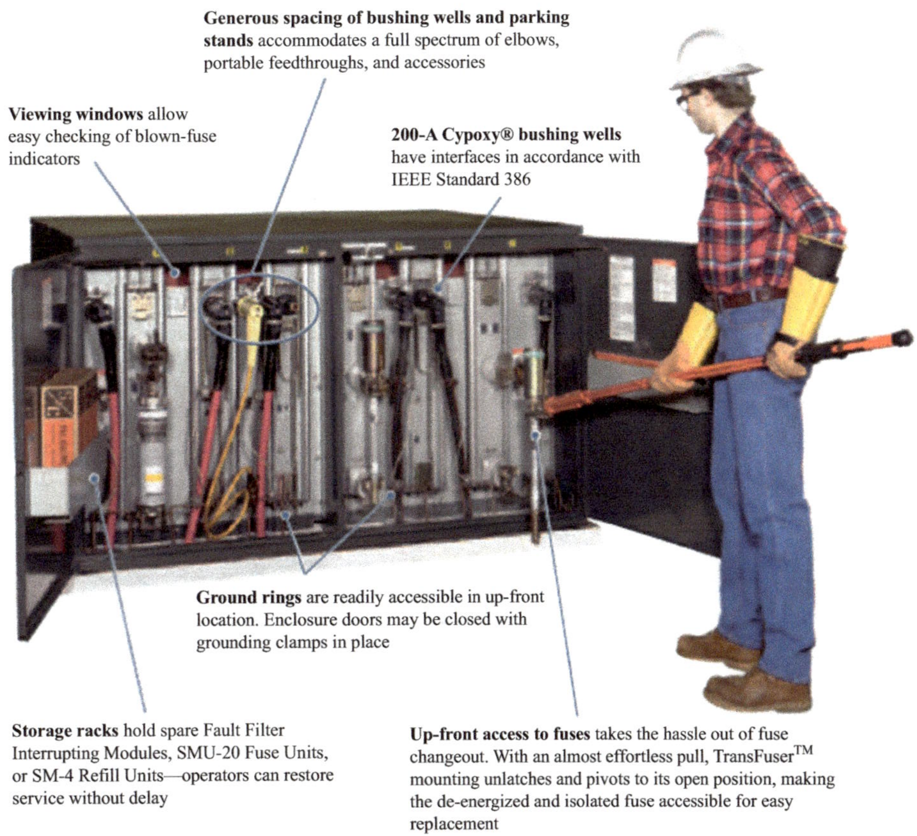

Figure 9.10 PME Pad-Mounted Gear. Reproduced by permission of S&C

9.3.1 Reclosers

9.3.1.1 General

A recloser is a device with the ability to detect phase and phase-to-earth overcurrent conditions, to interrupt the circuit if the overcurrent persists after a predetermined time and then to automatically reclose to reenergizer the line. If the fault that originated the operation still exists, then the recloser will stay open after a preset number of operations, thus isolating the faulted section from the rest of the system.

In an overhead distribution system, 80% to 95% of the faults are of a temporary nature and last, at the most, for a few cycles or seconds. Thus, the recloser, with its opening/closing characteristics, prevents a distribution circuit being left out of service for temporary faults. Typically, reclosers are designed to have up to three open-close operations and, after these, a final open operation to lock out the sequence. One further closing operation by manual means is usually allowed. The counting mechanisms register operations of the phase or earth-fault units that can also be initiated by externally controlled devices when appropriate communication means are available.

The operating TCC curves of reclosers normally incorporate three curves, one fast and two delayed, designated as A, B, and C, respectively. Figure 9.11 shows a typical set of time–current curves for reclosers. However, new reclosers with microprocessor-based controls may have keyboard-selectable time–current curves that enable an engineer to produce any curve to suit the coordination requirements for both phase and earth faults. This allows reprograming of the characteristics to tailor an arrangement to a customer's specific needs without the need to change components.

Coordination with other protection devices is important in order to ensure that, when a fault occurs, the smallest section of the circuit is disconnected to minimize disruption of supplies to customers. Generally, the time characteristic and the sequence of operation of the recloser are selected to coordinate with mechanisms upstream toward the source. After selecting the size and sequence of operation of the recloser, the devices downstream are adjusted in order to achieve correct coordination. A typical sequence of a recloser operation for a permanent fault is shown in Figure 9.12. The first shot is carried out in instantaneous mode to clear temporary faults before they cause damage to the lines. The three later ones operate in a timed manner with predetermined time settings. If the fault is permanent, the time-delay operation allows other protection devices nearer to the fault to open, limiting the amount of the network being disconnected.

Earth faults are less severe than phase faults and, therefore, it is important that the recloser has an appropriate sensitivity to detect them. One method is to use CTs connected residually so that the resultant residual current under normal conditions is approximately zero. The recloser should operate when the residual current exceeds the setting value, as would occur during earth faults.

Typically, reclosers are designed to have up to three open-close operations and, after these, a final open operation to lock out the sequence. One further closing

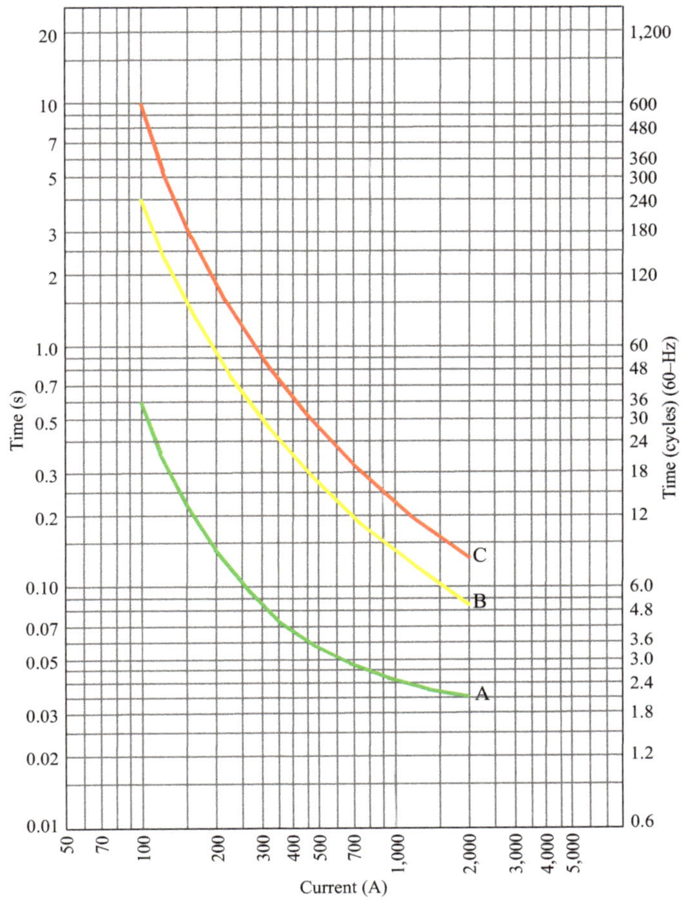

Figure 9.11 Time–current curves for reclosers

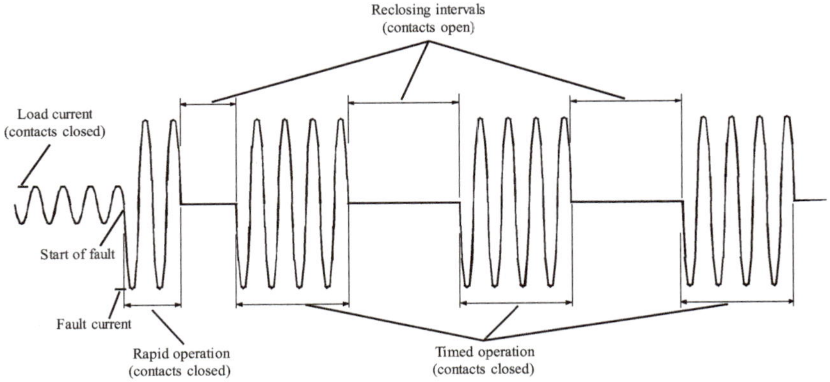

Figure 9.12 Typical sequence for recloser operation

operation by manual means is usually allowed. The counting mechanisms register operations of the phase or ground-fault units that can also be initiated by externally controlled devices when appropriate communication means are available.

9.3.1.2 Classification

Reclosers can be classified as follows:

- number of phases: single-phase and three-phase;
- arc-interrupting medium: oil or vacuum;
- type of insulation: oil, epoxy, or SF_6;
- type of control: hydraulic, electronic, or magnetic actuator.

Single-phase reclosers are used when the load is predominantly single phase. In such a case, when a single-phase fault occurs, the recloser should permanently disconnect the faulted phase so that supplies are maintained on the other phases. Three-phase reclosers are used when it is necessary to disconnect all three phases in order to prevent unbalanced loading on the system.

Reclosers with hydraulic operating mechanisms have a disconnecting coil in series with the line. When the current exceeds the setting value, the coil attracts a piston that opens the recloser main contacts and interrupts the circuit. The time characteristic and operating sequence of the recloser are dependent on the flow of oil in different chambers. The electronic type of control mechanism is normally located outside the recloser and receives current signals from a CT-type bushing. When the current exceeds the predetermined setting, a delayed shot is initiated which finally results in a tripping signal being transmitted to the recloser control mechanism. The control circuit determines the subsequent opening and closing of the mechanism, depending on its setting. Reclosers with electronic operating mechanisms use a coil or motor mechanism to close the contacts. Oil reclosers use the oil to extinguish the arc and also to act as the basic insulation. The same oil can be used in the control mechanism. Vacuum and SF_6 reclosers have the advantage of requiring less maintenance.

The following figures illustrate several features of reclosers. Figure 9.13 shows different types of single-phase reclosers: (a) NOJA OSM—vacuum interrupting, epoxy insulated, and electronically controlled; (b) COOPER NOVA—oil interrupting, epoxy insulated, and electronically controlled; (c) COOPER D—oil interrupting, oil insulated, and hydraulically controlled.

Figure 9.14 shows different types of three-phase reclosers: (a) G&W Viper-LT—vacuum interrupting, epoxy insulated, and electronically controlled; (b) SCHNEIDER U—vacuum interrupting, epoxy insulated, and electronically controlled; (c) ABB OVR—vacuum interrupting, epoxy insulated, and electronically controlled; (d) Hawker Siddeley Switchgear Ltd.'s GVR—vacuum interrupting, SF_6 insulated, and magnetic actuator.

Figure 9.15 presents the internal components of a three-phase recloser, air-insulated, vacuum interrupting, electronically controlled recloser, manufactured by Cooper Systems.

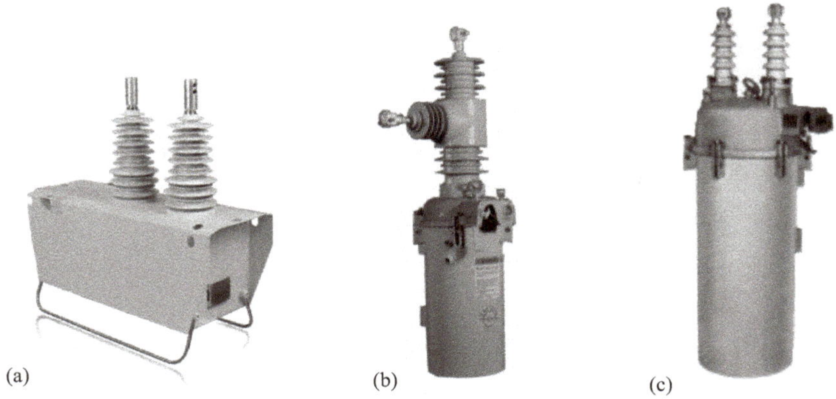

Figure 9.13 Types of single-phase reclosers: (a) NOJA OSM; (b) COOPER NOVA; and (c) COOPER D

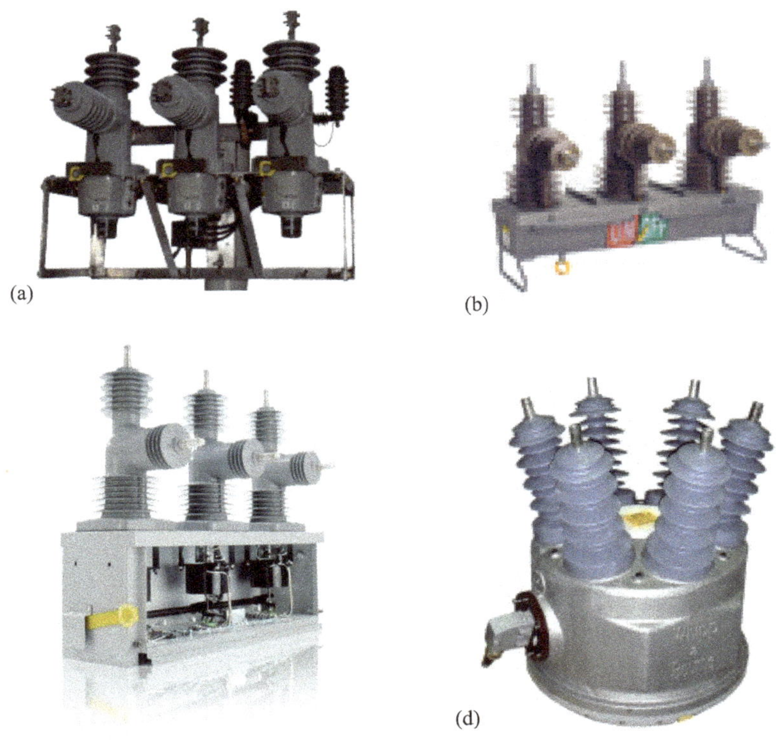

Figure 9.14 Types of three-phase reclosers: (a) G&W Viper-LT; (b) SCHNEIDER U; (c) ABB OVR; and (d) Hawker Siddeley Switchgear GVR

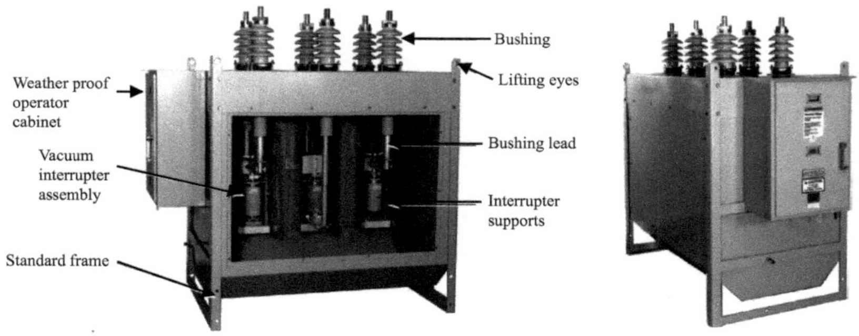

Figure 9.15 COOPER Kyle type VSA20A

The coordination margins with hydraulic reclosers depend upon the type of equipment used. In small reclosers, where the current coil and its piston produce the opening of the contacts, a separation greater than 12 cycles ensures non-simultaneous operation.

With large capacity reclosers, the piston associated with the current coil only actuates the opening mechanism. In such cases, a separation of more than eight cycles guarantees nonsimultaneous operation.

The principle of coordination between two large units in series is based on the time of separation between the operating characteristics, in the same way as for small units.

Electronically controlled reclosers can be coordinated more closely since there are no inherent errors such as those that exist with electromechanical mechanisms (due to overspeed, inertia, etc.). The downstream recloser must be faster than the upstream recloser, and the clearance time of the downstream recloser plus its tolerance should be lower than the upstream recloser clearance time less its tolerance.

Normally, the setting of the recloser at the substation is used to achieve at least one fast reclosure in order to clear temporary faults on the line between the substation and the load recloser. The latter should be set with the same, or a larger, number of rapid operations as the recloser at the substation.

It should be noted that the criteria of spacing between the TCCs of electronically controlled reclosers are different from those used for hydraulically controlled reclosers.

9.3.1.3 Applications

Reclosers are used at the following points on a distribution network:

- in substations, to provide primary protection for a circuit;
- in main feeder circuits, in order to permit the sectioning of long lines and thus prevent the loss of a complete circuit due to a fault toward the end of the circuit; and
- in branches or spurs, to prevent the tripping of the main circuit due to faults on the spurs.

Figure 9.16 illustrates several options to locate reclosers in a distribution system.

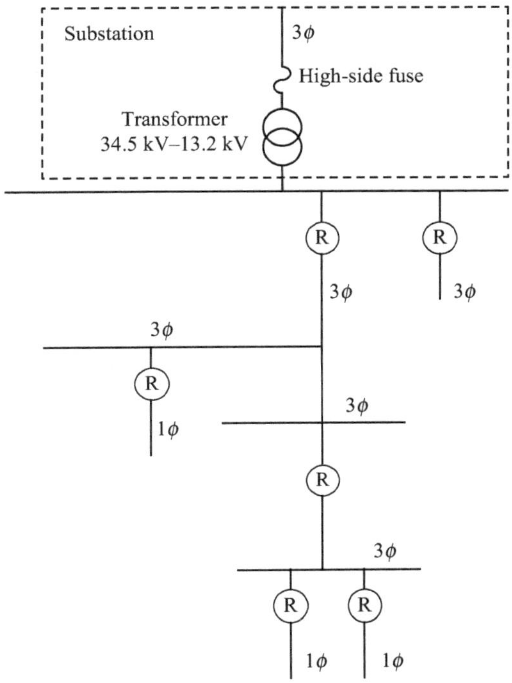

Figure 9.16 Option to locate reclosers

9.3.1.4 Specifications

The voltage rating and the short circuit capacity of the recloser should be equal to, or greater than, the values that exist at the point of installation. The same criteria should be applied to the current capability of the recloser in respect of the maximum load current to be carried by the circuit. It is also necessary to ensure that the fault current at the end of the line being protected is high enough to cause operation of the recloser.

To properly apply automatic circuit reclosers, five major factors must be considered:

1. System voltage: System voltage will be known. The recloser must have a voltage rating equal to (or greater than) system voltage.
2. Maximum fault current available at the recloser location: Maximum fault current will be known or can be calculated. The recloser interrupting must be equal to (or greater than) the maximum available fault current at the recloser location.
3. Maximum load current: The recloser continuous current rating must be equal to (or greater than) anticipated circuit load. For series-coil-type reclosers, the coil size can be selected to match the present load current, the anticipated future load current, or the substation transformer capacity. Minimum-trip

current is nominally twice the coil continuous-current rating. For electronically controlled reclosers, minimum-trip current must be greater than any anticipated peak load. Generally, a trip-current value of at least twice the expected load current is used.

4. Minimum-fault current within the zone to be protected: Minimum-fault current that might occur at the end of the line section must be checked to confirm that the recloser will sense and interrupt this current.
5. Coordination with the other protective devices on both the source and the load sides of the recloser.

After the first four application factors have been satisfied, coordination of the recloser with both the source and the load-side devices must be determined. Proper selection of time delays and sequences is vital to assure that any momentary interruption or longer term outage due to faults is restricted to the smallest possible section of the system.

Generally, recloser timing and sequences are selected to coordinate with the source-side devices. After the size and sequence of the required recloser has been determined, the protective equipment farther down the line is selected to coordinate with it.

9.3.2 Sectionalizers

9.3.2.1 General

A sectionalizer is a device that automatically isolates faulted sections of a distribution circuit once an upstream breaker or recloser has interrupted the fault current and is usually installed downstream of a recloser. Since sectionalizers have no capacity to break fault current, they must be used with a backup device that has fault current breaking capacity.

Sectionalizers will open to isolate a faulted area after a predefined number of "counts" with one count defined as seeing passage of fault current followed by a loss of voltage. They can be closed to pick up load, either remotely or via a local control panel or an operating handle actuated with a switching stick.

Sectionalizers count the number of operations of the recloser during fault conditions. After a preselected number of recloser openings, and while the recloser is open, the sectionalizer opens and isolates the faulty section of line. This permits the recloser to close and reestablish supplies to those areas free of faults. If the fault is temporary, the operating mechanism of the sectionalizer is reset. Sectionalizers are constructed in single- or three-phase arrangements with hydraulic or electronic operating mechanisms. A sectionalizer does not have a current–time operating characteristic and can be used between two protective devices, the operating curves of which are very close and where an additional step in coordination is not practicable.

Sectionalizers with hydraulic operating mechanisms have an operating coil in series with the line. Each time an overcurrent occurs the coil drives a piston that activates a counting mechanism when the circuit is opened and the current is zero by the displacement of oil across the chambers of the sectionalizer. After a prearranged number of circuit openings, the sectionalizer contacts are opened by means of pretensioned springs.

This type of sectionalizer can be closed manually. Sectionalizers with electronic operating mechanisms are more flexible in operation and easier to set. The load current is measured by means of CTs and the secondary current is fed to a control circuit that counts the number of operations of the recloser or the associated interrupter and then sends a tripping signal to the opening mechanism. This type of sectionalizer is constructed with manual or motor closing.

9.3.2.2 Classification

Sectionalizers can be classified as

- number of phases: single phase and three phase;
- type of insulation: oil, epoxy, or SF_6; and
- type of control: hydraulic, electronic, or magnetic actuator.

Figure 9.17 shows the different types of sectionalizers: (a) COOPER GH—single-phase oil insulated and hydraulic controlled; (b) Hawker Siddeley Switchgear Ltd.'s GVS 38—three-phase SF_6 insulated and magnetic actuator; (c) COOPER GN3VE—three-phase oil insulated and hydraulic controlled. Figure 9.18 shows the components of the untanked type GN3E three-phase sectionalizer from the Cooper Power Series.

9.3.2.3 Specifications

The nominal voltage and current of a sectionalizer should be equal or greater than the maximum values of voltage or load at the point of installation. The short circuit capacity (momentary rating) of a sectionalizer should be equal to or greater than the fault level at the point of installation. The maximum clearance time of the associated interrupter should not be permitted to exceed the short circuit rating of the sectionalizer. Coordination factors that need to be taken into account include the starting current setting and the number of operations of the associated interrupter before opening.

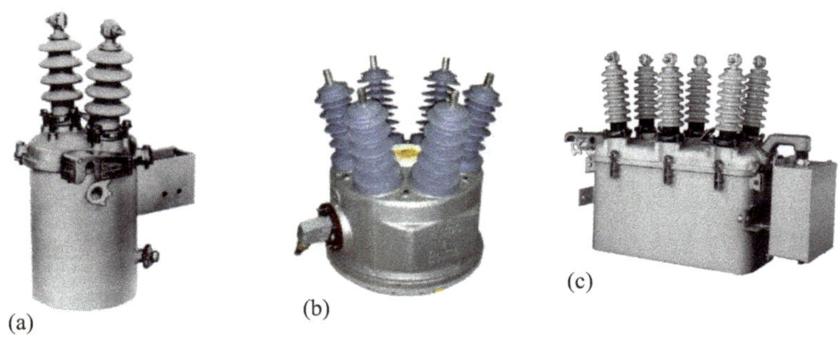

Figure 9.17 Types of sectionalizers: (a) COOPER GH; (b) Hawker Siddeley Switchgear; and (c) COOPER GN3VE

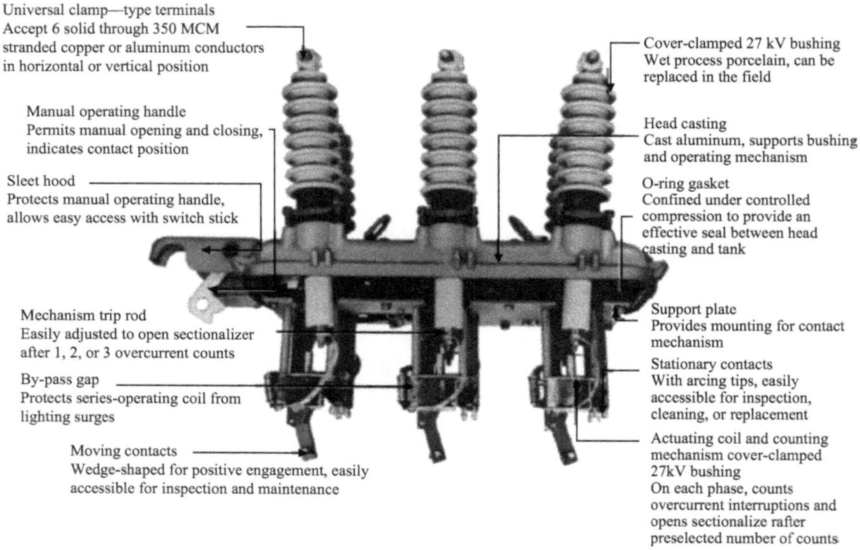

Figure 9.18 Untanked view of a sectionalizer

The following basic coordinating principles should be observed in the application of the electronically controlled sectionalizer:

1. The minimum actuating current setting of the sectionalizer should not be greater than 80% of the minimum trip current setting of the backup recloser or reclosing breaker for both phase and ground currents.
2. The counts-to-open setting of the sectionalizer must be at least one less than the number of operations to lock out of the backup device.
3. The count reset time of the sectionalizer control must be greater than the reset time of the backup protective device.
4. The minimum clearing time of the backup device must be greater than the charging time of the trip-energy storage capacitors to assure that the capacitors are fully charged before the backup trips.
5. Three-phase sectionalizers must be used with backup breakers or reclosers in which all three phases open simultaneously. The counting functions of the sectionalizers do not recognize a signal as originating in a particular phase but total the overcurrent interruptions in all three phases. Nonsimultaneous three-phase tripping of the backup could result in the sectionalizer interrupting fault current in one or more phases.
6. Application on multigrounded wye systems generally requires ground fault sensing and inrush current restraint. Setting the phase actuating level to the ground setting of the backup device may result in erroneous counts due to inrush currents and incorrect opening of the sectionalizer for source-side faults.

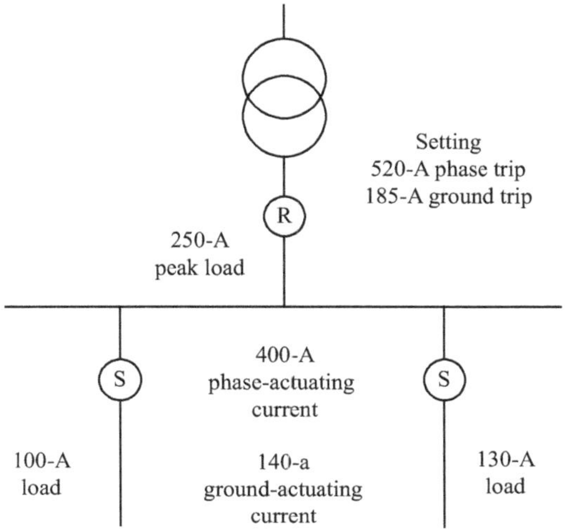

Figure 9.19 Sectionalizers application where two feeders of system are protected

7. Ground fault actuating settings should be set at or above peak loading at the sectionalizer location. This will prevent sectionalizer sensing system unbalance during upline phase-to-ground faults that generate system imbalances.

9.3.2.4 Applications

Sectionalizers can be applied where the main feeder is divided into two feeders close to the substation as shown in Figure 9.19.

Here, loads are likely to be reasonably equally divided. Sectionalizers may also be applied to protect an important branch that may be carrying only a small portion of the total load as shown in Figure 9.20.

The inrush current restraint has been designed to provide optimum protection from inrush currents. No settings are required as pickup values will be blocked during inrush conditions. This allows flexibility in the application without concern for increased connected loading on the system.

The sectionalizer is blocked for 3 s to allow for decay of magnetizing inrush, system imbalance, and initial inductive loading after long reclose intervals.

9.3.3 Fuses

9.3.3.1 General

A fuse is an overcurrent protection device; it possesses an element that is directly heated by the passage of current and that is destroyed when the current exceeds a predetermined value. A suitably selected fuse should open the circuit by the destruction of the fuse element, eliminate the arc established during the destruction of the element, and then maintain circuit conditions open with nominal voltage applied to its terminals (i.e., no arcing across the fuse element).

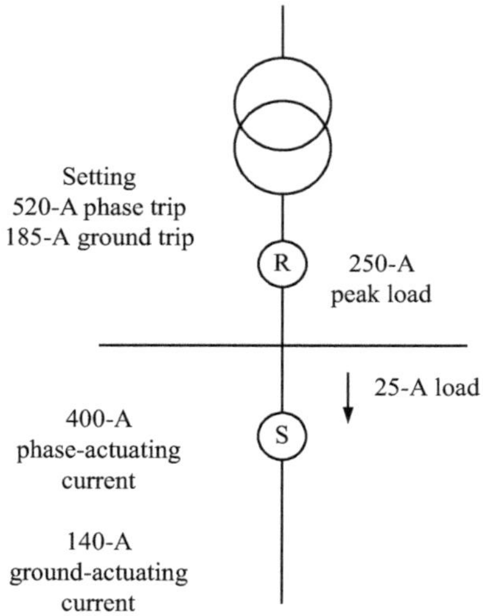

Figure 9.20 Sectionalizers application where one branch system is protected

The majority of fuses used in distribution systems operate on the expulsion principle, i.e., they have a tube to confine the arc, with the interior covered with deionizing fiber, and a fusible element. In the presence of a fault, the interior fiber is heated up when the fusible element melts and produces deionizing gases that accumulate in the tube. The arc is compressed and expelled out of the tube; in addition, the escape of gas from the ends of the tube causes the particles that sustain the arc to be expelled. In this way, the arc is extinguished when current zero is reached. The presence of deionizing gases, and the turbulence within the tube, ensures that the fault current is not reestablished after the current passes through zero point. The zone of operation is limited by two factors: the lower limit based on the minimum time required for the fusing of the element (minimum melting time) with the upper limit determined by the maximum total time that the fuse takes to clear the fault.

9.3.3.2 Applications

Fuses are the most popular protective device used in electrical systems. In distribution systems in particular, they are applied in most elements. Specific applications have been developed for the following elements:

- distribution transformers,
- capacitors, and
- feeders.

9.3.3.3 Type

Power fuses: Power fuses provide reliable and economical protection for transformers and capacitor banks. They are normally in outdoor substations served, and they incorporate silver or nickel–chrome fusible elements.

Fuse cutouts: A fuse cutout is a combination of a fuse and a switch, used in primary overhead feeder lines and distribution transformers to protect from current surges and overloads. The main components of fuse cutouts are the cutout body, the fuse holder, and the fuse element.

The cutout body, an open "C"-shaped frame that supports the "fuse holder" and a porcelain insulator that electrically isolates the conductive portions of the assembly from the support to which the insulator is fastened.

The fuse holder, often called the "fuse tube," contains the interchangeable fuse element and also acts as a simple knife switch. When the contained fuse operates or blows, the fuse holder drops open, disengages the knife switch, and hangs from a hinge assembly.

The fuse element, or "fuse link," is the replaceable portion of the assembly that operates due to high electrical currents.

Current-limiting fuse: A current-limiting fuse is a fuse that abruptly introduces a high resistance to reduce current magnitude and duration, resulting in subsequent current interruption. This type of fuse significantly reduces the current amplitude and the energy released in the event of a short circuit.

These fuses develop a positive internal gap of high dielectric strength after circuit interruption, thus precluding destructive reignitions when exposed to full system voltage—such as are experienced with current-limiting fuses after clearing under low recovery-voltage conditions.

Type SM Power Fuses have helically coiled silver fusible elements that are of solderless construction and are surrounded by air. Because of this construction, the fusible element is free from mechanical and thermal stress and confining support and therefore is not subject to damage—even by inrush currents that approach but do not exceed the fuse's minimum melting TCC curve. Current-limiting fuses, in contrast, have fusible elements that consist of a number of very fine diameter wires, or one or more perforated or notched ribbons, surrounded by, and in contact with, a filler material such as silica sand. Because of this construction, current-limiting fuses are susceptible to element damage caused by current surges that approach the fuse's minimum melting TCC curve. This damage may occur in one or more of the following ways: the fusible element may melt but not completely separate because the molten metal is constrained by the filler material—resulting, possibly, in resolidification of the element with a different cross-sectional area.

One or more, but not all, of the parallel wires or ribbons of the fusible element may melt and separate.

The fusible element may break as a result of fatigue brought about by current-cycling that can cause localized buckling from thermal expansion and contraction.

Damage to fusible elements of current-limiting fuses, as described above, may shift or alter their TCCs, resulting in a loss of complete coordination between the fuse and other downstream overcurrent protective devices. Moreover, a damage current-limiting fuse element may melt due to an otherwise harmless inrush current, but the fuse may fail to clear the circuit due to insufficient power flow—with the fuse continuing to arc and burn internally due to load-current flow. Because of the potential for damage to the fusible element from inrush currents, and because of the effects of loading and manufacturing tolerances, current-limiting fuse manufacturers typically require that when applying such fuses, adjustments be made to the minimum melting TCC curves. These adjustments are referred to as "safety zones" or "setback allowances" and range from 25% in terms of time to 25% in terms of current. The latter can result in an adjustment of 250% or more in terms of time, depending on the slope of the TCC curve at the point where the safety zone or setback allowance is measured. Furthermore, most current-limiting fuses inherently have steep, relatively straight TCC curves which, together with the required large safety-zone or setback-allowance adjustments, force the selection of a current-limiting fuse ampere rating substantially greater than the transformer full-load current in order to withstand combined transformer-magnetizing and load inrush currents, and also to coordinate with secondary-side protective devices. The selection of such large fuse ampere ratings results in reduced protection for the transformer and possible impairment of coordination with upstream protective devices. Also, since high-ampere-rated current-limiting fuses typically require the use of two or three lower ampere-rated fuses connected in parallel, increased cost and space requirements may be countered.

The value of the limited cutoff current is determined as a function of the prospective current for current values in the first half cycle where the short circuit current is limited as is shown in Figure 9.21.

The fuse elements or fuse links used in most distribution cutouts are tin or silver alloy fuse links that melt (or operate) when exposed to high current conditions. Ampere ratings of fuse elements vary from 1 A to 200 A.

In distribution systems, it is common to designate fuse links as K and T for fast and slow types, respectively, depending on the speed ratio (SR).

The SR is the ratio of minimum-melt current that causes fuse operation at 0.1 s to the minimum-melt current for 300 s operation.

For the K link, a SR of 6–8 is defined and, for a T link, 10–13.

9.3.3.4 Classification

There are a number of standards to classify fuses according to the rated voltages, rated currents, TCCs, manufacturing features, and other considerations. For example, there are several sections of ANSI/UL 198-1982 Standards that cover low-voltage fuses of 600 V or less. For medium- and high-voltage fuses within the range of 2.3 kV–138 kV, standards such as ANSI/IEEE C37.40, 41, 42, 46, 47, and 48 apply. Other organizations and countries have their own standards; in addition, fuse manufacturers have their own classifications and designations.

In distribution systems, the use of fuse links designated K and T for fast and slow types, respectively, depending on the SR, is very popular. The SR is the ratio of minimum-melt current that causes fuse operation at 0.1 s to the minimum-melt current for 300-s operation. For the K link, a SR of 6–8 is defined and for a T link, 10–13. Figure 9.22 shows the comparative operating characteristics of type 200K and 200T fuse links. For the 200K fuse a 4,400-A current is required for 0.1 s clearance time and 560 A for 300 s, giving an SR of 7.86. For the 200T fuse, 6,500 A is required for 0.1 s clearance, and 520 A for 300 s; for this case, the SR is 12.5.

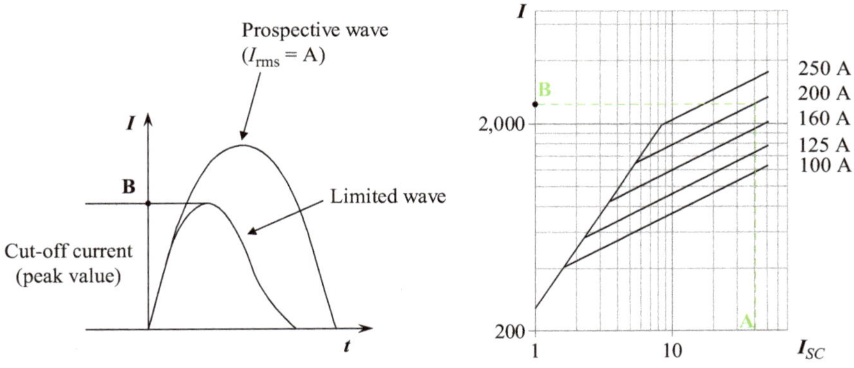

Figure 9.21 Sectionalizers application where one branch system is protected

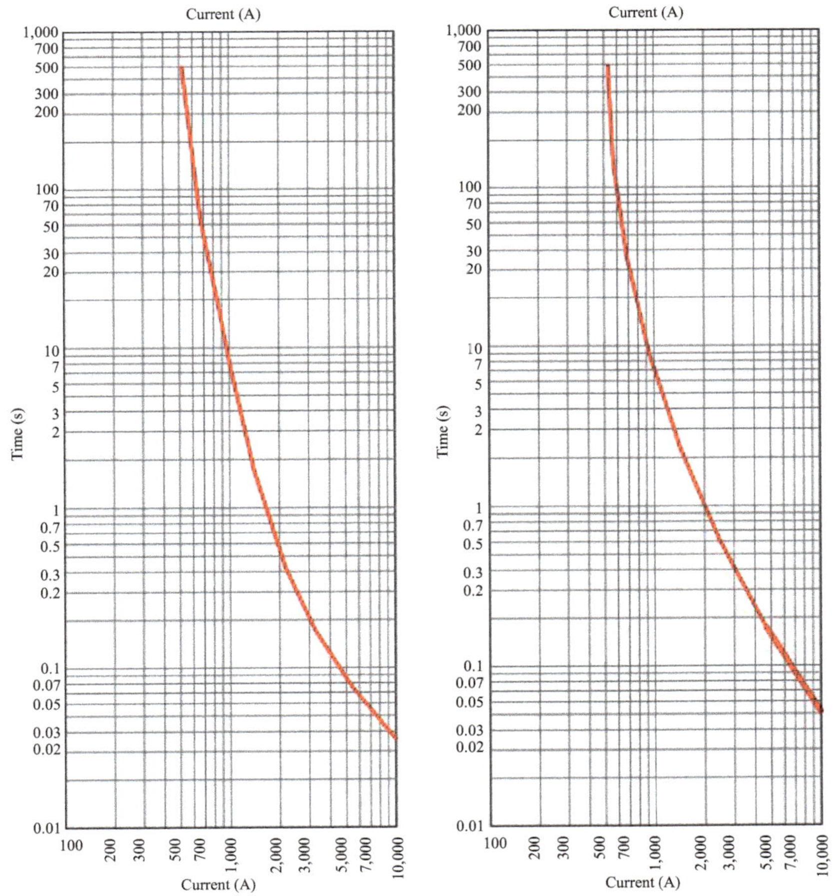

Figure 9.22 Characteristics of typical fuse links: (a) 200K fuse link and (b) 200T fuse link

9.3.3.5 Specifications

The following information is required in order to select a suitable fuse for use on the distribution system:

1. voltage and insulation level,
2. type of system,
3. maximum short circuit level,
4. load current.

The above four factors determine the fuse nominal current, voltage, and short circuit capability characteristics.

Selection of nominal current

The nominal current of the fuse should be greater than the maximum continuous load current at which the fuse will operate. An overload percentage should be allowed according to the protected-equipment conditions. In the case of power transformers, fuses should be selected such that the TCC is above the inrush curve of the transformer and below its thermal limit. Some manufacturers have produced tables to assist in the proper fuse selection for different ratings and connection arrangements.

Selection of nominal voltage

The system characteristics determine the voltage seen by the fuse at the moment when the fault current is interrupted. Such a voltage should be equal to, or less than, the nominal voltage of the fuse. Therefore, the following criteria should be used:

- In unearthed systems, the nominal voltage should be equal to, or greater than, the maximum phase-to-phase voltage.
- In three-phase earthed systems, for single-phase loads the nominal voltage should be equal to, or greater than, the maximum line-to-earth voltage, and for three-phase loads the nominal voltage is selected on the basis of the line-to-line voltage.

Selection of short circuit capacity

The symmetrical short circuit capacity of the fuse should be equal to, or greater than, the symmetrical fault current calculated for the point of installation of the fuse.

Fuse notation

When two or more fuses are used on a system, the device nearest to the load is called the main protection, and that upstream, toward the source, is called the backup. The criteria for coordinating them will be discussed later.

9.4 Setting criteria

The following basic criteria should be employed when coordinating time–current devices in distribution systems:

1. The main protection should clear a permanent or temporary fault before the backup protection operates or continue to operate until the circuit is disconnected. However, if the main protection is a fuse and the backup protection

is a recloser, it is normally acceptable to coordinate the fast operating curve or curves of the recloser to operate first, followed by the fuse, if the fault is not cleared (see Section 9.4.2).

2. Loss of supply caused by permanent faults should be restricted to the smallest part of the system for the shortest time possible.

Further in the section criteria and recommendations are given for the coordination of different devices used on distribution systems.

9.4.1 Fuse–fuse coordination

The essential criterion when using fuses is that the maximum clearance time for a main fuse should not exceed 75% of the minimum melting time of the backup fuse, as indicated in Figure 9.23. This ensures that the main fuse interrupts and clears the fault before the backup fuse is affected in any way. The factor of 75% compensates for effects such as load current and ambient temperature, or fatigue in the fuse element caused by the heating effect of fault currents that have passed through the fuse to a fault downstream but which were not sufficiently large enough to melt the fuse. Keeping the coordination between fuses is the idea as is shown in Figure 9.24.

The series fuse–fuse coordination is given by manufacturers based on information that is presented normally in tables that list maximum fault current values that represent the intersection of the total clearing TCC curve of the load-sided fuse link with the minimum melting TCC curve of the source-side fuse link.

Typically the tables present the maximum fault current with source-side fuse links arranged horizontally and the load-side fuse links arranged vertically. Coordination between the two fuses is guaranteed when the short circuit values are

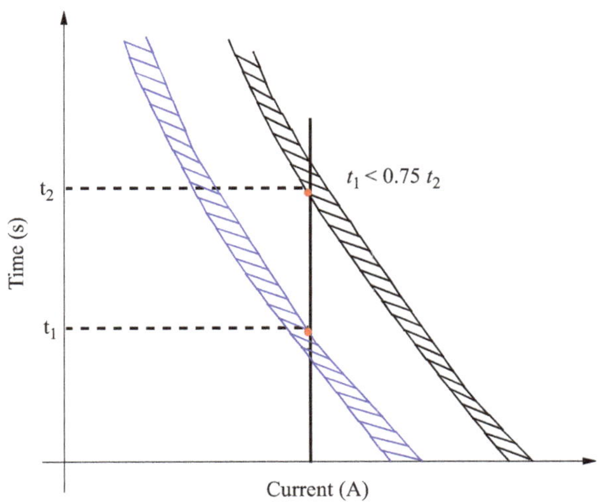

Figure 9.23 Criteria for fuse–fuse coordination

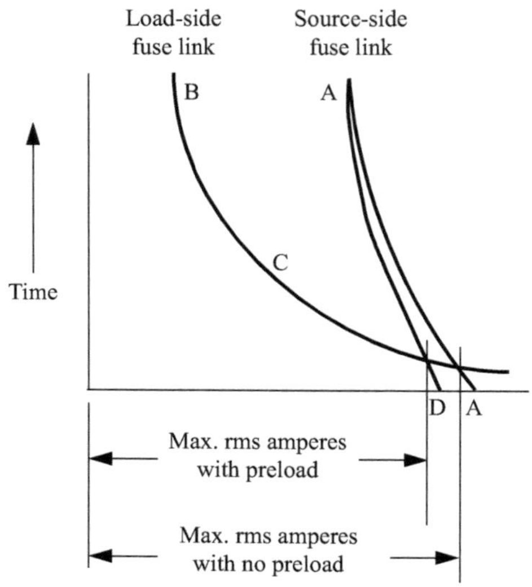

Figure 9.24 Time–current characteristics for fuse–fuse coordination

equal or lower to that given in the table for the intersection of the values corresponding to the two fuses.

Some manufacturers give the tables considering both preload and no preload. In the first case the curve is steeper as shown in the figure. Table 9.5 presents the coordination of S&C Standard Speed Fuse Links, based on preloading of source-side fuse links.

9.4.2 Recloser–fuse coordination

The criteria for determining recloser–fuse coordination depend on the relative locations of these devices, i.e., whether the fuse is at the source side and then backs up the operation of the recloser that is at the load side or vice versa. These possibilities are treated in the following paragraphs.

9.4.2.1 Fuse at the source side

When the fuse is at the source side, all the recloser operations should be faster than the minimum melting time of the fuse. This can be achieved through the use of multiplying factors on the recloser time–current curve to compensate for the fatigue of the fuse link produced by the cumulative heating effect generated by successive recloser operations. The recloser opening curve modified by the appropriate factor then becomes slower but, even so, should be faster than the fuse curve. This is illustrated in Figure 9.25.

The multiplying factors referred to above depend on the reclosing time in cycles and on the number of the reclosing attempts. Some values proposed by Cooper Power Systems are reproduced in Table 9.6.

Table 9.5 Maximum fault current in amperes, rms S&C Standard Speed Fuse Links

Load—side fuse link ampere rating	Source—side fuse link ampere rating																	
	5	7	10	15	20	25	30	40	50	65	80	100	101*	102*	103*	125	150	200
1	120	220	370	590	750	890	1100	1500	1850	2250	2800	3700	5200	8900	15000	4300	5500	7100
2	95	205	360	580	750	890	1100	1500	1850	2250	2800	3700	5200	8900	15000	4300	5500	7100
3		175	335	570	740	880	1100	1500	1850	2250	2800	3700	5200	8900	15000	4300	5500	7100
5		60	280	530	700	850	1050	1450	1800	2250	2800	3700	5200	8900	15000	4300	5500	7100
7			170	490	680	830	1050	1450	1800	2250	2750	3700	5200	8900	15000	4300	5500	7100
10				330	560	740	970	1400	1750	2200	2750	3700	5200	8900	15000	4250	5500	7100
15					120	480	780	1250	1650	2100	2700	3650	5100	8900	15000	4250	5500	7100
20							520	1100	1600	2050	2600	3600	5100	8900	15000	4200	5500	7000
25								920	1450	1950	2550	3500	5100	8900	15000	4150	5400	7000
30								560	1200	1750	2450	3450	5000	8800	15000	4050	5400	7000
40									285	1300	2100	3200	4800	8800	14500	3850	5200	6800
50										290	1550	2850	4650	8600	14500	3650	5100	6700
65											365	2400	4300	8500	14500	3250	4800	6500
80												1350	3750	8200	14000	2550	4350	6200
100													2300	7600	14000		3100	5200
101*															13000			1650
102*															6200			
125															13000			3150
150															11500			

*S&C Coordinating Speed Fuse Links, TCCNo. 172-6.

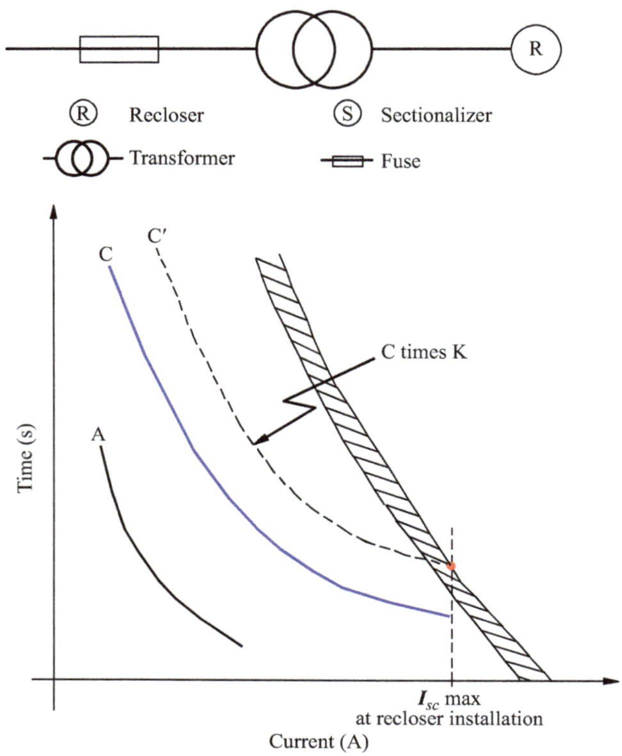

Figure 9.25 Criteria for source-side fuse and recloser coordination

Table 9.6 K factor for the source side fuse link

Reclosing time in cycles	Multipliers for		
	Two-fast, two-delayed sequence	One-fast, three-delayed sequence	Four-delayed sequence
25	2.70	3.20	3.70
30	2.60	3.10	3.50
50	2.10	2.50	2.70
90	1.85	2.10	2.20
120	1.70	1.80	1.90
240	1.40	1.40	1.45
600	1.35	1.35	1.35

It is convenient to mention that if the fuse is at the high-voltage side of a power transformer and the recloser at the low-voltage side, either the fuse or the recloser curve should be shifted horizontally on the current axis to allow for the transformer turns ratio. Normally it is easier to shift the fuse curve based on the transformer tap

Table 9.7 K factor for the load side fuse link

Reclosing time in cycles	Multipliers for	
	One-fast operations	**Two-fast operations**
25–30	1.25	1.80
60	1.25	1.35
90	1.25	1.35
120	1.25	1.35

that produces the highest current on the high-voltage side. On the other hand, if the transformer connection group is delta-star, the considerations given in Section 9.2 should be taken into account.

9.4.2.2 Fuses at the load side

The procedure to coordinate a recloser and a fuse, when the latter is at the load side, is carried out with the following rules:

- The minimum melting time of the fuse must be greater than the fast curve of the recloser times the multiplying factor, given in Table 9.7 and taken from the same reference as above.
- The maximum clearing time of the fuse must be smaller than the delayed curve of the recloser without any multiplying factor; the recloser should have at least two or more delayed operations to prevent loss of service in case the recloser trips when the fuse operates.

The application of the two rules is illustrated in Figure 9.26.

Better coordination between a recloser and fuses is obtained by setting the recloser to give two instantaneous operations followed by two timed operations. In general, the first opening of a recloser will clear 80% of the temporary faults, while the second will clear a further 10%. The load fuses are set to operate before the third opening, clearing permanent faults. A less effective coordination is obtained using one instantaneous operation followed by three timed operations.

9.4.3 Recloser–sectionalizer coordination

Since the sectionalizers have no time–current operating characteristic, their coordination does not require an analysis of these curves.

The coordination criteria in this case are based upon the number of operations of the backup recloser. These operations can be any combination of rapid or timed shots as mentioned previously, e.g., two fast and two delayed. The sectionalizer should be set for one shot less than those of the recloser, e.g., three disconnections in this case. If a permanent fault occurs beyond the sectionalizer, the sectionalizer will open and isolate the fault after the third opening of the recloser. The recloser will then reenergize the section to restore the circuit. If additional sectionalizers are

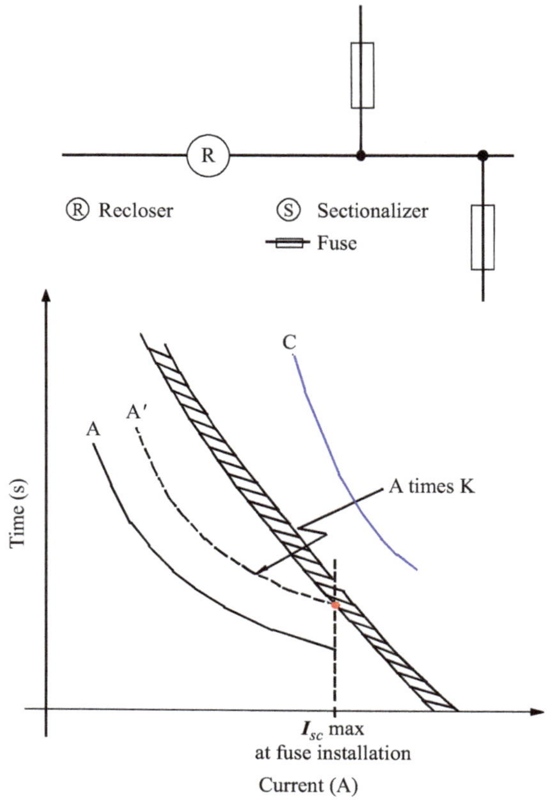

Figure 9.26 Criteria for load-side fuse and recloser coordination

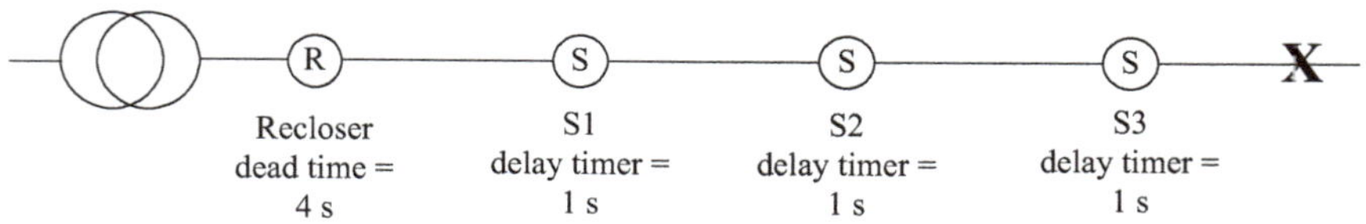

Figure 9.27 Coordination of one recloser with three sectionalizers

installed in series, the furthest recloser should be adjusted for a smaller number of counts. A fault beyond the last sectionalizer results in the operation of the recloser and the start of the counters in all the sectionalizers. Figure 9.27 shows an example of coordination between three sectionalizers and their setting.

9.4.4 Recloser–sectionalizer–fuse coordination

Each one of the devices should be adjusted in order to coordinate with the recloser. In turn, the sequence of operation of the recloser should be adjusted in order to obtain the appropriate coordination for faults beyond the fuse by following the criteria already mentioned.

Figure 9.28 shows a portion of a 13.2-kV distribution feeder that is protected by a set of overcurrent relays at the substation location. A recloser and a sectionalizer have been installed downstream to improve the reliability of supply to customers.

The recloser chosen has two fast and two delayed operations with 90 cycles intervals.

The time–current curves for the transformer and branch fuses, the recloser, and the relays are shown in Figure 9.28. For a fault at the distribution transformer, its fuse should operate first, being backed up by the recloser fast operating shots. If the fault is still not cleared, then the branch fuse should operate next followed by the delayed opening shots of the recloser and finally by the operation of the feeder relay. The sectionalizer will isolate the faulted section of the network after the full number of counts has elapsed, leaving that part of the feeder upstream still in service.

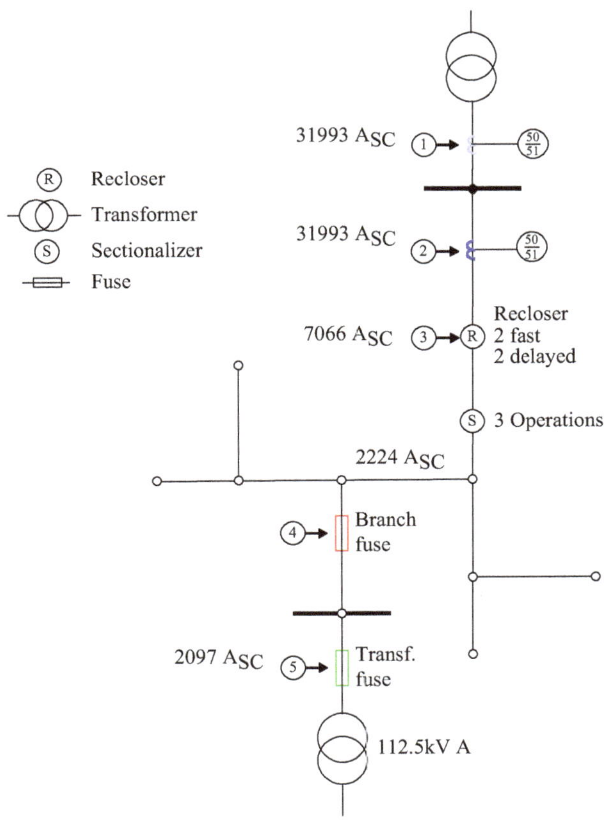

Figure 9.28 Portion of a distribution feeder

As the nominal current of the 112.5 kV A distribution transformer at 13.2 kV is 4.9 A, a 6T fuse was selected on the basis of allowing a 20% overload. The fast curve of the recloser was chosen with the help of the following expression based on the criteria already given, which guarantees that it lies in between the curves of both fuses:

$$t_{recloser} \cdot k \leq t_{MMT\ of\ branch\ fuse} \cdot 0.75$$

where $t_{MMT\ of\ branch\ fuse}$ is the minimum melting time. The 0.75 factor is used in order to guarantee the coordination of the branch and transformer fuses, as indicated in Section 9.4.1.

At the branch fuse location the short circuit current is 2,224 A, which results in operation of the branch fuse in 0.02 s. From Table 9.7, the *K* factor for the two fast operations and a reclosing time of 90 cycles is 1.35. With these values, the maximum time for the recloser operation is (0.02 0.75/1.35) = 0.011 s.

The time and the pickup current of the recloser determine the fast curve of the recloser.

The feeder relay curve is selected so that it is above that of the delayed curve of the recloser, and so that the relay reset time is considered. The curves of Figure 9.29 show that adequate coordination has been achieved.

9.4.5 Recloser–recloser coordination

The coordination between reclosers is obtained by appropriately selecting the amperes setting of the trip coil in the hydraulic reclosers, or of the pickups in electronic reclosers.

9.4.6 Recloser–relay coordination

Two factors should be taken into account for the coordination of these devices; the interrupter opens the circuit some cycles after the associated relay trips, and the relay has to integrate the clearance time of the recloser. The reset time of the relay is normally long and, if the fault current is reapplied before the relay has completely reset, the relay will move toward its operating point from this partially reset position.

For example, consider a recloser with two fast and two delayed sequence with reclosing intervals of 2 s, which is required to coordinate with an inverse time-delay overcurrent relay that takes 0.6 s to close its contacts at the fault level under question and 16 s to completely reset. The impulse margin time of the relay is neglected for the sake of this illustration. The rapid operating time of the recloser is 0.030 s, and the delayed operating time is 0.30 s. The percentage of the relay operation during which each of the two rapid recloser openings takes place is (0.03 s/0.6 s) × 100% = 5%. The percentage of the relay reset that takes place during the recloser interval is (2 s/16 s) × 100% = 12.5%. Therefore, the relay completely resets after both of the two rapid openings of the recloser.

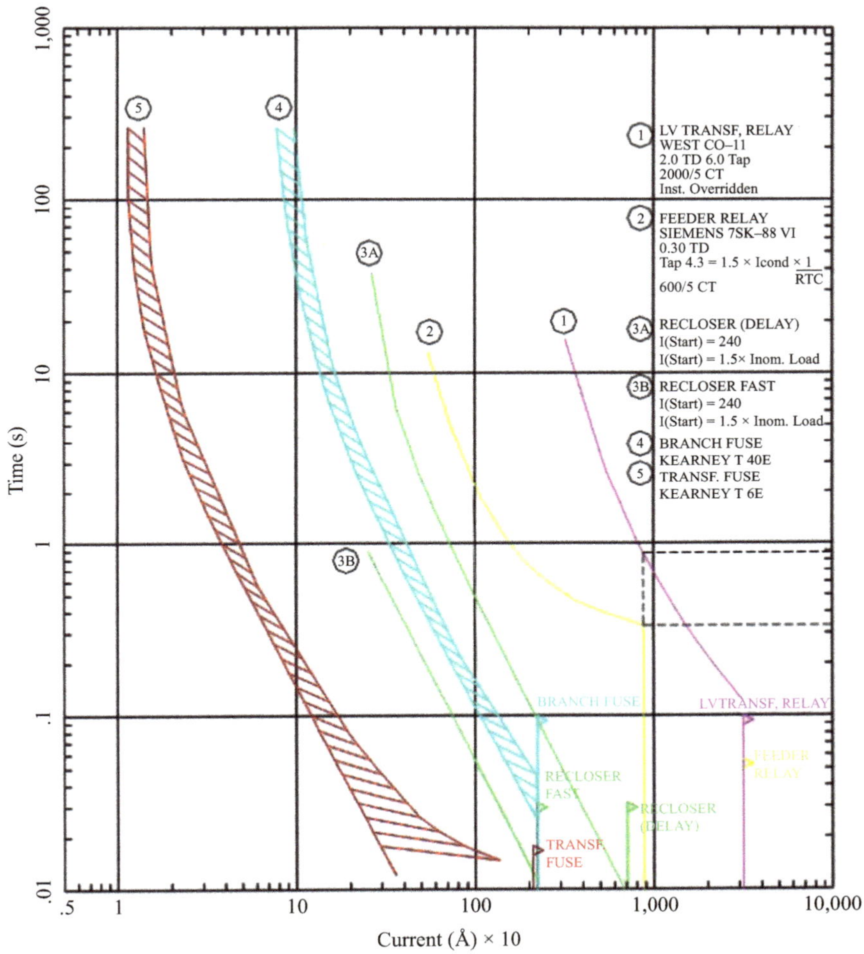

Figure 9.29 Phase–current curves

The percentage of the relay operation during the first time-delay opening of the recloser is (0.3 s/0.6 s) × 100% = 50%. The relay reset for the third opening of the recloser = 12.5%, as previously, so that the net percentage of relay operation after the third opening of the recloser = 50% − 12.5% = 37.5%. The percentage of the relay operation during the second time-delay opening of the recloser takes place = (0.3 s/0.6 s) × 100% = 50%, and the total percentage of the relay operation after the fourth opening of the recloser = 37.5% + 50% = 87.5%.

From the above analysis it can be concluded that the relay does not reach 100% operation by the time the final opening shot starts, and therefore coordination is guaranteed.

9.5 Protection considerations when distributed generation is available

As previously indicated, distributed generation (DG) brings about important benefits to the operation of distribution systems. The main benefit, of course, is the possibility of having generation at the user level that increases service reliability. If the source comes from green power, not only the price is lowered but also the pollution emission. However, important considerations have to be done to make sure that the generation is properly handled as follows.

9.5.1 Short circuit levels

Short circuit levels increase along the feeders with the DG that increase the withstand capabilities of breakers, sectionalizers, reclosers, capacitors, etc. In particular, extra care has to be exercised with breakers and reclosers to make sure that their duties are above the maximum total short circuit currents.

9.5.2 Synchronization

The reclosing functionality at the substation that feeds distribution lines should be disabled if there is not a proper way to open the generators connected along the feeders upon the occurrence and clarification of a fault. This prevents the possibility of energizing a line with generation at the other end without following an appropriate synchronization procedure. This of course reduces the flexibility of the operation but is required to avoid accidents that could be fatal.

9.5.3 Overcurrent protection

The use of overcurrent relays should be examined as the short circuit currents can flow in and out of the substation if distributed generators are present. In this case it could be convenient to replace them with directional overcurrent protection to achieve a better coordination.

9.5.4 Adaptive protection

Due to the inherent nature of DG, machines can be on and off during normal operation. This is even more possible if the generation is from solar or wind power. In these cases, the relays should accommodate to different topologies that make their operation risky or too slow unless adaptive protection is used. Most numerical relays have four or more setting groups that could be selected according to the equal number of operating scenarios.

Microgrids or grids that contain secondary local generation are a great case where adaptive protection should be applied. From Figure 9.30, you can see that two sources are present. During normal operation the utility grid that operates the relays requires one set of protection settings. When the utility grid is interrupted, the diesel generator starts. In order to protect the system the relays must change settings in order to be coordinated correctly. Having adaptive protection will allow for continuous protection of the distribution system regardless of the source (Figure 9.30).

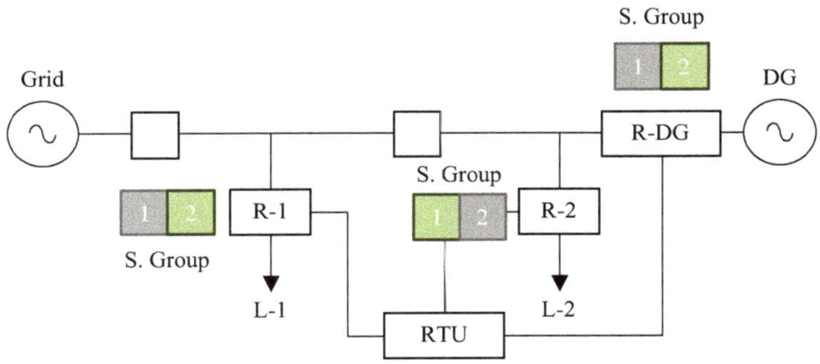

Figure 9.30 Example of adaptive protection setting with a RTU device

9.6 Proposed exercises

1. Calculate the pickup setting, time dial setting, and the instantaneous setting of the phase relays installed in the high- and low-voltage sides of the 115 kV/13.2 kV transformers T1 and T3, in the substation illustrated in Figure 9.31. The short circuit levels, CT ratios, and other data are shown in the same diagram.
2. For the system shown in Figure 9.32, carry out the following calculations:
   (i) The maximum values of short circuit current for three-phase faults at busbars A–C, taking into account that busbar D has a fault level of 12,906.89 A rms symmetrical (2,570.87 MV A).
      (a) The maximum peak values to which breakers 1, 5, and 8 can be subjected.
      (b) The rms asymmetrical values that breakers 1, 5, and 8 can withstand for five cycles for guarantee purposes.

   For these calculations assume that the *L*/*R* ratio is 0.2.
   (ii) The turns ratios of the CTs associated with breakers 1–8. The CT in breaker 6 is 100/5. Take into account that the secondaries are rated at 5 A and that the ratios available in the primaries are multiples of 50 up to 400 and from then on are in multiples of 100.
   (iii) The instantaneous, pickup, and time dial settings for the phase relays in order to guarantee a coordinated protection system, allowing a time discrimination margin of 0.4 s.
   (iv) The percentage of the 34.5 kV line that is protected by the instantaneous element of the overcurrent relay associated with breaker 5.

   Bear in mind the following additional information:

- The settings of relay 6 are as follows: pickup 7 A, time dial 5, instantaneous setting 1,000 A primary current.

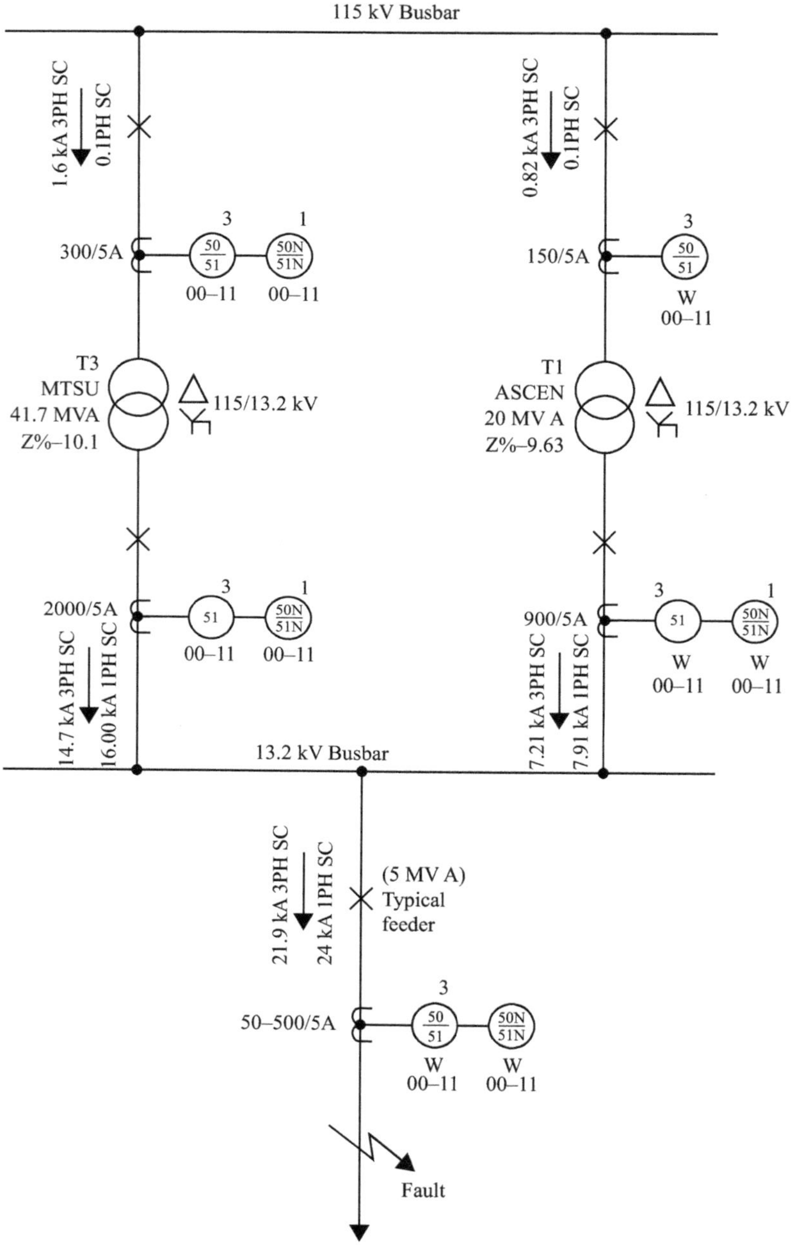

Figure 9.31 Single-line diagram for exercise 1

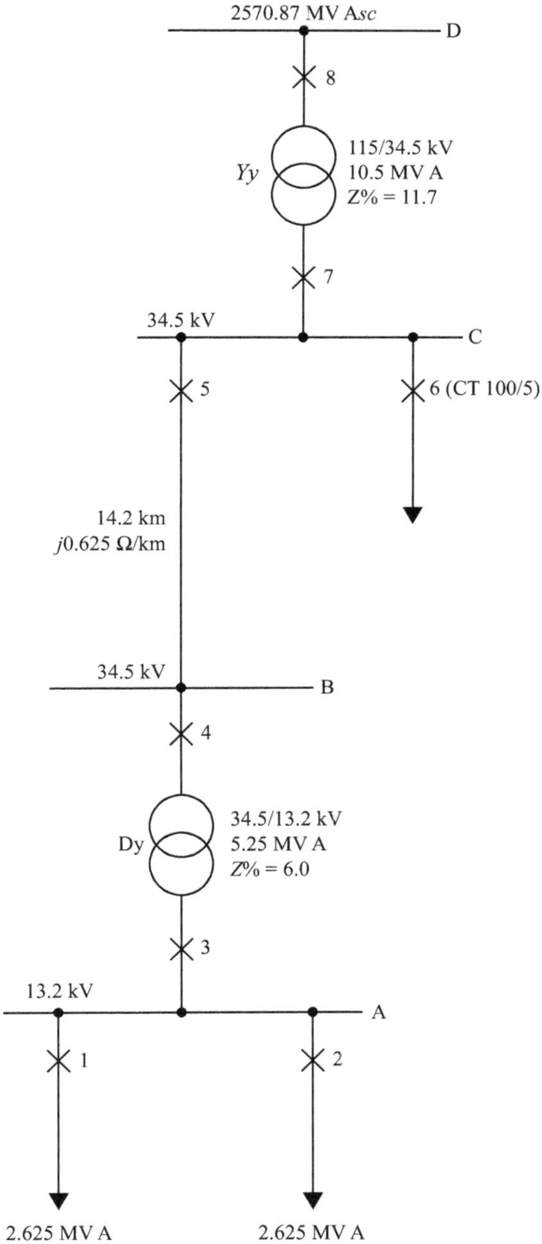

Figure 9.32 Single-line diagram for exercise 2

- All the relays are inverse time type, with the following characteristics:
  - pickup: 1–12 in steps of 2 A
  - time dial: as in Figure 5.15
  - instantaneous element: 6–144 in steps of 1 A.

Calculate the setting of the instantaneous elements of the relays associated with the feeders assuming $0.5I_{sc}$ on busbar A.

Further reading

[1] ABB. "Distribution protection," ABB buyer's guide, vols. I, II and III, 1991–1992.
[2] ALSTOM. Network Protection & Automation, 1st ed. Cayfosa, Barcelona, Spain, July 2002.
[3] ANSI/UL 198M-2018. Standard for Mine-Duty Fuses, 5th ed., Underwriters Laboratories, 2018.
[4] Basler Electric. "Protection and control devices standards, dimensions and accessories," Product Bulletin SDA-5 8-01, August 2001.
[5] Beckwith Electric Co. "Instruction book M-7651 feeder protection relay," Publication 800-7651-SP-00, June 2010.
[6] Cooper Power Systems. Electrical Distribution System Protection Manual, 3rd ed., Cooper Power Systems, 1990.
[7] Davies, T. Protection of Industrial Power Systems. Pergamon Press, Cleveland, OH, USA, 1983.
[8] Electricity Training Association. Power System Protection. The Institution of Electrical Engineers, London, UK, 1997.
[9] Feenan, J. "The versatility of high-voltage fuses in the protection of distribution systems," Power Engineering Journal, vol. 1, pp. 109–115, 1987.
[10] Gers, J. M. and Holmes, E. J. Protection of Electricity Distribution Networks, 3rd ed. London, United Kingdom, 2011.
[11] Holbach, J. and Dufaure, T. "Comparison of IEC61850 GOOSE Messages and Control Wiring Between Protection Relays," in 62th Annual Georgia Tech Protective Relaying Conf., 2008.
[12] IEC. IEC 60255 – Electrical Relays.
[13] IEE Conference Publication. Developments in Power System Protection, 7th International Conference, IEE Conference Publication, Amsterdam, Netherlands, 2001, vol. 479.
[14] IEEE. IEEE Std. 242-2001 (Revision of IEEE Std. 242-1986) – IEEE Recommended Practice for Protection and Coordination of Industrial and Commercial Power Systems (IEEE Buff Book), 2001.
[15] IEEE. IEEE Std. C37.112-2018 – IEEE Standard for Inverse-Time Characteristics Equations for Overcurrent Relays.
[16] IEEE. IEEE Std. C37.40-2003 – IEEE Standard Service Conditions and Definitions for High-Voltage Fuses, Distribution Enclosed Single-Pole Air Switches, Fuse Disconnecting Switches, and Accessories.

[17] IEEE. IEEE Std. C37.41-2016 – IEEE Standard Design Tests for High-Voltage (>1000 V) Fuses and Accessories.
[18] IEEE. IEEE Std. C37.42-2016 – IEEE Standard Specifications for High-Voltage (>1000 V) Fuses and Accessories.
[19] IEEE. IEEE Std. C37.47-2011 – IEEE Standard Specifications for High-Voltage (>1000 V) Distribution Class Current-Limiting Type Fuses and Fuse Disconnecting Switches.
[20] IEEE. IEEE Std. C37.48-2005 – IEEE Guide for Application, Operation, and Maintenance of High-Voltage Fuses, Distribution Enclosed Single-Pole Air Switches, Fuse Disconnecting Switches, and Accessories.
[21] IEEE. "Microprocessor relays and protection systems," Tutorial Course, 88EH0269-1-PWR, 1987.
[22] Laycock, W. J. "Management of protection," Power Engineering Journal, vol. 5, pp. 201–207, 1991.
[23] Momoh, J. Electric Power Distribution, Automation, Protection, and Control, 1st ed. CRC Press, Boca Raton, FL, USA, September 2008.
[24] SIEMENS. Switching, Protection and Distribution in Low Voltage Networks, 2nd ed. Publicist MCD Verlag, Munich, Germany, 1994.
[25] Wright, A. and Newberry, P. G. Electric Fuses, 2nd ed. Peter Peregrinus, London, UK, 1994.
[26] Wright, A. "Application of fuses to power networks," Power Engineering Journal, vol. 4, pp. 293–296, 1991, Wright, A. "Application of fuses to power networks," Power Engineering Journal, vol. 5, pp. 129–134, 1991.

Chapter 10

Distributed generation and energy storage systems

In this chapter, several technologies applied in generation using DERs are going to be presented. The renewable energy resources or technologies that will be discussed in terms of basic operational principles include solar energy, wind generation, small hydropower plants, and energy storage.

Distributed energy resource (DER) is the name given to any generation units that are based in renewable energies of low-power capacity, and usually connected to a feeder, near to the user. Larger units known as "distributed generation (DG) units" are often connected to the transmission system at the point of interconnection.

10.1 Current situation of renewable generation

Most of the global energy in the world is based on fossil fuels. There is a small participation of nuclear energy and renewables. However, the generation of electricity with renewable and alternative sources has increased gradually in the world. According to the "Renewable Energy Policy Network for the 21st Century (REN21)," the global capacity of generation units based on renewables has increased from 1,320 GW in 2010 to 2,378 GW in 2018 [1].

There is an increased interest in connecting these generation units to the electric grid. The problem with renewable energy is its uncertain nature and the location of the resources. Therefore, it is hard to define accurately where these generation units are going to be connected. This has prompted the possibility to using both the transmission and distribution systems in order to integrate them properly.

Only small generators may be connected to the lowest voltage networks, but large installations of hundreds of megawatts are connected to the busbars of high-voltage distribution systems.

10.2 Solar plants

The total energy that the Earth receives from the Sun represents an important amount that can be used to supply energy demand requirements. The total amount of energy received from the Sun can vary between locations and the atmospheric conditions. For instance, the solar radiation is higher in locations near the equator than near poles with an environment with low cloudiness.

The solar energy reaching 1 m^2 at the Earth's surface may vary between 1,000 and 2,000 kW h/m^2 [2]. It is required to optimize the performance of the system that takes advantage of this sunlight energy and uses it into other useful forms of energy. This process needs several amounts of data about the sunlight properties, atmospheric effects, and sun position, among others.

There are many applications in which the energy from the Sun can be transformed; for instance, large solar facilities that are based on concentrating sunlight to boil water. This high-pressure vapor powers a turbine in a similar way as done in a conventional thermal power station [3].

The most common way of using solar radiation is by using solar panels, which are installed in the roofs or in large photovoltaic (PV) solar farms [2]. PV systems use the sunlight and convert it into electricity. This type of technology is designed around the PV cell. The connection of these cells is called PV module, which is connected into different arrays depending of the desired voltage or current at the system output. An advantage of this technology is the power produced close to the load because electric losses of transporting the energy around the system are reduced considerably. In the following sections, more details about this technology and its components are developed.

10.2.1 PV cell model

A PV cell is formed by a p–n junction, which is layer of combined *p-type* and *n-type* semiconductor materials. When the photons in the sunlight reach the PV cell, an electric potential difference across the junction terminals is produced. This excites the electrons in both the layers. The incident light creates electron–hole pairs in the junction, which produce a diffusion of electrons through the cell and the outer contacts become electrodes (the *n*-type side becomes positive electrode, while the *p*-type side becomes negative). A current will flow around the electric circuit once that a load is connected between both the electrodes [2,4] as shown in Figure 10.1.

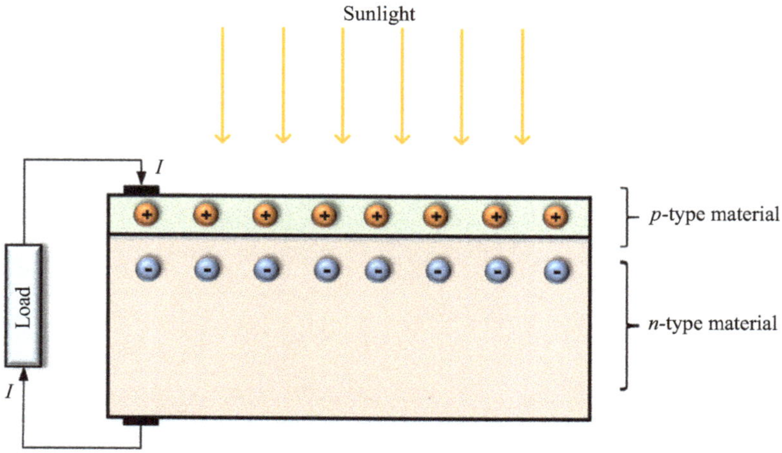

Figure 10.1 A typical PV cell

The ideal electrical behavior of a PV cell is described by the *I–V* characteristic curve of a diode [5], where both current and voltage depends on the cell irradiation level and temperature of exposure.

$$I = I_{Ph} - I_0 \cdot (e^{qV/kT} - 1) \tag{10.1}$$

$$V = \frac{kT}{q}\ln\left(1 - \frac{I - I_{Ph}}{I_0}\right) \tag{10.2}$$

Then,

$$V_{OC} = \frac{kT}{q}\ln\left(\frac{I_{Ph}}{I_0}\right)$$

where I is the total current flowing through the cell terminals in amps, A; V is the cell voltage across terminals in volts, V; V_{OC} is the open-circuit voltage across terminals; I_{Ph} is the cell current due to the incident photons or photocurrent; I_0 is the saturation current of the diode; q is the electron electric charge in coulombs, $q = 1.6021 \cdot 10^{-19}$ C; k is the Boltzmann constant, $1.3806 \cdot 10^{-23}$ J/K; T is the cell temperature in Kelvin, K.

In order to determine the short circuit current of a PV cell, the voltage should be set on $V = 0$ for the exponent. This leads to $I_{SC} = I_{Ph}$. Typically, a PV cell produces an open-circuit voltage around 0.6 V at 25°C. Figure 10.2 shows the *I–V* curve of a PV cell for different values of short circuit currents and cell temperatures.

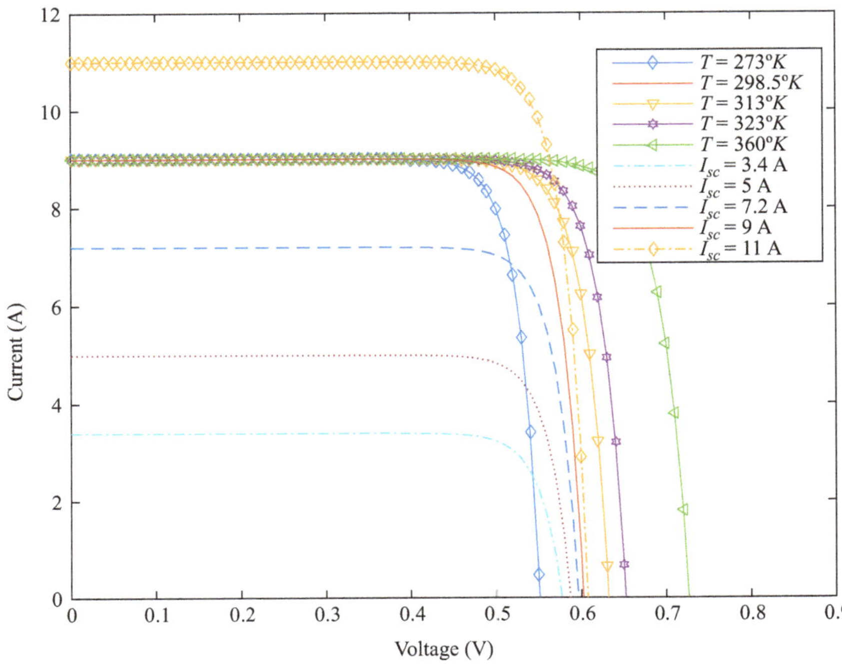

Figure 10.2 I–V characteristic of a PV cell

Figure 10.3 shows the power output of a PV cell, which corresponds to a non-linear expression that relates the current and voltage across the cell terminals. The maximum power point can be determined as the peak of the curve, which is located around the inflection point in the *I–V* curve. This point can be expressed in terms of the short circuit current, the open-circuit voltage and a fill factor (*FF*), which indicates the overall behavior or quality of the PV cell, as indicated in the following expression:

$$P_{\max} = I_{\max} \cdot V_{\max} = FF \cdot I_{SC} V_{OC} \tag{10.3}$$

A large internal resistance and higher saturation will turn into smaller FFs and vice versa. FF for PV cells may vary from 0.5 to 0.82 [5,6]. From (10.3), the FF can be determined as follows:

$$FF = \frac{P_{\max}}{I_{SC} \cdot V_{OC}} = \frac{I_{\max} \cdot V_{\max}}{I_{SC} \cdot V_{OC}} \tag{10.4}$$

PV cells can be connected in series configuration to obtain more voltage at the output, and in parallel configuration for more current. These connections together produce a PV module that provides more power. Several modules can achieve open-circuit voltages around 600 V_{DC}. They are connected to an inverter that has a maximum power point tracking (MPPT) algorithm to extract power [7].

Figure 10.4 illustrates the equivalent circuit of a PV cell. The dependency and proportionality of the surface area and the incident irradiance are represented by a current source. The shunt resistors represent the losses associated with leakage

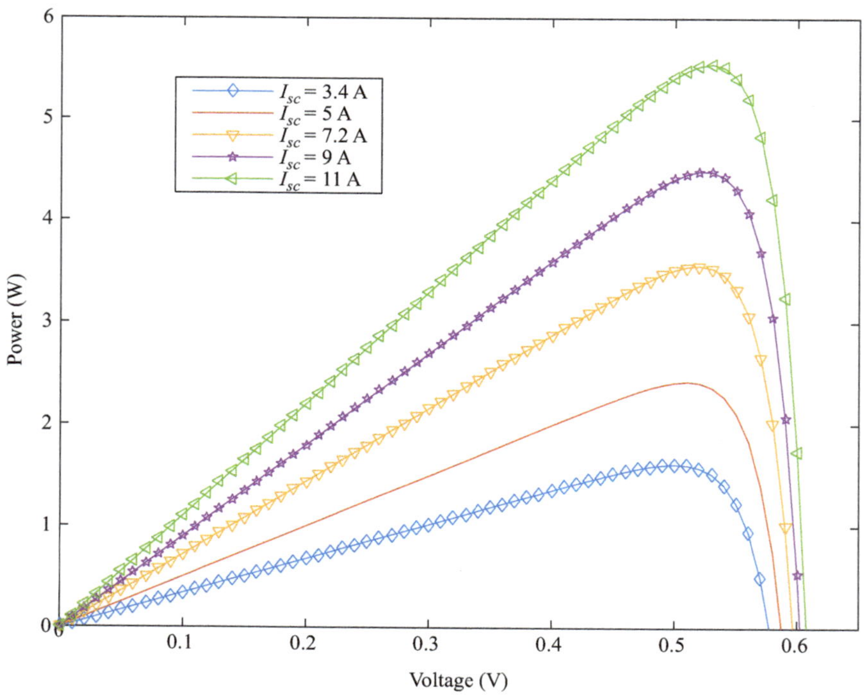

Figure 10.3 P–V characteristic of a PV cell

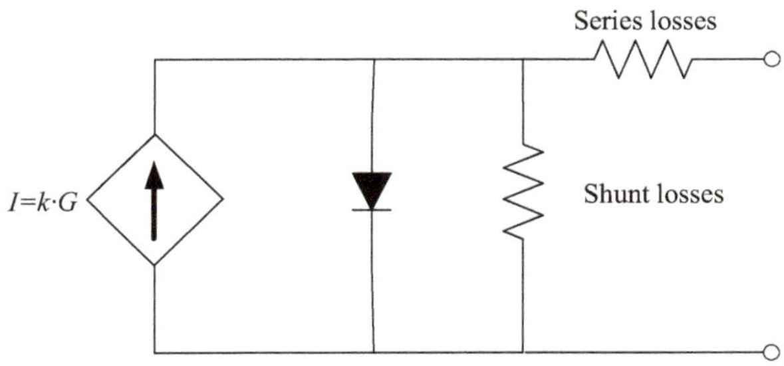

Figure 10.4 Real PV cell equivalent circuit [2]

Figure 10.5 Power inverter operation scheme

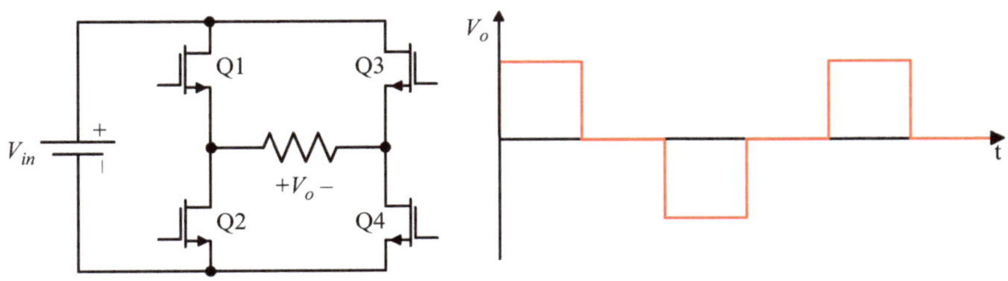

Figure 10.6 One-phase square wave inverter

currents through the edges and structure of the cell, while the series correspond to the resistance of terminals, temperature effects, and semiconductors materials.

10.2.2 Inverters

The electric signals that come out the PV system are DC signals and they must be converted into AC. A power inverter is a single-phase or three-phase solid-state electronic device formed by transistors or thyristors that operate as switches, converting the DC input to an AC symmetrical output with a desire amplitude and frequency. The basic operation of an inverter is shown in Figure 10.5.

The simplest inverter configuration is a single-phase square wave inverter shown in Figure 10.6. This configuration has a significant amount of harmonic content and may cause the electric loads to overheat or malfunction. With pulse-width modulation (PWM), a sine wave can be obtained by controlling the on-and-off switching of a pulse waveform as seen in Figure 10.7. This produces less harmonic content and more efficiency in the resultant waveform.

Figure 10.8 illustrates a common configuration for a PV system as a DG unit, which is *formed by* a PV array or PV system, an MPPT algorithm, an energy

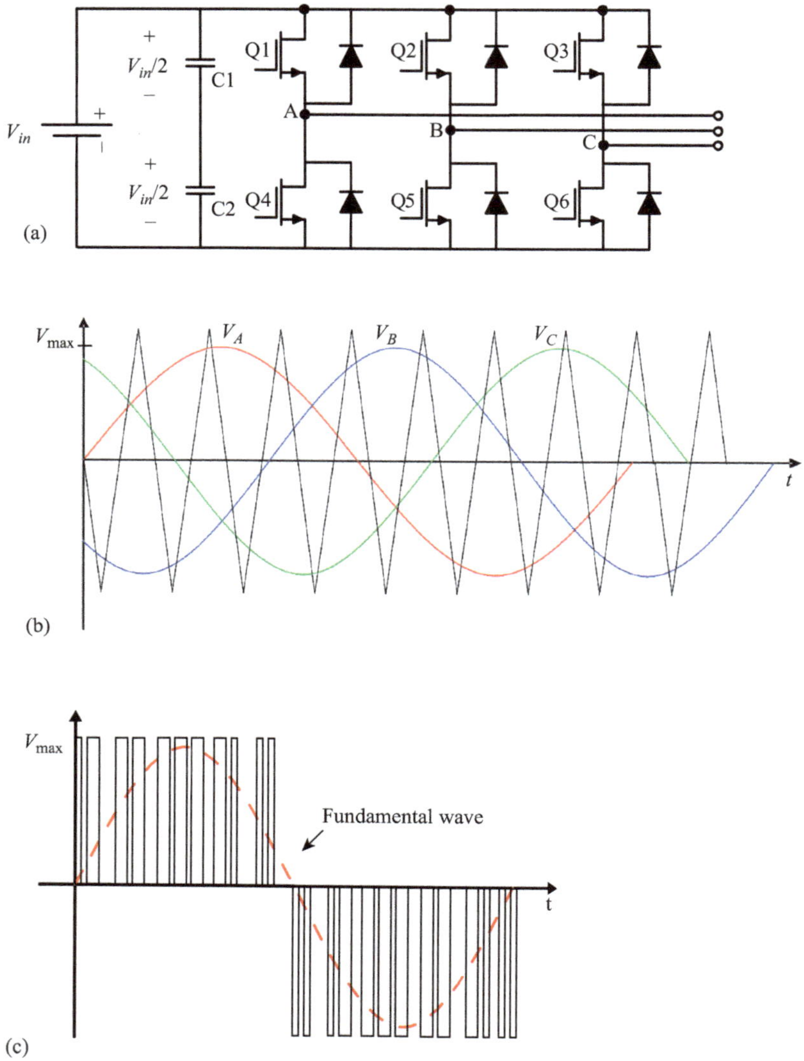

Figure 10.7 *Three-phase sine wave PWM inverter: (a) equivalent circuit, (b) resultant waveform, and (c) fundamental wave modulation [8,9]*

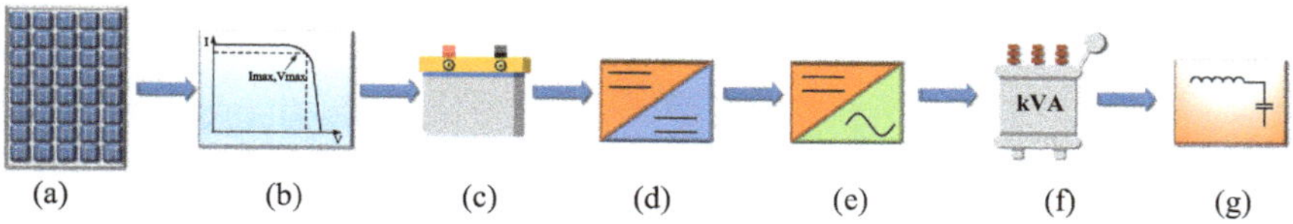

Figure 10.8 *Inverter system for a PV module as a DG unit: (a) PV module, (b) MPPT, (c) energy storage, (d) DC:DC converter, (e) DC:AC converter, (f) isolation, and (g) output filter*

storage device, DC–DC converter, DC–AC inverter, an isolation transformer, and an output filter to avoid DC and harmonic injection.

There are several commercial manufacturers such as SMA and Siemens (Germany), ABB (Switzerland), Fronius (Austria), among others. They offer different power ranges up to 3,000 kW at unit power factor ($pf = 1.0$) for solar string applications [10].

10.2.3 Grid-connected and stand-alone systems

A common application of the PV technologies includes exporting energy to the main grid, also referred to as grid-connected systems. There is also stand-alone operation, in which the system is disconnected from the electric system. This system is equipped with an energy storage system that stores the energy surplus and releases it at times of low-power generation. Finally, there are hybrid systems that are coupled to loads and surplus is exported to the grid or stored if storage system is installed.

PV integration into the power system can be either centralized or decentralized [11]. A centralized system is a large-scale installation that produces a significant amount of power. In order to dispatch this power, electrical substations and transmission lines are required. Decentralized systems produce energy on-site or close to the consumers who are owners of the system. The energy surplus can be sold to the local utility according to the market regulations. Decentralized configuration has more freedom to be operated. Figure 10.9 shows the typical components of each configuration.

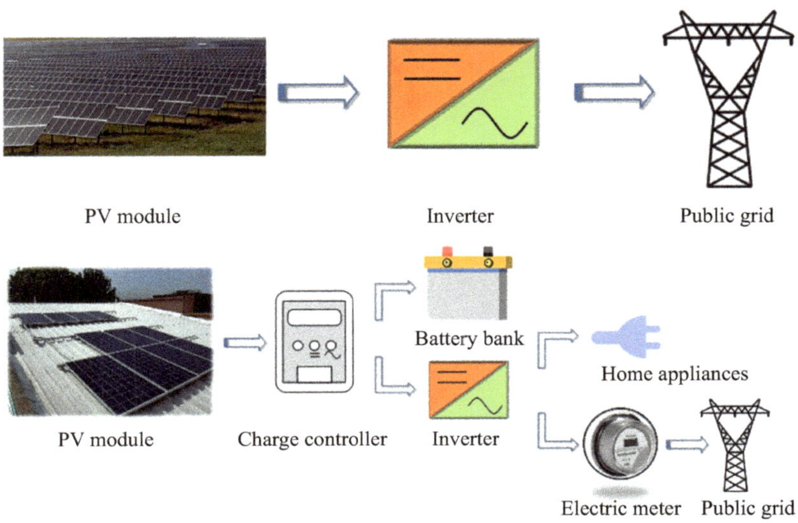

Figure 10.9 Centralized and decentralized PV systems

Example 10.1 To illustrate the performance of a grid-connected PV system on a distribution network, a 2 MV A PV system is connected to a 0.4 kV bus as shown later. The power consumed by the load is 1.5 MV A with unity power factor. Both the load and the PV system are modeled with a load profile that indicates the variation of the load and sunlight throughout a day. The example has been modeled and run with a PSAT and the results are depicted in Figures 10.10 and 10.11. By running

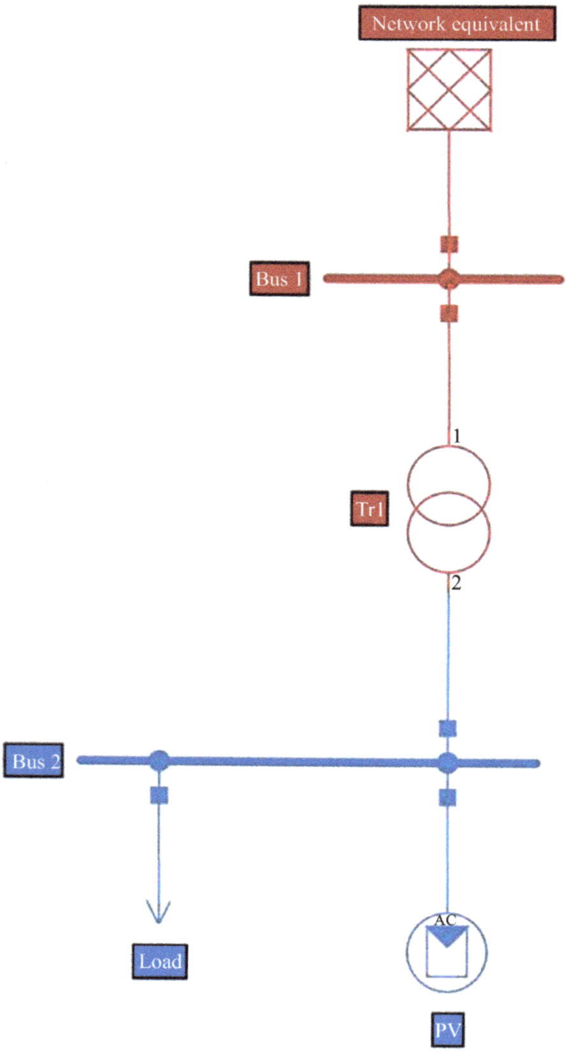

Figure 10.10 Grid-connected PV system example

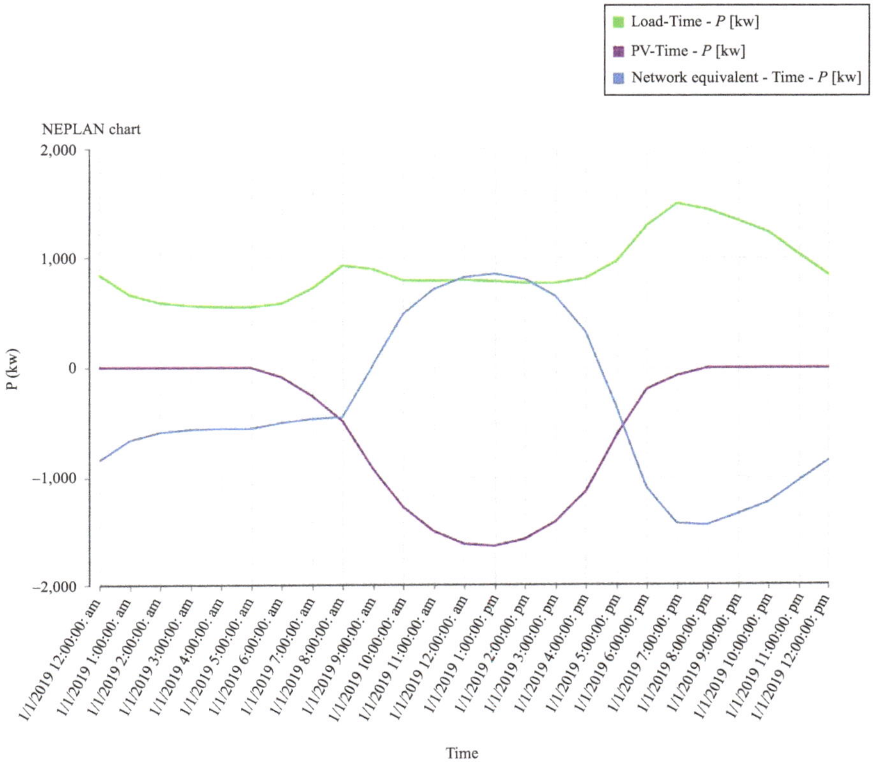

Figure 10.11 Load flow of the PV system through 1 day

the load flow throughout a day it can be seen how the surplus energy, that is, the energy that is not being consumed by the load is delivered to the system.

Example 10.2 In order to size a PV system for a stand-alone (off-grid) application, some basic requirements must be met.

1. energy demand estimation
2. solar resource availability
3. PV system sizing
4. charge regulator
5. battery bank
6. solar inverter

A brief explanation is given in the next sections in order to size a PV system correctly.

Step 1: Energy demand estimation

As the PV system is thought to supply the 100% of the load demand, the total energy demand must be estimated. Table 10.1 indicates some common appliances with their respective power demand and average hours of usage per day.

By computing the total energy per appliance (Table 10.2)

$$\text{Total demand(W h/day)} = (P \cdot \text{time} \cdot \#\text{units})$$

The total demand is

$$\text{Total demand(W h/day)} = \sum_{i=1}^{n=8} (P_i \cdot \text{time}_i \cdot \#\text{units}_i)$$

$$\text{Total demand(W h/day)} = (400 + 960 + 600 + 300 + 480 + 257 + 60 + 107)\text{W h/day}$$

$$\text{Total demand} = 3{,}164 \text{ W h/day}$$

Table 10.1 Basic home appliances and typical power consumption for Example 10.2

Description	**Type**	**Units**	**Power (W)**	**h/day**	**Energy (W h/day)**
Lightning	AC	8	10	5	
Fridge	DC	1	80	12	
TV	AC	1	120	5	
Laptop	AC	1	75	4	
Fan	AC	1	40	12	
Stereo system	AC	1	100	6	
Mobile charger	AC	4	5	3	
Washing machine	AC	1	500	1.5	

Table 10.2 Energy demand per home appliance for Example 10.2

Description	**Type**	**Units**	**Power (W)**	**h/day**	**Energy (W h/day)**
Lightning	AC	8	10	5	400
Fridge	DC	1	80	12	960
TV	AC	1	120	5	600
Laptop	AC	1	75	4	300
Fan	AC	1	40	12	480
Stereo system	AC	1	100	6	257
Mobile charger	AC	4	5	3	60
Washing machine	AC	1	500	1.5	107

The demand per type of appliance, considering a safety factor of 15% and a power factor of 0.95 lagging

$$L_{DC} = E_{DC} \cdot 1.15$$

$$L_{DC} = 960 \text{ W h/day} \cdot 1.15 = 1{,}104 \text{ W h/day}$$

$$L_{AC} = \frac{E_{AC} \cdot 1.15}{pf}$$

$$L_{AC} = \frac{2{,}204 \text{ W h/day} \cdot 1.15}{0.95} = 2{,}668 \text{ W h/day}$$

The total load of the PV system is then

$$L_T = L_{DC} + L_{AC} = 3,772 \text{ W h/day}$$

$$PVS_{demand} = 3,772 \text{ W h/day}$$

Step 2: Solar resource availability

It is also important to take into account the availability of the energy resource and its variability through the year according to the facility location. Such information can be obtained from websites with meteorology and solar energy data. The main relevant information for this example is the daily solar radiation and Earth's temperature of each month as shown in Figure 10.12.

As the PV system performance is dependent on the environmental conditions, months with the higher and lower temperatures must be identified. Also, demand can vary with months and seasons; thus, months with higher demand/irradiation ratio are classified as critical months (Table 10.3).

In this case, November is the most critical month as the ratio demand/irradiation is higher.

Month	**Daily solar radiation horizontal**	**Earth temperature**
	kW h/m^2/d	°C
January	4.06	20.9
February	4.28	21.4
March	4.37	21.6
April	4.21	21.6
May	4.10	21.4
June	4.05	21.3
July	4.34	22.3
August	4.31	23.6
September	4.26	23.4
October	3.99	22.1
November	3.89	20.9
December	3.82	20.7
Annual	4.14	21.8

Figure 10.12 Daily solar radiation per month for Example 10.2

Table 10.3 Demand/irradiation ratio per month for Example 10.2

Month	Irradiation (kW h/m^2/day)	PSH (h)			Demand/ Irradiation
			Demand (%)	Demand (kW h/day)	
January	4.06	4.06	45	1.7	0.42
February	4.28	4.28	100	3.7	0.88
March	4.37	4.37	100	3.7	0.86
April	4.21	4.21	100	3.7	0.90
May	4.10	4.10	100	3.7	0.92
June	4.05	4.05	45	1.7	0.42
July	4.34	4.34	45	1.7	0.39
August	4.31	4.31	100	3.7	0.88
September	4.26	4.26	100	3.7	0.89
October	3.99	3.99	45	1.7	0.43
November	3.89	**3.89**	100	**3.7**	**0.97**
December	3.82	3.82	45	1.7	0.44

Note: The bold values represent those for the critical month.

Table 10.4 Typical PV module datasheet for Example 10.2

	STC	NOCT
Maximum power ($P_{\max}$)	270 W_p	202 W_p
Maximum power voltage (V_{mp})	31.7 V	29.0 V
Maximum power current (I_{mp})	8.52 A	6.97 A
Open-circuit voltage (V_{oc})	38.8 V	35.6 V
Short circuit current (I_{SC})	9.09 A	7.35 A
Module efficiency STC (%)	16.50%	
Operating temperature (°C)	−40°C + 85°C	
Maximum system voltage	1,000 V_{DC} (UL)	
Maximum series fuse rating	15 A	
Temperature coefficients of $P_{\max}$	−0.40%/°C	
Temperature coefficients of V_{oc}	−0.30%/°C	
Temperature coefficients of I_{SC}	0.06%/°C	
Nominal operating cell temperature (NOCT)	45 ± 2°C	

Step 3: PV system design

Table 10.4 shows the technical specifications of a PV poly-crystalline module. Each module has a maximum power capacity of 270 W and open-circuit voltage of 38.8 V.

From the modules datasheet, information such as operating temperature conditions, short circuit and open voltages characteristics are very useful for the system specification.

The number of modules is determined as the ratio of the energy demand from the PV system and the energy considering the power output of each module, the peak sun-hours of the critical month and a module performance factor, assumed as 75%.

For this example, five modules are needed to fulfill the energy requirements, the total power of the PV system is equal to 1.35 kW. These modules can be arranged in series and parallel configuration depending on the voltage and current limitations of the inverter and regulator involved.

$$\#Modules = \frac{L_T}{P_{module} \cdot PSH_{critical\ month} \cdot PR}$$

$$\#Modules = \frac{3,772 \text{ W h/day}}{270 \text{ W} \cdot 3.89 \text{ h/day} \cdot 0.75} = 4.78 \approx 5\ modules$$

$$P_{PVS} = 5\ modules \cdot 270 \text{ W} = 1.35 \text{ kW}$$

As mentioned, temperature plays an important role in the PV modules performance. The maximum and minimum temperature, at which the modules are exposed to, defines the number of series and parallel elements of the array according to the voltage and current limits, respectively. This number of units also depends on the charge regulator parameter ranges and the operating cell temperature conditions. Figure 10.13 shows the variation of series and parallel modules with temperature.

For the maximum number of elements connected in series, the minimum operating cell temperature must be determined. If the temperature is reduced, the open-circuit voltage is then increased.

For the lower temperature scenario from the site data, that is 20.7°C, a maximum number of three series units are obtained. As can be seen, the open-circuit voltage of the cell rises from 38.8 V to 41.21 V due to a reduction in the operating temperature.

$$T_{p\ \min} = T_{a\ \min} + \frac{T_{NOC} - 20}{800 \text{ W/m}^2} \times I_a$$

$$T_{p\ \min} = 20.7°\text{C} + \frac{45°\text{C} - 20}{800 \text{ W/m}^2} \times \frac{3,820 \text{ W h/m}^2}{4.5 \text{ h}} = 47.23°\text{C}$$

$$V_{oc(T_{p\ \min})} = \left(1 + \frac{(T_{P\ \min} - 25°\text{C}) \times \Delta V_{oc}\%}{100}\right) \times V_{oc(STC)}$$

$$V_{oc(T_{p\ \min})} = \left(1 + \frac{(47.23°\text{C} - 25°\text{C}) \times 0.3}{100}\right) \times 38.8 \text{ V}_{DC} = 41.39 \text{ V}_{DC}$$

Hence,

$$n_{series\_max} = \frac{V_{\max-Equip}}{V_{oc(T_{p\ \min})}}$$

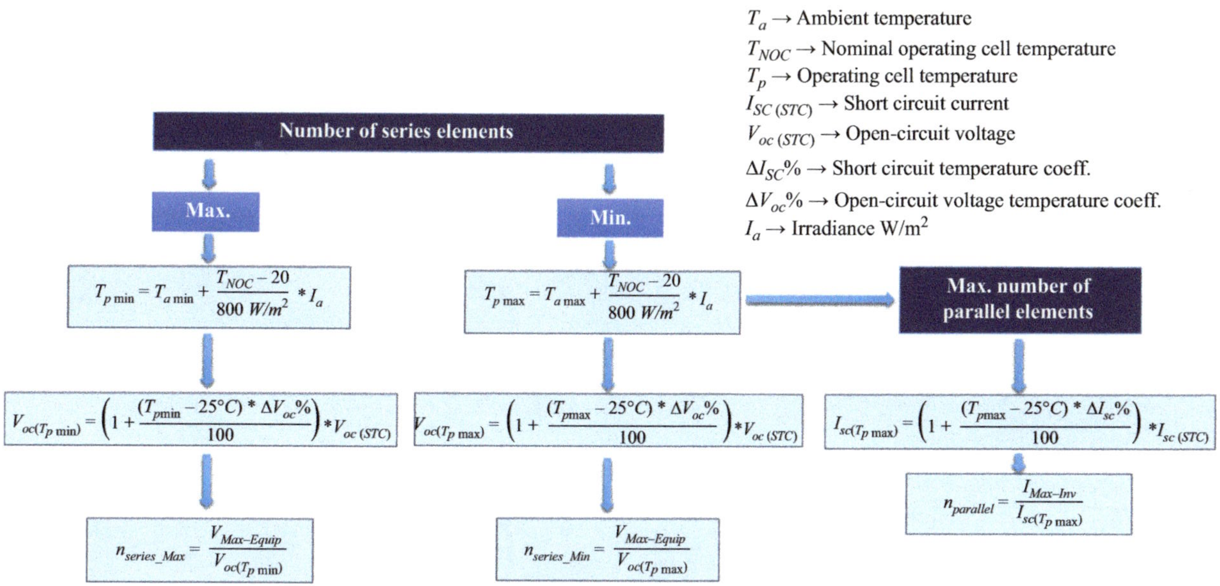

Figure 10.13 Variation of series and parallel modules with temperature

$$n_{series\_max} = \frac{145\ \mathrm{V}_{DC}}{41.2\ \mathrm{V}_{DC}} = 3.503 \approx 3 \text{ series units, maximum}$$

For the minimum number of elements connected in series, the maximum operating cell temperature must be determined. If the temperature is increased, the open-circuit voltage then decreases.

On the other hand, for the higher temperature scenario, that is, 23.6°C, a minimum number of two series units are obtained. The open-circuit voltage of the cell falls from 38.8 V to 35.04 V due to an increase on the cell-operating temperature.

$$T_{p\ \max} = T_{a\ \max} + \frac{T_{NOC} - 20}{800\ \mathrm{W/m^2}} \times I_a$$

$$T_{p\ \max} = 23.6°\mathrm{C} + \frac{45°\mathrm{C} - 20}{800\ \mathrm{W/m^2}} \times \frac{4,310\ \mathrm{W\,h/m^2}}{4.0\ \mathrm{h}} = 57.27°\mathrm{C}$$

$$V_{oc(T_{p\ \max})} = \left(1 + \frac{(T_{P\ \max} - 25°\mathrm{C}) \times \Delta V_{oc}\%}{100}\right) \times V_{oc(STC)}$$

$$V_{oc(T_{p\ \max})} = \left(1 + \frac{(57.27°\mathrm{C} - 25°\mathrm{C}) \times 0.3}{100}\right) \times 38.8\ \mathrm{V}_{DC} = 35.04\ \mathrm{V}_{DC}$$

$$n_{series\_Min} = \frac{V_{\min-Equip}}{V_{oc(T_{p\ \max})}}$$

$$n_{series\_Min} = \frac{60\ \mathrm{V}_{DC}}{35.04\ \mathrm{V}_{DC}} = 1.71 \approx 2 \text{ series units, minimum}$$

Real PV cells or modules can be modeled as dependent current sources, representing the dependency and proportionality of the cell to the incident irradiance. Therefore, as the irradiance increases, the short circuit current also increases. As can be noticed, the short circuit current increases from 9.09 A to 9.27 A due to the irradiance.

This current capacity yields to a maximum of five parallel units connected to the regulator.

$$T_{p\ \max} = T_{a\ \max} + \frac{T_{NOC} - 20}{800\ \mathrm{W/m^2}} \times I_a$$

$$T_{p\ \max} = 23.6°\mathrm{C} + \frac{45°\mathrm{C} - 20}{800\ \mathrm{W/m^2}} \times \frac{4,310\ \mathrm{W\,h/m^2}}{4.0\ \mathrm{h}} = 57.27°\mathrm{C}$$

$$I_{sc(T_{p\ \max})} = \left(1 + \frac{(T_{P\ \max} - 25°\mathrm{C}) \times \Delta I_{sc}\%}{100}\right) \times I_{sc(STC)}$$

$$I_{sc(T_{p\ \max})} = \left(1 + \frac{(57.27°\mathrm{C} - 25°\mathrm{C}) \times 0.06}{100}\right) \times 9.09\ A_{DC} = 9.27\ \mathrm{A}_{DC}$$

$$n_{parallel} = \frac{I_{\max-Equip}}{I_{sc(T_p \max)}}$$

$$n_{parallel} = \frac{50\ \mathrm{A}_{DC}}{9.27\ \mathrm{A}_{dc}} = 5.39 \approx 5\ \text{Parallel strings}$$

Step 4: Charge regulator

A charge regulator is chosen as a way to supply a battery energy storage system if needed, also, the regulator is useful for DC/DC voltage control. Table 10.5 shows the charge regulator datasheet for this example.

The 1.6 kW regulator has DC voltage ranges from 60 V to 115 V with a maximum PV array open-circuit voltage of 145 V, and a maximum operating current of 50 A. Hence, the PV system design must be in accordance with these parameters.

For this, a brief verification must be done:

$$P_{PVS} = 5\ \mathit{Modules} \times 270\ \mathrm{W}_p = 1.35\ \mathrm{kW}_p$$

$$I_{in} = I_{SC\_Panel} \times \#\text{Parallel Strings}(\mathrm{A})$$

$$I_{in} = 9.09\ \mathrm{A}_{DC} \times 2 = 18.18\ \mathrm{A}$$

$$I_{out} = \frac{P_{DC} + P_{AC}/0.95}{V_{bus\_DC}}(\mathrm{A})$$

$$I_{out} = \frac{80\ \mathrm{W} + 850\ \mathrm{W}/0.95}{48\ V_{DC}} = 20.31\ \mathrm{A}$$

It is important to validate that the array design accomplishes the regulator requirements. A configuration of two parallel strings is chosen. The total regulator input current from the PV system is 18.18 A and the output current is equal to 20.31 A, which is in accordance with the regulator specifications as shown in the datasheet.

Table 10.5 Charge regulator datasheet for Example 10.2

Input	**Values**
Operating voltage	60 V_{DC}–115 V_{DC}
Maximum PV array open-circuit voltage	145 V_{DC}
Maximum PV array power	1,600 W
Maximum current	50 A
Output	**Values**
Nominal battery voltage	48 V_{DC}
Maximum charging current	60 A

Table 10.6 Inverter datasheet for Example 10.2

Output	**Values**
Output voltage	120 $V_{AC} \pm 3\%$
Maximum output current	12.5 A
Output frequency	60 Hz $\pm$ 1%
Total harmonic distortion	<3%
Continuous output power (at $pf = 1$)	1,500 W
Input	**Values**
Nominal DC input voltage	48 V_{DC}
Maximum input current	50 A

Step 5: Battery bank

The charge of the battery storage system is calculated with a safety factor of 10%.

$$Q_{\mathrm{A\,h}} = \frac{L_t \times D_{aut}}{V_{bus\_DC} \times P_d} \times 1.1(\mathrm{A\,h})$$

$$Q_{\mathrm{A\,h}} = \frac{3,772\ \mathrm{W\,h/day} \times 2}{48\ \mathrm{V}_{DC} \times 0.5} \times 1.1(\mathrm{A\,h})$$

$$Q_{\mathrm{A\,h}} = 346\ \mathrm{A\,h}; 48\ \mathrm{V}_{DC}$$

where L_t is the total demand, D_{aut} is the days of autonomy, V_{bus_DC} is the DC bus voltage, and P_d is the discharge depth.

Step 6: Solar inverter

The solar inverter is chosen considering the voltage level and the power output oversized by 25% of the PV system capacity.

$$P_{Inverter} = 1.25 \times P_T(\mathrm{V\,A})$$

$$P_{Inverter} = 1.25 \times (850/0.95 + 80)(\mathrm{V\,A})$$

$$P_{Inverter} = 1,218\ \mathrm{V\,A}$$

Table 10.6 shows the selected solar inverter.

The final arrangement of the PV system is shown in Figure 10.14.

10.3 Wind generation

The energy available from the wind drafts is obtained from kinetic energy due to large masses of air moving over the Earth's atmosphere [12]. The blades of a wind turbine allow for transforming this kinetic energy into mechanical energy and then, electrical energy through a system of gearboxes, a turbine, and the generator. The total system efficiency depends greatly on the efficiency with which the rotor interacts with the wind stream.

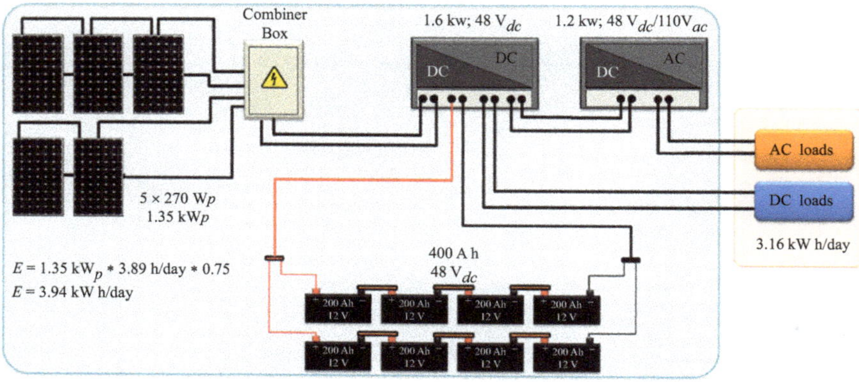

Figure 10.14 PV arrangement for Example 10.2

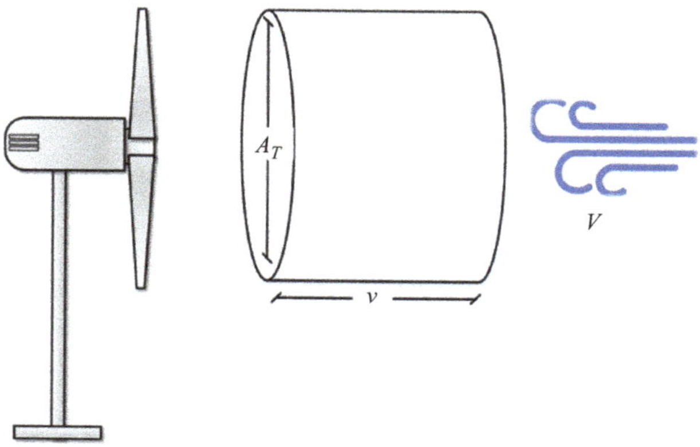

Figure 10.15 Air parcel interacting with a wind turbine rotor [12]

The energy of a wind stream can be estimated with the total air mass (m) and the moving velocity (V) as follows:

$$E_K = \frac{1}{2} m V^2 \tag{10.5}$$

The total amount of kinetic energy available for a wind turbine can be determined considering the wind rotor cross-sectional area A exposed to the wind stream or parcel. This is illustrated in Figure 10.15.

The total air interacting with the rotor per unit time can be defined in terms of its cross-sectional area, which is equal to the area covered by the rotor (A_T) and thickness in function of the wind velocity (V). The air density and volume available to the rotor are denoted as ρ_a and v, respectively. Therefore, the expression for kinetic energy can be defined as follows:

$$E_K = \frac{1}{2}\rho_a v \cdot V^2 \tag{10.6}$$

Hence, the power, which is energy per unit time, is expressed as follows:

$$P = \frac{1}{2}\rho_a A_T V^3 \tag{10.7}$$

These expressions show that the power of the turbines is defined by air density, the area of the wind rotor, and the wind velocity. The geographical location and the rotor size are relevant to the design. Wind velocity is the most relevant design factor, since this term has a cubic effect in the previous equation and this helps to reduce the cost of the wind turbine. Therefore, rotors dimensions can be reduced 8 times to obtain the same power if wind power is doubled.

The efficiency of a wind turbine is defined in terms of the power coefficient (C_P), which represents the part of the kinetic energy from the air mass that is transferred to the rotor and turned into electric energy (the rest of the air mass does not interact with the blades). Therefore, the power coefficient is the ratio of actual power developed by the rotor to the theoretical power available in the wind parcel, as indicated in the following expression:

$$C_P = \frac{2P_T}{\rho_a A_T V^3} \tag{10.8}$$

where P_T is the total power developed by the turbine. The power coefficient of a turbine depends on many factors such as the profile of the rotor blades, blade arrangement and setting, etc. [13].

The power developed by a rotor depends on the relative speed between the blade and the wind speed. For instance, if the blades are moving slow, a portion of the air parcel may pass without interacting with the turbine. Similarly, if the rotor is rotating fast and the wind is very low, the wind may be deflected from the turbine and turbulence will produce a poor energy conversion. This is known as the tip–speed ratio (TSR or λ), which is the ratio between the velocity of the rotor tip and the wind velocity, as shown in the following expression:

$$\lambda = \frac{R\Omega}{V} = \frac{2\pi NR}{V} \tag{10.9}$$

where R is the radius, N is the rotational speed, and Ω is the angular velocity of the rotor.

The power coefficient can be obtained through experimental tests and simulations of the wind turbine performance. Nonetheless, some numerical approximations can be done as shown in [4] by expressing the C_P as a nonlinear function

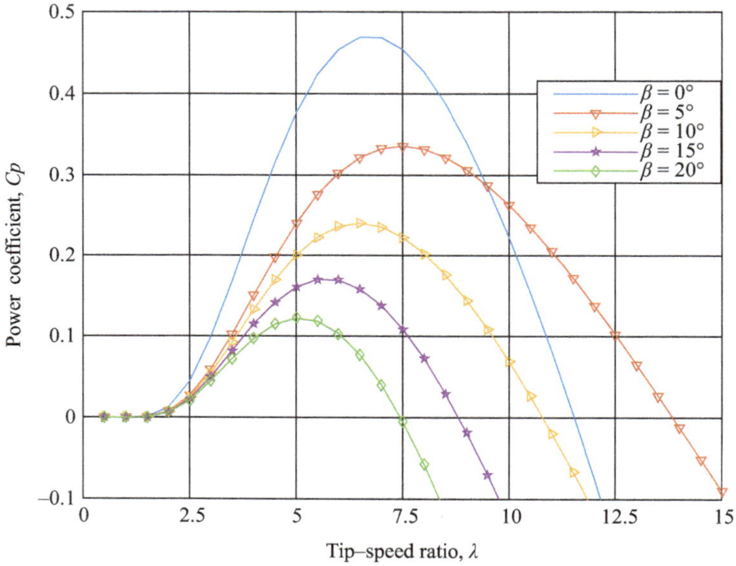

Figure 10.16 Power coefficient as a function of the TSR

of the TSR and pitch angle of the blades, both dependent on the wind speed. This is shown in the following equations:

$$C_P(\lambda, \beta) = \frac{1}{2}\left(\frac{98}{\lambda_i} - 0.4\beta - 5\right) \cdot e^{-16.5/\lambda_i} \tag{10.10}$$

$$\lambda_i = \left[\frac{1}{\lambda + 0.089} - \frac{0.035}{(\beta^3 + 1)}\right]^{-1} \tag{10.11}$$

where β is the blade pitch angle in degrees and λ is the tip–speed ratio.

Typical modern wind turbines operate at TSR range between 5 and 10, close to the maximum value of C_P. The maximum limit is known as the "Betz's limit," a theoretical maximum that states that only wind power up to 59.3% can be achieved.

Figure 10.16 shows several C_P curves considering the pitch angle for a 0.5 kW three-blade wind turbine. As can be seen, the maximum C_P obtained is near 0.47.

Wind turbines are categorized according to the way the blades interact with the wind, the orientation of the rotor axis, and the type of generator. The following sections discuss in more detail their characteristics [12,14].

10.3.1 Drag and lift blades

Drag blades are characterized by the air pushing the blade itself, forcing the rotor to turn on its axis. This type of blades is very limited as its efficiency, since the blades cannot turn faster than the wind speed.

Lift-bladed turbines use airfoils, which have the advantage of moving faster than the wind speed and being more efficient due to its aerodynamic construction.

10.3.2 Rotor axis orientation

Wind turbines can be also classified as horizontal axis or vertical axis turbines. Horizontal-axis wind turbines (HAWT) rotate with their axis parallel to the wind stream and the ground. This is the most common and commercial type of wind turbine. The components of this type of turbine are shown in Figure 10.17.

The main advantages of the HAWT include a low cut-in wind speed, access to high wind streams because of their tower height, relatively high-power coefficient, and high efficiency. Their disadvantages include a complex design and maintenance, massive structure, and complex control systems. Also, their generators, gearboxes, and power transformers are placed over the tower. In some new applications, the power transformer is located outside the nacelle to reduce space and material.

The other turbine type according to its axis orientations is the vertical-axis wind turbines (VAWT). These turbines rotate with their axis perpendicular to the wind stream and to the ground. The advantages of the VAWT include receiving wind from any direction, simpler and more economical designs, maintenance can be done at ground level, and complex yaw mechanisms are no longer needed. Disadvantages of the VAWT include a lower wind speed because the height of towers, which makes less power available and airflow around the ground can create turbulent flows. Also, their inertia is higher, which requires more power to start up the turbine.

10.3.3 Number of blades

Horizontal axis turbines can also be classified considering the number of blades. Construction of single-, two-, three-, and multibladed systems can be implemented. These configurations are illustrated in Figure 10.18.

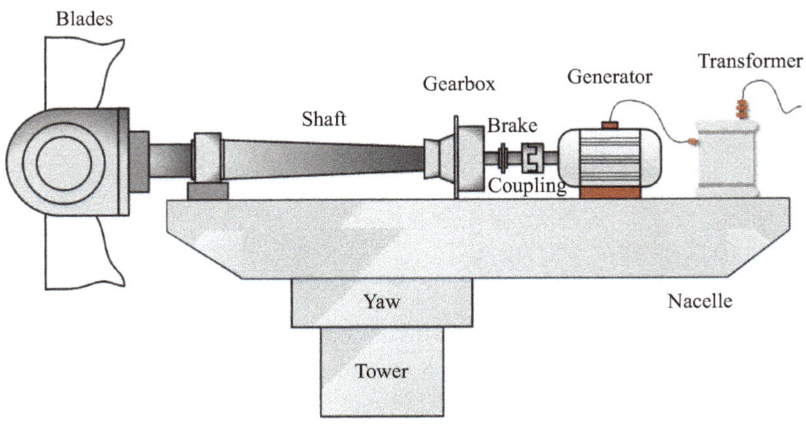

Figure 10.17 Mechanical structure of a HAWT [13]

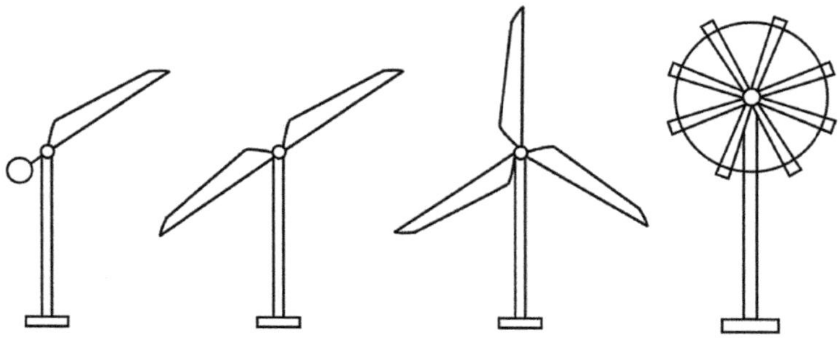

Figure 10.18 Single-bladed, two-bladed, three-bladed, and multibladed wind turbines [12]

Single-bladed turbines use less material, which makes them cheaper. However, a counterweight is needed to balance the blade. Two-bladed turbines also need a counterweight and drag losses can be minimum. Most of the commercial wind turbines used for electricity generation are three-bladed. Multibladed systems are often used for water-pumping applications because of the need of high starting torque.

10.3.4 Speed of rotation

According to the speed of rotation of the rotor, wind turbines are classified into fixed-speed wind turbines (FSWT) or variable-speed wind turbines (VSWTs). FSWT are simpler to operate because they generate electricity when the wind speed makes the generator shaft speed higher than the synchronous speed. These turbines are limited in the generated power, although they require less maintenance and are cheaper in comparison with the other type. FSWT operate almost linear, as illustrated in Figure 10.19. When the speed is outside the linear region, the turbine enters into an unstable nonlinear region. This type is known as "constant" or "fixed-speed" because the narrow speed ranges.

VSWT systems are more complex than FSWT because of the power converters that allow to generate electricity in a wide range of turbine speeds. A voltage is injected in the rotor circuit of the generator to reach different torque-speed characteristics.

10.3.5 Generator types

Wind turbine generators may have different types according to the generator configurations, which are shown in the following section [15,16].

10.3.5.1 Type 1—squirrel-cage induction generator

This type of wing turbine generator uses a fixed-speed turbine with a squirrel-cage induction generator. The induction generator generates electricity when it is driven above synchronous speed. A negative slip indicates that the wind turbine operates in generating mode, in which the operating slips for an induction generator are

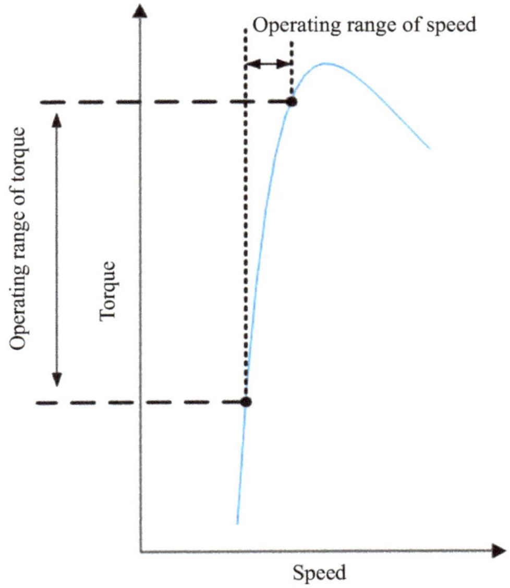

Figure 10.19 Torque-speed characteristic of an FSWT [13]

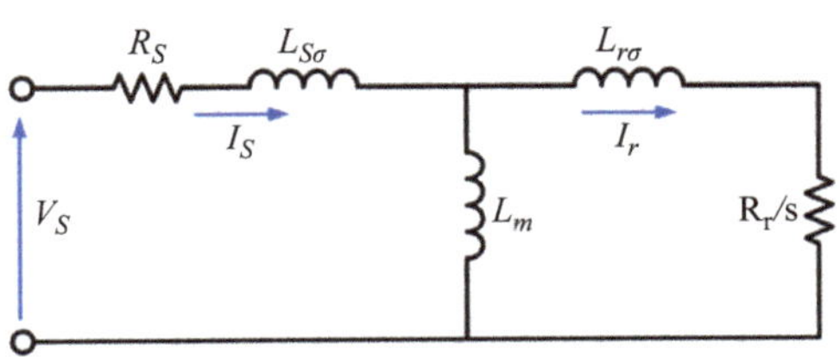

Figure 10.20 Equivalent circuit of a Type 1 generator [15]

normally between 0% and −1%. Figure 10.20 shows the simplified single-phase equivalent circuit of a squirrel-cage induction machine, while Figure 10.21 shows an example single-line connection diagram.

Figure 10.20 is referred to the stator where R_S and R_r are the stator and rotor resistances, $L_{S\sigma}$ and $L_{r\sigma}$ are the stator and rotor leakage inductances, L_m is the magnetizing reactance, and *S* is the rotor slip. In the case of voltage fault, the inertia of the wind rotor drives the generator after the voltage drops at the generator terminals. The rotor flux may not change instantaneously right after the voltage drop due to fault. Therefore, voltage is produced at the generator terminals causing fault current flow into the fault until the rotor flux decays to zero. This process takes a few electrical cycles.

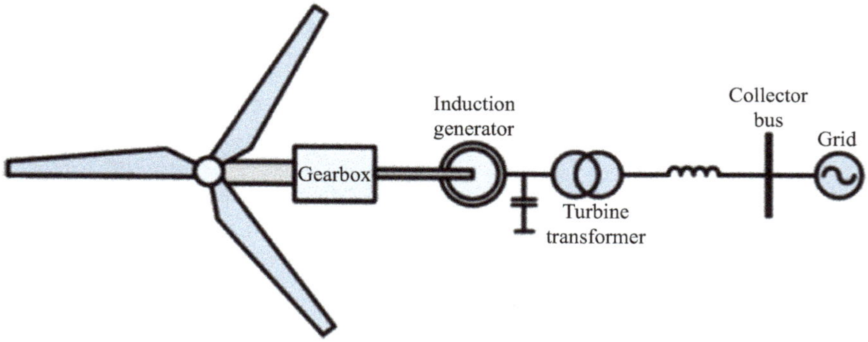

Figure 10.21 Circuit diagram of a Type 1 generator [15]

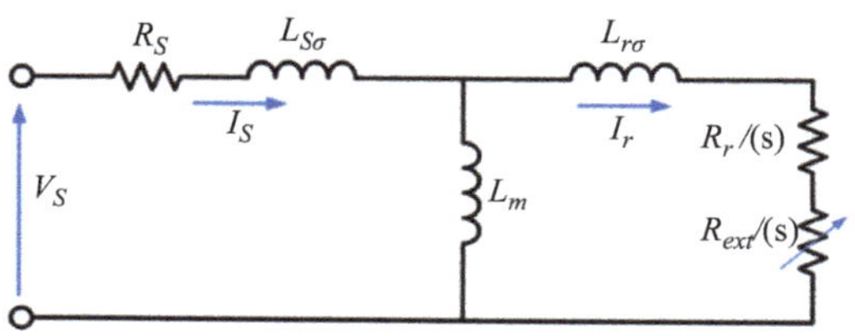

Figure 10.22 Equivalent circuit of a Type 2 generator [15]

10.3.5.2 Type 2—wound-rotor induction generator with variable external rotor resistance

A three-phase rotor winding is connected to a power electronic component and three-phase external resistance. The external rotor-resistance controller (ERRC) is a very fast controller that allows the effective rotor resistance to vary; thus, the torque speed characteristic of this type of generator can be shaped accordingly. The equivalent circuit for a Type 2 generator is illustrated in Figure 10.22, where Figure 10.23 shows an example single-line connection diagram.

10.3.5.3 Type 3—doubly fed induction generator

This type of wind turbine generator is implemented by a doubly fed induction generator. The rotor speed is allowed to vary between 0.3 and −0.3 slip; thus, the power converter can be sized to about 30% of rated power (partial rating). Maximum energy yield is accomplished for low-to-medium wind speeds. Above rated wind speeds, the aerodynamic power is controlled by pitch to limit rotor speed and to minimize mechanical loads. The equivalent circuit for a Type 3 generator is illustrated in Figure 10.24, where Figure 10.25 shows an example single-line connection diagram.

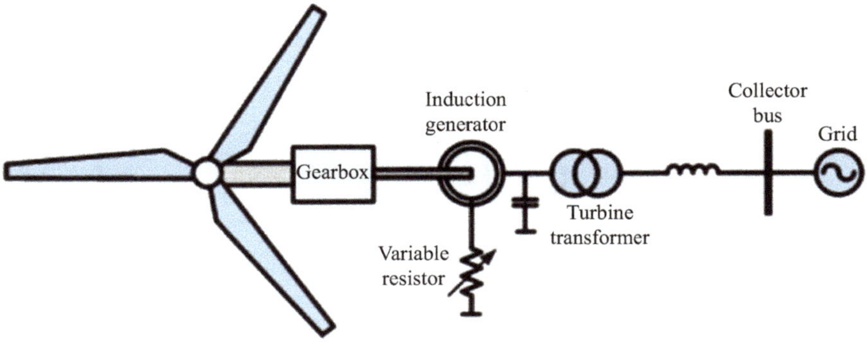

Figure 10.23 Circuit diagram of a Type 2 generator [15]

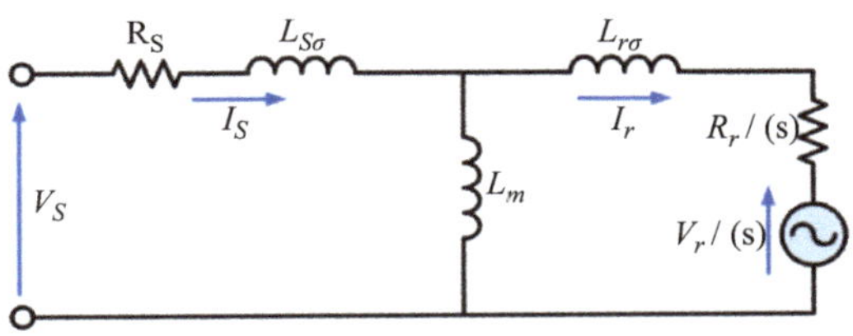

Figure 10.24 Equivalent circuit of a Type 3 generator [15]

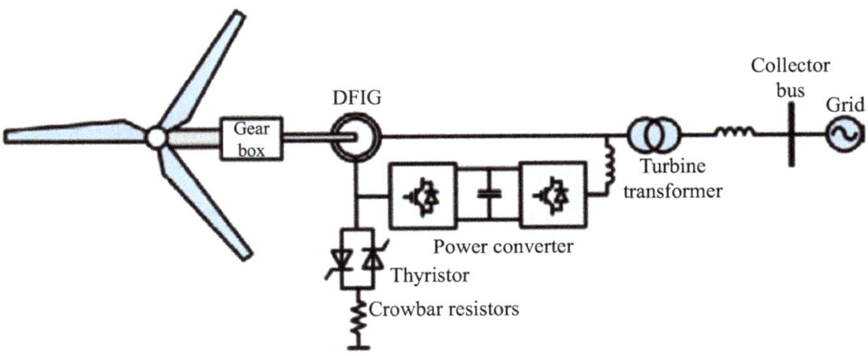

Figure 10.25 Circuit diagram of a Type 3 generator [15]

10.3.5.4 Type 4—full-converter wind turbine generator

This is a VSWT generator implemented with full power conversion. Recent advances and lower cost of power electronics make it feasible to build VSWTs with power converters with the same rating as the turbines. Maximum energy yield is

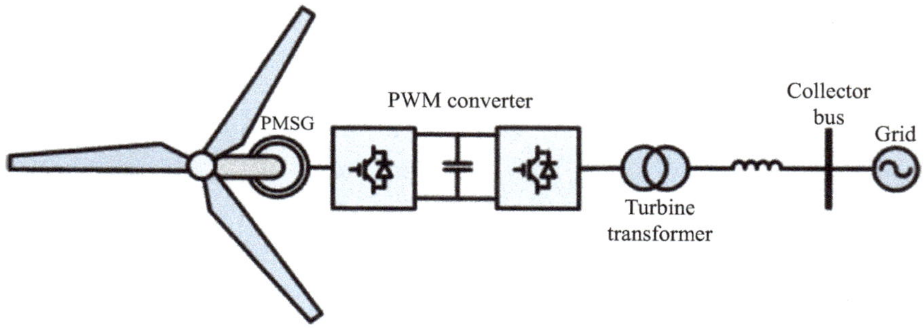

Figure 10.26 Circuit diagram of a Type 4 generator [15]

accomplished for low-to-medium wind speeds. Above rated wind speeds, the aerodynamic power is controlled by pitch to limit rotor speed and to minimize mechanical loads. With the use of a power converter, the real and reactive power can be controlled independently and instantaneously within design limits. Figure 10.26 shows an example of a single-line connection diagram.

The short circuit current contribution for a three-phase fault is limited to its rated current or a little above its rated current. An overload capability of 10% above rated power is a common practice to design a power converter for a Type 4 wind turbine. The generator stays connected to the power converter in any fault condition and is isolated from the faulted lines on the grid. Although there is a fault on the grid, the generator output current is controlled to stay within the current limit (e.g., 1.1 pu).

10.3.6 Control systems

Wind turbines have several types of control mechanisms [4,13,17]. The most common are soft starting, generation control, pitch control, and yaw control. Other control strategies also implemented are feathering, reactive power control, stability, low and high voltage ride through, and ramp control. The formers will be discussed in this section.

10.3.6.1 Soft starting

The soft starting of a wind turbine is an electrical device used in fixed-speed turbines during the grid connection process, which reduces the inrush-current and to avoid voltage disturbances into the grid.

10.3.6.2 Stall and pitch control

Most of the modern turbines implement this type of control to regulate the output power. The blades are pitched as the wind speed varies in order to obtain a better performance. As the wind speed increases, the blades turn into the opposite direction of the wind to increase the attack angle and to control its power at higher wind speeds without any damaging. This type of control depends on the operation region of the wind turbine by modifying the angle of the blades. This is done in order to achieve a higher aerodynamic efficiency.

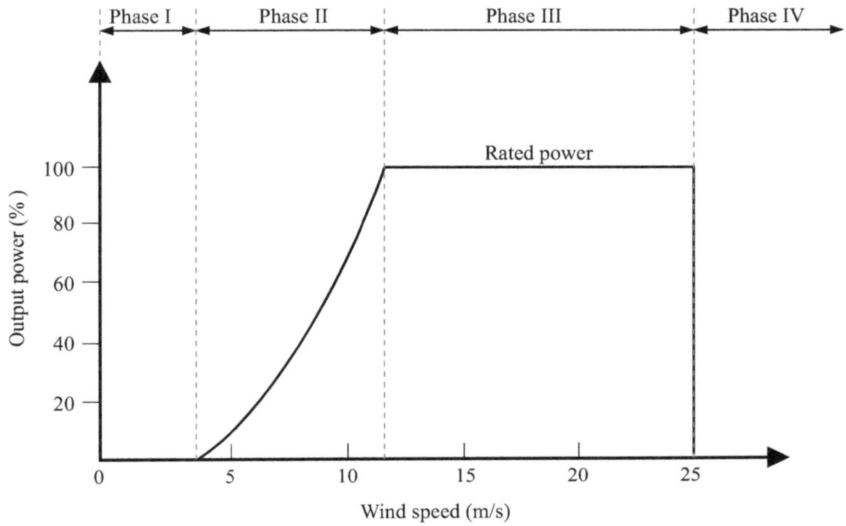

Figure 10.27 Phases of wind power generation [13]

The operation region is shown in Figure 10.27. Phase I (wind speed below 5 m/s) indicates that the generator is not generating any power, since the wind speed is not high enough. In Phase II, the output power is increased gradually with the wind speed up to its rated power (wind speed around 12 m/s). In this phase, the pitch angle maximizes the output power by tracking the maximum C_P, until the output power remains constant at its rated value at Phase III. Control mechanisms are applied to avoid overcurrent and/or overloading. When the wind speed exceeds the cut-out speed in Phase IV (wind speed around 25 m/s), the blades are adjusted to limit the output power until they are stopped.

10.3.6.3 Yaw control

Yaw control is useful to face the wind turbine to the wind stream. It is used to turn the rotor partly away at higher wind speeds. When the wind speed exceeds the cutoff limit of the wind turbine, the axis position is adjusted to a nearly perpendicular position with respect to the wind; thus the generated power is reduced.

10.3.7 Wind farms

When large amounts of power are required, wind turbines can be gathered in a cluster, forming what is known as a wind farm [13]. Some of the advantages of clustering wind turbines are in terms of an easier installation than several scattered units, operation and maintenance, and power transmission efficiency.

The wind turbines in a wind farm are arranged in rows near each other. Factors like wind direction, topographic and environmental conditions, available area, and existing electrical network must be considered.

To avoid shadowing and wind turbulence due to nearby units, safety distances are needed. In [12], a spacing of $3D_T$–$4D_T$ within the rows is suggested, where D_T is the rotor diameter. Between rows, a spacing of $10D_T$ is recommended in order to restore the wind stream before it interacts with the turbines in next row. This is illustrated in Figure 10.28.

Alternatively, the number of turbines per row (N_{TR}) and the number of row (N_R) can be calculated as follows:

$$N_{TR} = \frac{L_R}{S_R} + 1 \tag{10.12}$$

where L_R is the row length and S_R is the row spacing.

Considering

$$N_T = \frac{P_F}{P_T} \tag{10.13}$$

where N_T is the total number of turbines in the farm, P_F is the total capacity of the wind farm, and P_T is the rated power of a single turbine.

Hence, the total number of rows is

$$N_R = \frac{N_T}{N_{TR}} \tag{10.14}$$

In some cases, areas with high wind speeds are located far from the habitable zones and existing electrical networks. Therefore, a small network must be installed to take advantage of the wind resource. Offshore wind farms are an example of the latter. Offshore farms often offer higher efficiency and productivity than onshore wind farms.

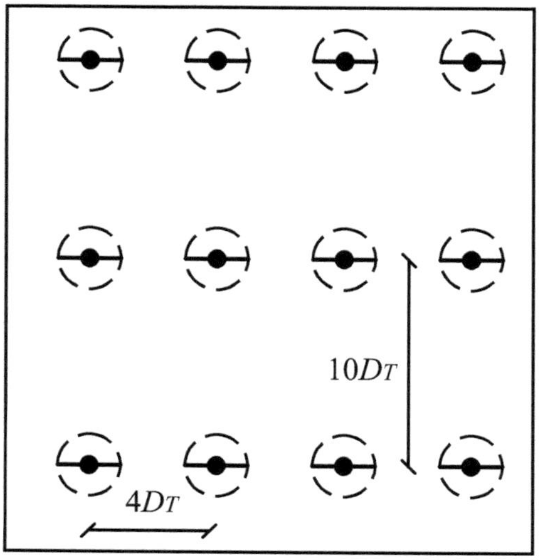

Figure 10.28 Wind farm layout

10.4 Small hydroelectric plants

Hydropower plants represent a mature technology with a well-known operational scheme [2,18]. A dam is constructed to collect water that powers the turbines, as shown in Figure 10.29. Some hydro schemes can experience large variation of water according to dam's capacity and seasonal rainfalls. Small hydropower plants become an interesting option for customers close to mountainous areas, in which constructing large hydropower plants is not possible. They can be constructed fast with a technology that offers over 80% of efficiency. They have automatic operating systems and low operational and maintenance costs.

Most of these plants operate in grid-connected mode, but several units operate in isolated operation mode, which requires an exact match between the demand for and the supply of electrical energy. The generation can be controlled considering that criteria, in which the dam is designed to match the required power demand. In some scenarios, in which the water capacity surpasses the required amount, it has to be bypassed by the turbines. The available power from a hydro plant is calculated by using the following expression:

$$P_H = \eta \cdot g \cdot Q \cdot H \ (kW) \tag{10.15}$$

where η is the efficiency of the turbine, g is the gravity (9.81 m/s^2), Q is the total flow through the turbine, and H is the head [6].

Hydropower plants are classified according to the head of the installation, which is the vertical distance through which the waterfalls and by linking it with the flow. These are low-head, middle-head, and high-head hydropower plants. Low-head hydro plants are rivers and channels with a small downward gradient. They have the water supplied to the turbine practically without storage. Middle-head hydropower plants are built exclusively in connection with dams (control

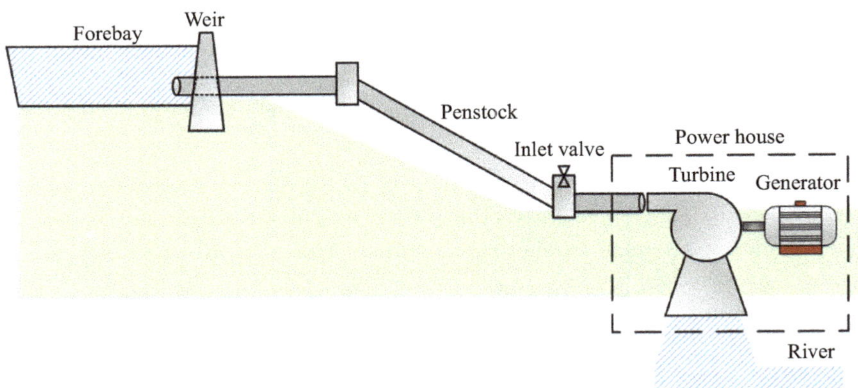

Figure 10.29 Small-scale hydropower plant

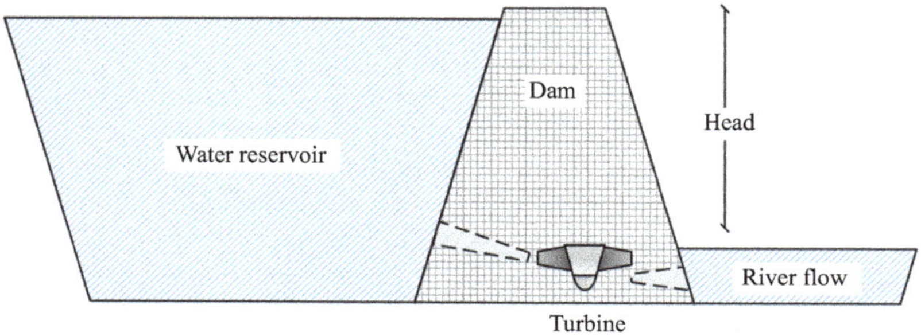

Figure 10.30 Schematic diagram of a hydroelectric power plant with water reservoir

works), which have heights up to 200 m. The sole of the dam uses pressure pipelines or penstocks to supply water to the powerhouse. High-head hydropower plants are in central and high mountains with small flow rates heads up to 2,000 m, from where the energy comes from rather than from the flow rate.

In some applications, small hydro plants with water reservoir can be used as a storage system for wind energy systems, which have the storage capacity able to support the total wind energy supply. Figure 10.30 shows the schematic diagram of a hydroelectric power plant with water reservoir.

10.5 Energy storage systems

Due to the uncertainty nature associated with some sources such as wind or solar energy, it is required to support their generation capacity by storage units. The energy storage systems are becoming very important in the increased utilization of the intermittent and periodic alternative energy sources, such as wind, solar, tidal, and wave energy, among others [19–22]. These storage systems get the energy when there is a surplus after attending the local demand.

Some of the most relevant advantages of using energy storage devices on a distribution network are the following [19,23–26]:

- Capacity firming: Storage units can help to maintain a committed level of supplied energy for a period of time. They smooth the output and control the ramp rate (MW/min) to eliminate rapid voltage and power swings on the electrical grid.
- Peak shaving: This helps to respond to high variability load demands, which stress system requirements and make the operation more expensive.
- Voltage support: These units help to maintain the grid voltage by injecting or absorbing both active and reactive power when the system presents operational voltage values outside the allowed ranges.

- Frequency regulation: This approach for regulating frequency is an attractive option due to its rapid response time, which is charged or discharged according to the system demand.
- Power quality: In power quality applications, an energy storage system helps one to protect downstream loads against short-duration events that affect the quality indices.

Example 10.3 An energy storage device is connected to the PV system shown in Figure 10.31. The system is the same as Figure 10.10 where the energy storage device has been added. The example has been modeled and run with a PSAT and the results are depicted in Figure 10.32 that shows the load flow profile when the

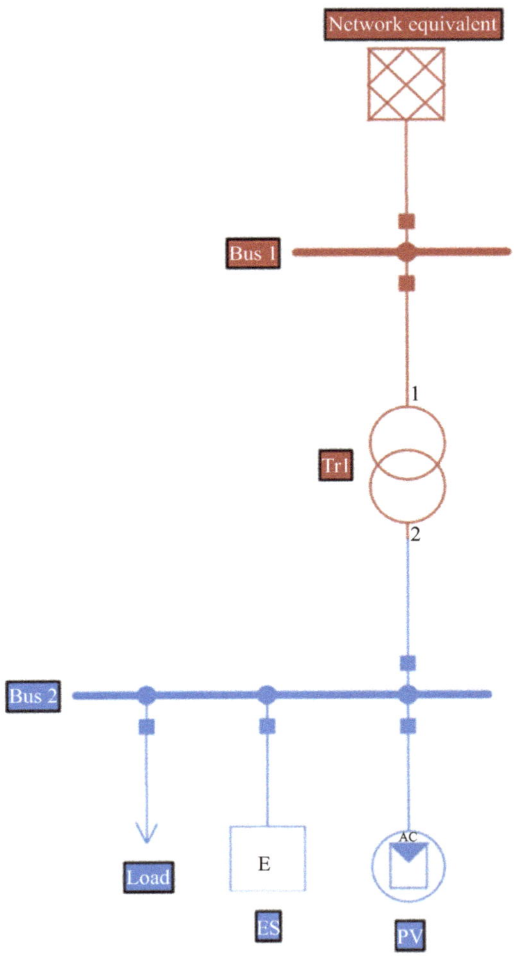

Figure 10.31 Integration of a PV system with energy storage in the power system

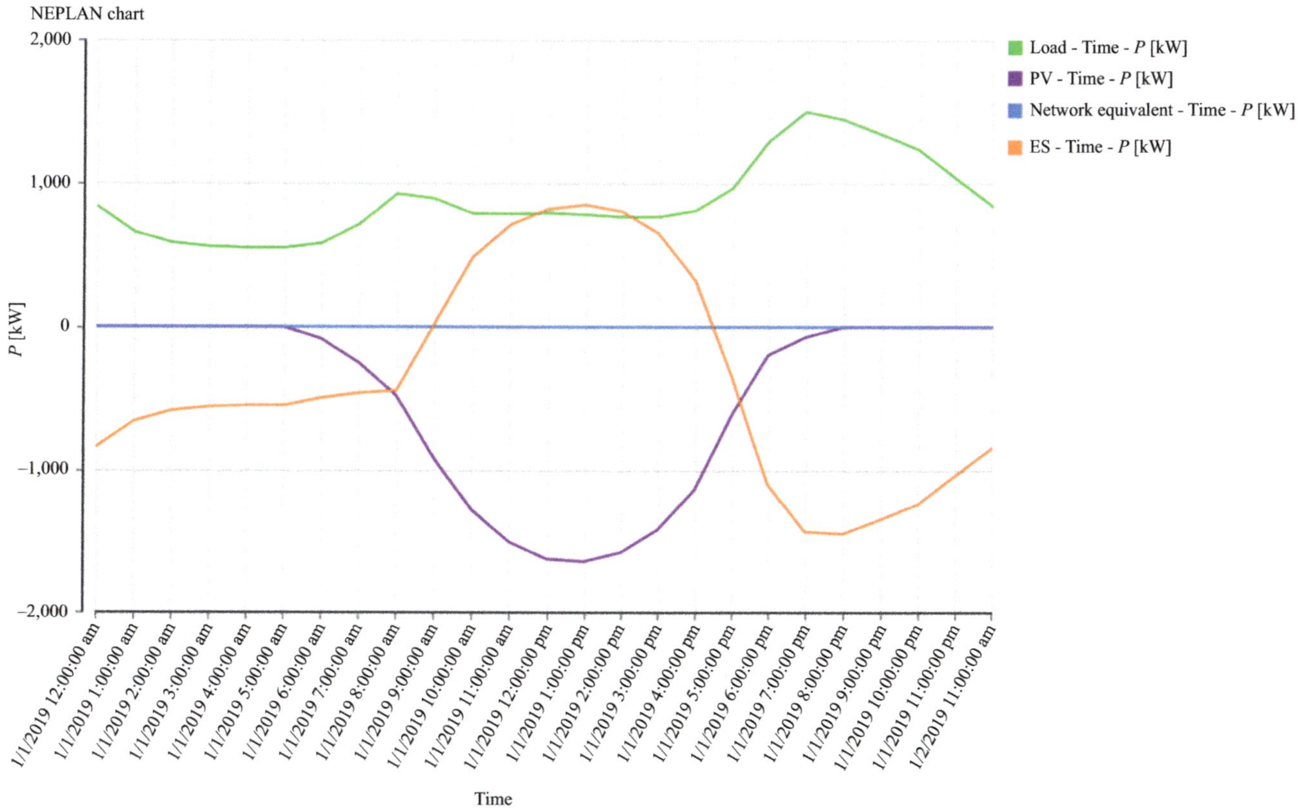

Figure 10.32 Example of a load flow profile through one typical day

power output from the PV system is reduced. No power from the grid is then needed.

The most frequently used systems for the storage of energy are presented in this section [6,20,22]. There are some based on electromechanical principles, such as pump systems or flywheels, and some are based on thermal and chemical systems such as batteries, fuel cells, and supercapacitors.

10.5.1 Electromechanical storage

There are two types of energy storage that result from the application of forces upon materials systems. One of these involves changes in potential energy (such as pumped-hydro storage and compressed-air), and the other involves changes in the kinetic energy (such as flywheel energy storage).

10.5.2 Electrochemical storage

Chemical storage is a very desirable method for energy storage, which has very high energy density (in kJ/kg) and is easy to be used by combustion or direct conversion to electricity. Additionally, an energy storage device with desirable properties must charge and discharge quickly. This allows providing large quantities of energy and high power when it is required. Some examples of the electrochemical energy storage devices are fuel cells, supercapacitors, superconducting magnetic energy storage and batteries.

One of the most common technologies used for power system applications is batteries that can be used for a very wide range of applications, from assisting the very large-scale electrical grid down to tiny portable devices used for many purposes. There are standard batteries such as lead acid or alkaline batteries, lithium batteries. Also, there are high temperature or thermal batteries, which are activated at high temperatures by the application of heat from an external source to liquefy the electrolyte.

The system illustrated in Figure 10.33 is taken from NEPLAN [27] and represents a 200 V–6.5 A h NiMH battery connected to a constant DC load of 50 A. The DC machine (in shunt excitation) is connected in parallel with the load and operates freely at no-load torque. When the state of charge (SOC) of the battery goes below 40%, a negative load torque of 200 N m is applied to the machine so it acts as a generator, recharging the battery and supplying the load. When the SOC goes above 80%, the load torque is removed so only the battery supplies the load. This control scheme is implemented in the control circuit block (CCT) of the corresponding battery that has been modeled in the referred software.

As illustrated previously, the SOC determines the available capacity (in A h) as a percentage of its rated capacity. Accurate SOC estimation is one of the main tasks of battery-management systems, which will help improve the system performance and reliability, and will also lengthen the lifetime of the battery. In fact, precise

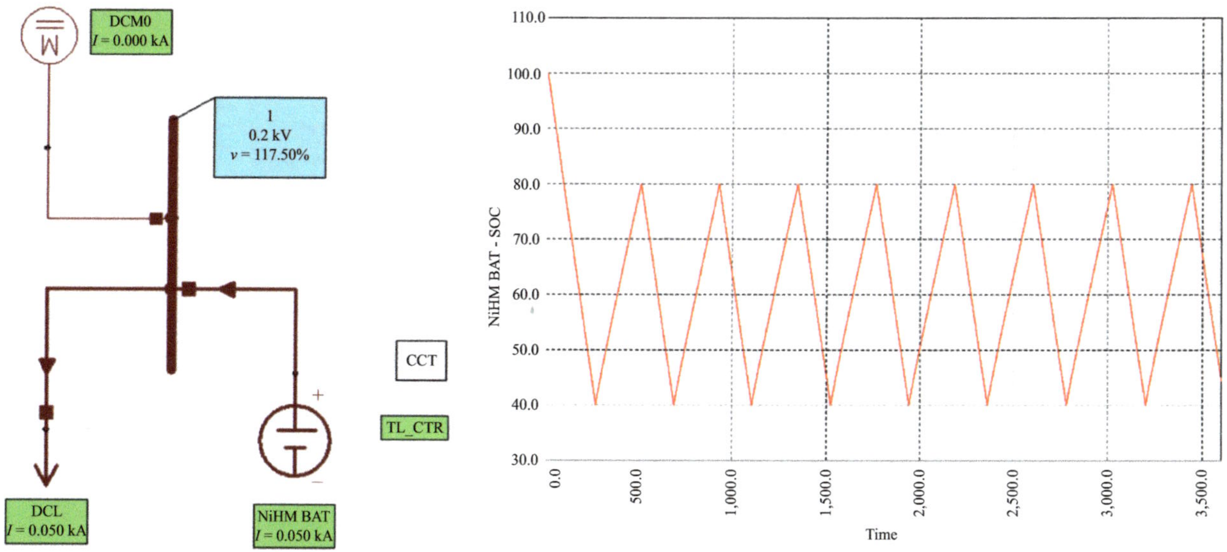

Figure 10.33 NiMH battery state of charge operation

SOC estimation of the battery can avoid unpredicted system interruption and prevent the batteries from being overcharged and overdischarged [28].

10.6 Proposed exercises

1. Determine the open-circuit voltage of a 60-cell silicon module when the ambient temperature rises to 30°C and the total irradiation varies to 1.0 kW/m^2. The module has a NOCT of 45°C and a $V_{oc} = 35$ V.

$$T_{Cell} = T_A + \left(\frac{NOCT - 20°\mathrm{C}}{0.8\ \mathrm{kW/m^2}}\right) \cdot G$$

2. Determine the maximum power considering a FF of 0.8. Considering the following *I–V* curve of a PV module (Figure 10.34).

$$P_{\max} = FF \cdot I_{SC} \cdot V_{OC}$$

3. Determine the FF of the PV module of the PV module, based on the following *I–V* and *P–V* curves (Figure 10.35).
4. Write a script in MATLAB® that plots the *I–V* curve for a PV cell as shown in Figure 10.3. Assume the saturation current I_o as 6.0×10^{-10} A. Consider the values listed next for the photocurrent and different temperatures.

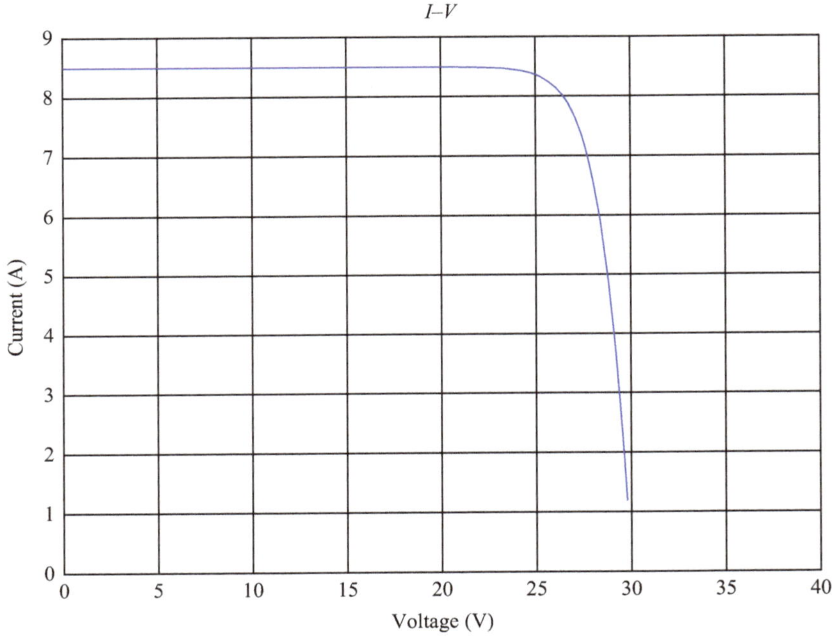

Figure 10.34 I–V curve for exercise 2

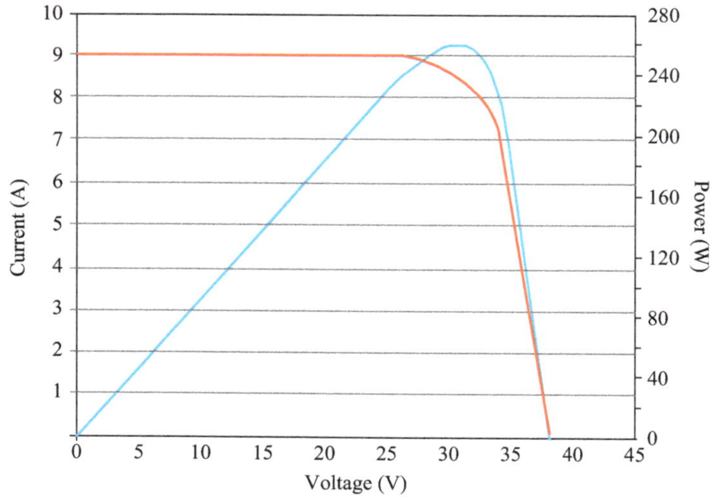

Figure 10.35 Curve for exercise 3

(i) I_{ph} (A): 3.4, 5.0, 7.2, 9.1
(ii) T (K): 273, 313, 323, 360

5. A utility acquires a field of 169 km^2 (13 km × 13 km) and wants to install wind turbines of 66.5 m blades length. Determine the number of wind turbines allowed to achieve a row spacing of seven times the turbine's diameter (D_T).
6. For a wind turbine with a swept area of 14,540 m^2, compute the total power of a wind stream when the wind speed is equal to 13 m/s and the air density is 1 kg/m^3. Then assume a power coefficient of 0.35 in order to determine the power converted from the wind into rotational energy.

$$P = \frac{1}{2}\rho_a A_T V^3$$

$$\rho_a = 1.0\,\text{kg/m}^3$$

$$A_T = 14{,}540\text{m}^2$$

$$V = 13.0\,\text{m/s}$$

7. Based in Figure 10.36, find the optimal angular velocity for a wind turbine with a 66.5 m blade length and wind speed of 13 m/s. Assume the wind turbine is operating for the maximum performance coefficient.
8. Write a script in MATLAB that plots power coefficient curve as shown in Figure 10.16. Consider the following values:
   (i) Pitch angle: 0°,5°,10°,15°, and 20°
   (ii) TSR: 0–15

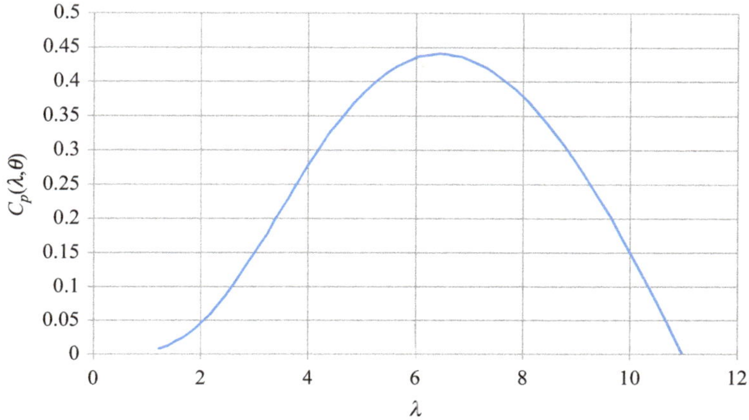

Figure 10.36 *C_P vs TSR curve for exercise 7*

9. Determine the total power delivered by a pumped-hydro storage plant when the water flow rate is 135 m^3/s and the conversion efficiency is 0.75. Assume an elevation change of 24 m and the density of water as 1,000 kg/m^3.

$$E_{stored} = g\rho_w V(H_{upper} - H_{lower})$$

10. Determine the total amount of charge in A h of a battery bank in order to supply the total demand of an off-grid PV system. Consider the following:
    (i) Total demand: 7,890 W h/day
    (ii) Battery bank days of autonomy: 2
    (iii) Bus voltage: 48 V
    (iv) Battery bank discharge depth: 75%

References

[1] REN21. "Renewable Energy Policy Network for the 21st Century (REN21)," Renewables 2019 Global Status Report. REN21, 2019.

[2] Jenkins, N., Ekanayake, J., and Goran, S. Distributed Generation. IET, London, UK, 2010.

[3] Sen, Z. Solar Energy Fundamentals and Modeling Techniques. Springer, London, UK, 2008.

[4] Zambroni de Souza, A. and Castilla, M. Microgrids Design and Implementation. Springer, Cham, Switzerland, 2019.

[5] Messenger, R. and Ventre, J. Photovoltaic Systems Engineering, 4th ed. CRC Press, Boca Raton, FL, USA, 2010.

[6] Kaltschmitt, M., Streicher, W., and Wiese, A. Renewable Energy: Technology, Economics and Environment. Springer, Berlin, Heidelberg, 2007.

[7] Technische Universität München, TUM. Renewable Energy. Technische Universität München, TUM Munich, Germany, 2014.
[8] Rashid, M. Power Electronics: Circuits, Devices, and Applications, 2nd ed. Prentice Hall, Upper Saddle River, NJ, USA, 1993.
[9] Mohan, N., Undeland, T., and Robbins, W. Power Electronics: Converters, Applications and Design, 3rd ed. Wiley, Hoboken, NJ, USA, 2002.
[10] Clean Energy Reviews (CER). "Best Solar Inverters 2018." May 12, 2018. [Online]. Available: https://www.cleanenergyreviews.info/blog/best-grid-connect-solar-inverters-sma-fronius-solaredge-abb. [Accessed: 21-Jun-2018].
[11] Mustafa, F., Ahmed, I., Malek, A., and MobasSharin, A.-Y. Centralized and Decentralized Solar Power Control and Distribution: A Comparative Analysis of Operational and Economic Aspects. BRAC University, Dhaka, Bangladesh, 2017.
[12] Mathew, S. Wind Energy. Fundamentals, Resource Analysis and Economics. Springer, Berlin, Heidelberg, 2006.
[13] El-Sharkawi, M. Wind Energy: An Introduction. CRC Press, Boca Raton, FL, USA, 2015.
[14] Nelson, V. and Starcher, K. Introduction to Renewable Energy, 2nd ed. CRC Press, Boca Raton, FL, USA, 2016.
[15] Gevorgian, V. and Muljadi, E. "Wind Power Plant Short Circuit Current Contribution for Different Fault and Wind Turbine Topologies," in The 9th Annual International Workshop on Large-Scale Integration of Wind Power into Power Systems as well as on Transmission Networks for Offshore Wind Power Plants, National Renewable Energy Laboratory, 2010.
[16] Gers, J. M. and Viggiano, C. "Protective Relay Setting Criteria Considering DERs and Distributed Automation," 13th International Conference on Development in Power System Protection 2016 (DPSP), Edinburgh, UK, 2016, p. 8.
[17] Gharedaghi, F., Deysi, M., Jamali, H., and Khalili, A. "Soft starter investigation on grid connection of wind turbines," Australian Journal of Basic and Applied Sciences, vol. 5, no. 10, pp. 1146–1153, 2011.
[18] Bollen, M. and Hassan, F. Integration of Distributed Generation in the Power System. Wiley, Hoboken, NJ, USA, 2011.
[19] Efstathios, E. Alternative Energy Sources. Springer, Berlin, Heidelberg, 2012.
[20] Akinyele, D. O. and Rayudu, R. K. "Review of energy storage technologies for sustainable power networks," Sustainable Energy Technologies and Assessments, vol. 8, pp. 74–91, 2014.
[21] Oberhofer, A. and Meisen, P. Energy-Storage Technologies, Global Energy Network Institute, July, 2012.
[22] Huggins, R. A. Energy Storage: Fundamentals, Materials and Applications. 2nd ed. Springer, Cham, Switzerland, 2015.
[23] ABB. "Frequency Regulation." [Online]. Available: https://new.abb.com/distributed-energy-microgrids/applications/energy-storage-applications/frequency-regulation. [Accessed: 12-Apr-2018].

[24] ABB. "Peak Shaving." [Online]. Available: https://new.abb.com/distributed-energy-microgrids/applications/energy-storage-applications/peak-shaving. [Accessed: 12-Apr-2018].
[25] ABB. "Load Leveling." [Online]. Available: https://new.abb.com/distributed-energy-microgrids/applications/energy-storage-applications/load-leveling. [Accessed: 12-Apr-2018].
[26] ABB. "Spinning Reserve." [Online]. Available: https://new.abb.com/distributed-energy-microgrids/applications/energy-storage-applications/spinning-reserve. [Accessed: 12-Apr-2018].
[27] NEPLAN. NiHM Battery With DC Machine Shunt Excited. NEPLAN AG, 2018.
[28] Murnane, M. and Ghazel, A. A Closer Look at State of Charge (SOC) and State of Health (SOH) Estimation Techniques for Batteries. Analog Devices, 2017. [Online]. Available: https://www.analog.com/media/en/technical-documentation/technical-articles/A-Closer-Look-at-State-Of-Charge-and-State-Health-Estimation-Techniques-....pdf. [Accessed: 18-Jun-2019].

Chapter 11
Fundamentals on microgrid technology

In this chapter, the fundamentals of a microgrid will be presented. It is important to understand the main characteristics of a microgrid as a subsystem and an active component of the distribution network, as well as its operation modes.

Some general concepts of control strategies and protection solutions in microgrids are also explained, considering the impacts of high penetration of distributed generation (DG) units and energy resources into the distribution network.

11.1 Introduction to microgrids

The incorporation of microgrids in the electric system has taken place since the growing concern for energy availability and cost of kW h, in some portions of electrical systems. Also, there are some infrastructures of current electrical transmission and distribution systems that have been deteriorated by aging. One efficient way to improve the performance of the energy distribution is through the use of microgrids [1].

A microgrid can be described as a modern component of the electric network in which different components such as loads, distributed energy resources (DERs), and energy storage (ES) systems interact together through control devices. This makes that microgrids are seen as a single entity by the main grid. An illustration of this concept is shown in Figure 11.1 [2].

Generation technologies such as conventional DG units (micro turbines, fuel cells, and diesel), renewable energy units (wind energy, solar photovoltaics, and others), ES (thermal storage, batteries), and combined heat and power units are common components of microgrids.

The main motivations to build microgrids can be summarized as follows:

- Improve reliability since the values of quality indices are every day more stringent and in particular those of SAIDI, SAIFI, and CAIDI.
- Increase resiliency: It is the ability to withstand challenges and to continue in operation. A clear example is the value of microgrids during storms by maintaining the power supply.
- Improve economic and in particular the infrastructure cost savings by avoiding investments for the replacement and/or expansion of old power plants.
- Promote sustainability/emissions reduction.

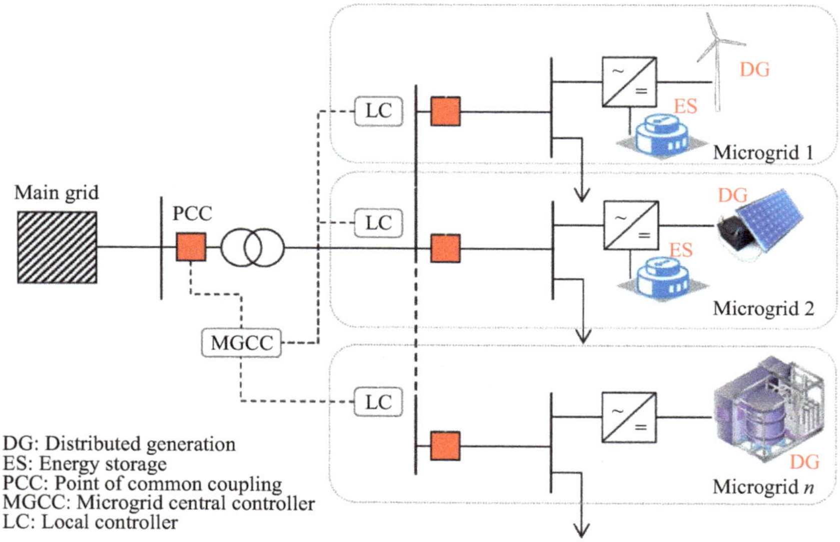

Figure 11.1 A general scheme of a microgrid

There is no standard term to define a microgrid; however, there are some international standards that have incorporated related specifications for microgrids, including *IEEE Std. 446-1995. Recommended Practice for Emergency and Standby Power Systems for Industrial and Commercial Applications* [3]; *IEEE Std. 519-2014. Recommended Practices and Requirements for Harmonic Control in Electric Power Systems* [4]; *IEEE Std. 1547-2018. Standard for Interconnection and Interoperability of Distributed Energy Resources with Associated Electric Power Systems Interfaces* [5]; IEC TS 62257-2015. *Recommendations for Renewable Energy and Hybrid Systems for Rural Electrification* [6]; and IEC TS 62898, *Microgrids – Parts 1 & 2* [7,8].

11.2 Microgrid components

According to the microgrid definitions given earlier, the main components as suggested in are presented as follows [1,9,10]:

- DG;
- ES;
- DERs;
- loads (traditional loads, plug-in electric vehicles, and charging stations);
- automatic meter reading (AMR)/advanced metering infrastructure (AMI);
- control devices;
- point of common coupling (PCC); and
- protection devices.

Additionally, the information and communication technology plays a relevant role in modern electric power networks. Several technologies have been used or tested in distribution systems and microgrids such as AMI and smart meters, which can use wireless, optical communication, and power line communication. Some common requirements on the communication system are defined to achieve bidirectional, real-time, and efficient communication. The communication system should be established on the basis of open and common standards, which may support high-speed and accurate communication between sensors, advanced electronics and applications [9].

11.3 Classification of microgrids

The previous definition of a microgrid described a group of interconnected loads with DERs and control devices. Nevertheless, it does not describe the way how these components are related to or how they interact with the main grid or among them. Different types of microgrids should be established according to their configuration, capacity, and type of DG technologies. Therefore, microgrids can be classified by their configuration, AC/DC type of connection, and their main feeder location and operation mode [1,9,11].

11.3.1 Classification by configuration

Microgrid can be classified according to the type of configuration and the load associated as consumer microgrids, community microgrids, and utility microgrids [9,12].

- Consumer microgrids are composed by few DGs, simple control functions, and design for a single consumer. Normally, they are managed by the same customers.
- Community microgrids require a more complex design and operation than the consumer microgrid. This type has multiple-consumer side, and it is normally consumer owned. Some loads are defined as "sheddable" in the case of a frequency issue to maintain power balance.
- Finally, utility microgrids are those which loads are prioritized based on user's requirements on reliability, and high-priority loads will be powered preferentially in an emergency. The supply resources are provided on the utility side with consumer interactions. They are organized according to the utility objectives.

11.3.2 Classification by AC/DC type

According to the type of connection, microgrids are divided into DC, AC, and AC/DC or hybrid microgrids [9]. DC microgrid, the generation, and consumption units are connected to a DC bus by using converters. An AC microgrid is connected to the distribution network via an AC bus, which is connected to the distribution system through the PCC. A hybrid microgrid combines AC buses and DC buses according to the nature of loads connected. This integrates the advantages of both AC and DC microgrids.

On the one hand, DC microgrids depend on DG voltage, which makes a coordinated operation of the DGs easier. Also, DC microgrid does not involve any synchronization process with the electric grid, which simplifies the control process. However, most of the loads in the current system are AC loads, which implies the use of more converters.

On the other hand, AC microgrids require no inverter for power supply to AC loads, which makes this type of microgrid the most common used around the world. Since it operates with the electric grid, this type requires the inclusion of a synchronization process in the control process.

11.3.3 Classification by modes of operation

A microgrid, in general, can operate either in grid-connected mode or islanded mode [9]. The first mode of operation indicates that the microgrid is connected with the main power grid, while the other one refers to a completely independent operation from the rest of the electric grid. There is an intermediate operation mode stage, in which the microgrid changes from the grid-connected mode to the islanded mode of operation and vice versa. This is done considering a series of steps to achieve the transition of operational modes correctly and without affecting the components of the microgrids.

Some of the advantages of grid-connected mode include economic optimal dispatch by using resources from the distribution network, automatic voltage control, and reactive power control. Also, it makes use of the ES to minimize total feeder maximum demand and to minimize the ratio between total feeder maximum and minimum demand.

The islanded mode is used intentionally or unintentionally. For the latter, predetermined control and unintentional islanding strategies are defined in order to prevent unplanned or uncontrollable behaviors.

11.3.4 Classification by feeder location

Based on the characteristics and properties of feeders, a microgrid can be categorized in urban, rural, and off-grid microgrids [11].

Some of the urban microgrid characteristics include a high-density balance load structure around the system. The short circuit ratio between the system grid short circuit capacity at the PCC and the total DER generation capacity of the microgrid is regularly above 25.

Another type of microgrids is the rural microgrids, in which load is scattered around the system. The short circuit ratio is not as high as it is for the urban microgrid. However, it presents voltage imbalance and fluctuation. Therefore, DER units can be used to regulate the feeder voltage.

Finally, an off-grid microgrid operates in islanded mode since it is located far from the main grid or is surrounded by a difficult environment for transmission line connection. This makes that it cannot be strictly considered as “microgrid.” Nevertheless, strategies for off-grid microgrids are fully applicable to islanded mode.

11.4 Microgrid control

As was stated in the previous section, there are many possible configurations for microgrids, where the main objective is the coordinated operation of both energy supply and demand in order to guarantee voltage and frequency stability. Therefore, microgrid control has an important role in order to guarantee operational conditions. There is no general structure of control architecture for microgrids because it depends on the type of microgrid or the existing infrastructure. Typical control actions are network reconfigurations by switching operation and voltage control via capacitors switching or transformers tap changing.

A typical distribution network with a high penetration of DGs is shown in Figure 11.2 [1]. The distribution management system (DMS) defined the way how the system is controlled and managed, which monitors the HV/MV sections. This control system is not in charge of controlling the DGs or loads. However, it can control large installations in certain specific scenarios. The other required element for controlling is the AMR/AMI system, which is responsible for the collection of electronic meter readings. When there are integrated smart meters, the data management capabilities are increased, which also enables a two-way communication between utilities and customers. The distribution system operator is the entity that manages and controls the distribution system and collects the energy metering data. It sends the measured data to the Energy Service Company (ESCO), which is responsible for billing.

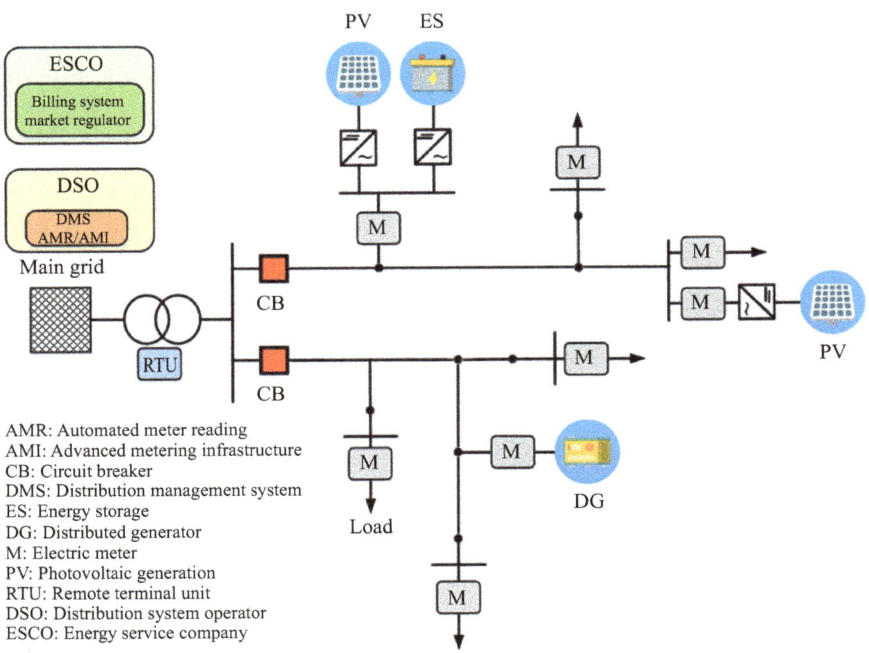

Figure 11.2 Typical distribution system management structure

The previous structure just mentioned one layer required for microgrid control. Local control levels at DGs and loads are required in order to enable advanced market participation. Also, scalability capacities are expected to integrate a large number of users. Finally, plug-and-play characteristics are expected to enable new participants and integrate new functionalities and business cases.

Two control structures can be identified for microgrid control: centralized and decentralized.

11.4.1 Centralized control

In a centralized control, the microgrid has a three-layer control scheme as indicated in Figure 11.3 [1,9,13]. The first layer called *local control layer* is the bottom layer composed of a local controller, which is in charge of primary voltage and frequency

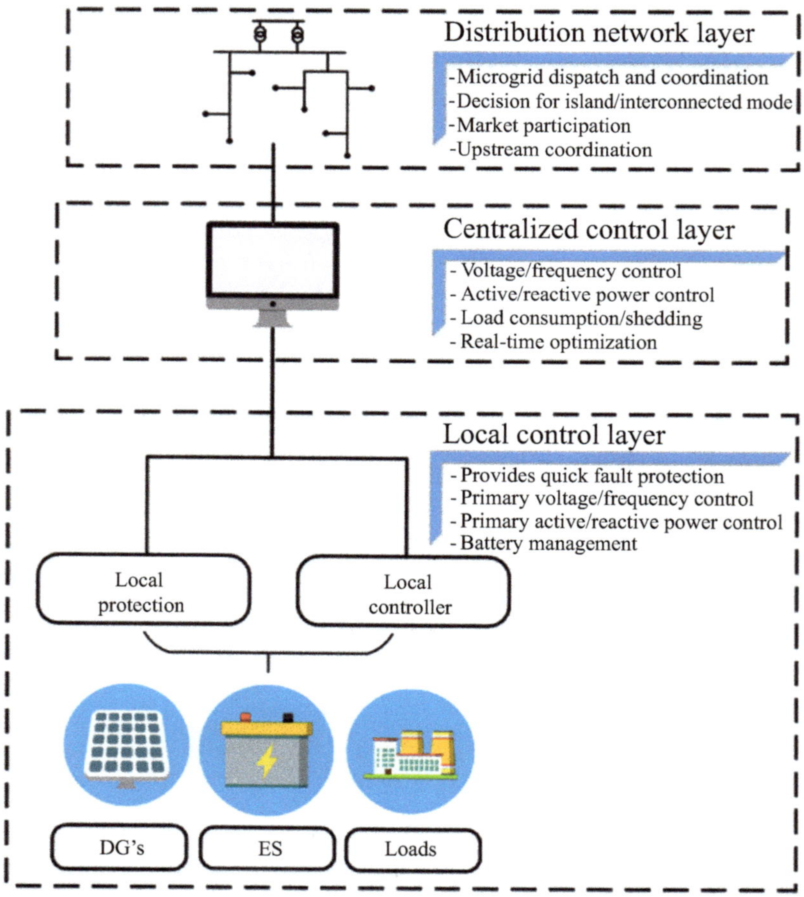

Figure 11.3 Centralized control structure

regulations. It also has a local protection that provides fault protection for the microgrid. The middle layer called *centralized control layer* is the microgrid control center and the core of the microgrid control system. During grid-connected operation, this control layer tries to bring the best performance of the microgrid. When the microgrid operates in the islanded mode, it gives priority to stability and safe operation of the microgrid. Finally, the *distribution network layer* is the last layer, which is coordinated by the distribution network, and it is in charge of dispatching the microgrid in order to maximize its profit and provide ancillary services to the main grid. In order to achieve this, this layer relates microgrids to the market according to their participation.

The advantage of a centralized control system is that all data is centralized and allows defining global objective in the control approach, which results in high controllability of the system. However, the computational burden is high, as the optimization is computed based on a large amount of information. Moreover, there is a single point of failure, and redundancy of the central controller is expensive. The loss of communication may cause a shutdown of the overall system.

11.4.2 Decentralized control

In a decentralized control, the microgrid demand is defined in the control goal by the collaboration of different controllers [1,13]. Aspects like the market prices and status of the microgrid elements are considered in the definition of the control approach, but the control divides the problem into local controllers or agents. They together achieve an optimal state through a hierarchical system coordination of all controllers. There is less information required at the central controller of different DG units of the microgrid, which results in a more scalable system. A central agent communicates global constraints to the other agents, such as transfer power limits or voltage limits. The information can be flowing around different agents when a single communication channel fails.

Table 11.1 compares the characteristics of both types of control.

11.5 Microgrid protection

A microgrid is also required to operate securely and stably. Its relay protection should provide reliability, response speed, flexibility, and selectivity, as is shown in Figure 11.4. When the microgrid is in grid-connected mode, power can flow bidirectionally between the distribution network and the microgrid. In addition, being different from traditional rotating generators, the microgrid also uses power electronic technologies, which brings a difference to the relay protection settings [9].

A traditional distribution system is a radial one-end-source system, which requires no directional element for feeder protection and is usually provided with traditional current protections. In microgrids, this configuration can change since multiple power sources and no-radial configurations are available. Islanding protection, overcurrent protection, and recloser for synchronous grid connection are the main protections provided at the PCC.

Table 11.1 Microgrid control methods comparison [1]

	Centralized control	**Decentralized control**
DG ownership	Single owner	Multiple owners
Goals	A clear, single task, e.g., minimization of energy costs	Uncertainty over what each owner wants at any particular moment
Availability of operating personnel (monitoring, low-level management, special switching operations, etc.)	Available	Not available
Market participation	Implementation of complicated algorithms	Owners unlikely to use complex algorithms
Installation of new equipment	Requirements of specialized software	Should be plug-and-play
Optimality	Optimal solutions	Mostly suboptimal solutions
Communication requirements	High	Modest
Market participation	All units collaborate	Some units may be competitive
Microgrid operation is attached to a larger and more critical operation	Possible	Not possible

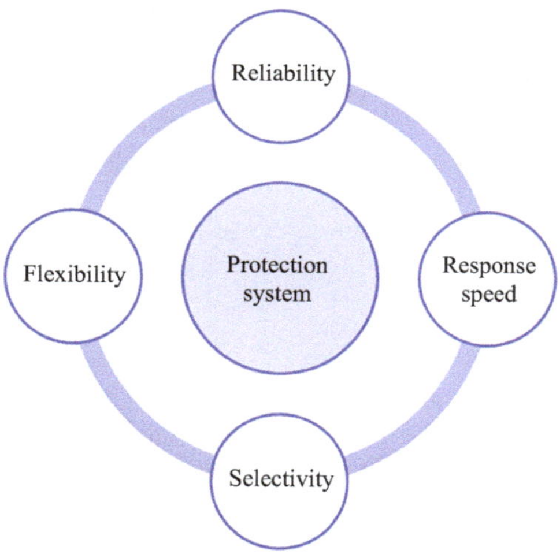

Figure 11.4 Protection system features

After the event of a fault inside a microgrid, the connected DGs contribute to the fault current, which makes varying the short circuit level on each node. The type, location, and capacity of the DGs affect the operation of the relays in the conventional distribution networks and the microgrid.

When the microgrid is in island mode, the short circuit level is reduced due to the low short circuit contribution from the inverter-based DGs. The following example illustrates the short circuit level variation due to the presence of DGs.

Example 11.1 A short circuit analysis was performed for the microgrid shown in Figure 11.5 where the following scenarios were considered:

Scenario 1: All the generation units are turned off.
Scenario 2: All the generation units are turned on.
Scenario 3: All the generation units are turned on and the system is operating disconnected from the main grid.

The analyses were carried out with PSAT software and the results are summarized in Figures 11.6 and 11.7, which show that the short circuit level increases with the integration of distribution generators. As the fault current contribution depends on the generation unit technology and its capacity, the short circuit levels of system operating disconnected from the main substation are significantly less than the system operating with all the components connected. These variations can lead to a misoperation of the protection devices.

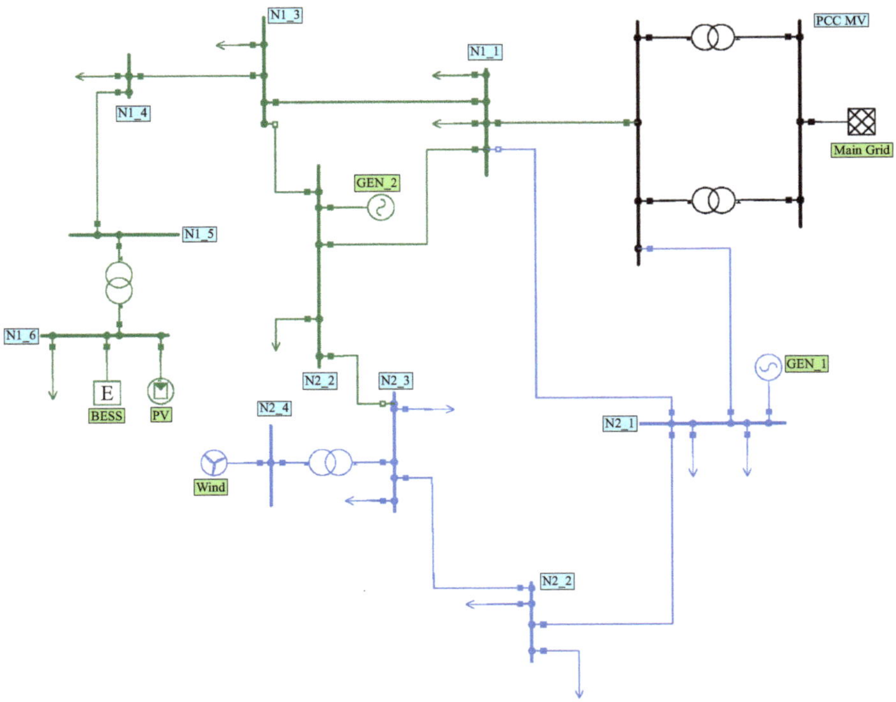

Figure 11.5 Microgrid for Example 11.1

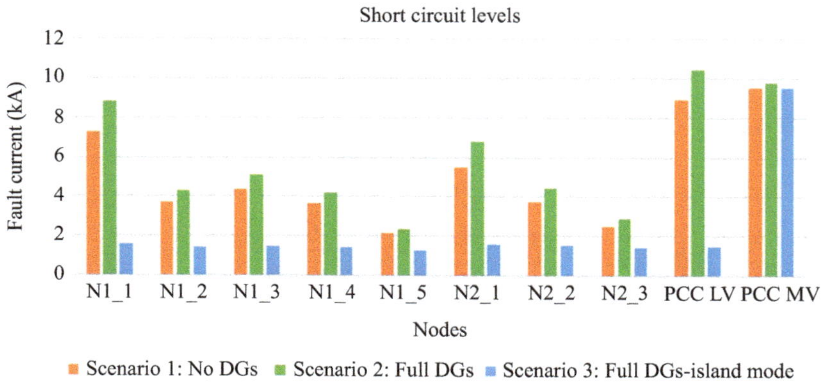

Figure 11.6 Short circuit analysis for three different scenarios for Example 11.1

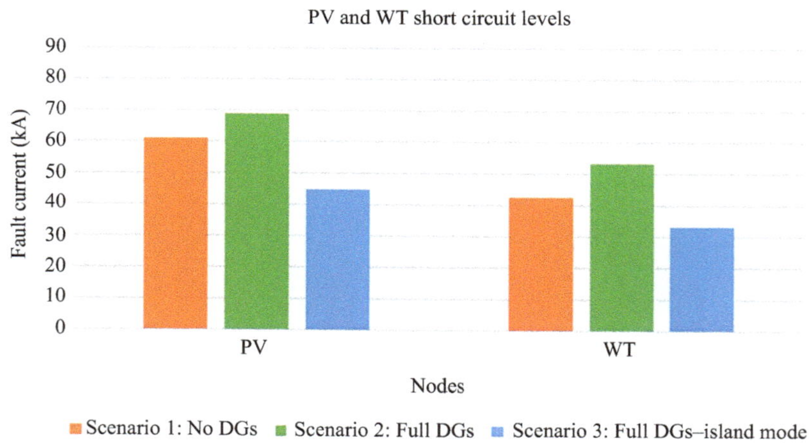

Figure 11.7 Short circuit analysis for three different scenarios for the PV and wind turbine nodes for Example 11.1

The protection equipment will operate incorrectly due to the short circuit levels variations. To overcome this, it is essential to use adaptive protection, which can be implemented by using the multiple setting groups feature included in most numerical relays [14].

11.6 Benefits of microgrids

The optimal operation scheduling in microgrids can have economic, technical, and environmental objectives [1]. Figure 11.8 shows the different microgrid operation strategies.

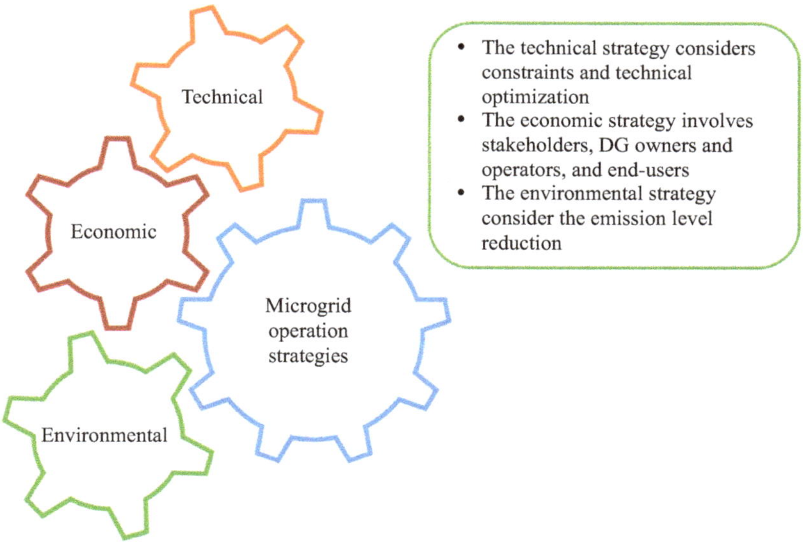

Figure 11.8 Microgrid operation strategies

In the planning or operation process, four different microgrid operational objectives can be identified: economic, technical, environmental, and combined objectives.

The economic objective focuses on minimizing total costs regardless of network impact/performance. This option may be envisaged by DG owners or operators. DGs are operated without concern for grid or emission obligations.

The technical objective focuses on optimizing network operation without any consideration of DG production costs and revenues. This means minimizing power losses, voltage variation, and device loading. This option might be preferred by system operators.

The environmental objective focuses on dispatching DG units with lower specific emission levels with higher priority, disregarding financial or technical aspects. This is preferred for meeting environmental targets and is currently mainly supported by regulatory schemes.

Finally, the combined objectives solve a multi-objective DG optimal dispatch problem, considering some of the objectives defined before. It converts technical and environmental criteria into economic equivalents, considering constraints from both network and DG physical limits.

Some of the most relevant benefits associated with each option will be discussed in the following sections.

11.6.1 Economic benefits of a microgrid

The economic values created by a microgrid can be roughly categorized into providing locality and selectivity benefits. Locality benefit is mainly attributed to the creation of an internal "over-the-grid" energy market within the microgrid, where micro source units could sell at prices higher than the prices at wholesale level and end consumers could buy at prices lower than the retail level. This is equivalent to a market policy of a good citizen behavior. Selectivity benefit is associated with the optimization of real-time dispatch decisions that minimize opportunity cost of the whole microgrid with a consideration of technical and environmental constraints. This is equivalent to a market policy of an ideal citizen behavior.

Both locality and selectivity benefits can be attributed to either the consumer side or the micro source side, where a proper market design/interest allocation mechanism is needed to ensure a reasonable split of both benefits between consumer and micro source owner. From a macro-economic point of view, the locality benefits of a microgrid can be grouped into two aspects:

- Microgrid can act as an initiator of local retail and service markets.
- Microgrid can act as a hedging tool against potential risks of price volatility, outage, load growth, and so on.

The selectivity benefit of a microgrid can also be further categorized into the following two aspects:

- Microgrid can act as an aggregator of both supply- and demand-side players.
- Microgrid can serve as an interest arbitrator for different stakeholders.

11.6.2 Technical benefits of a microgrid

A microgrid can potentially improve the technical performance of the local distribution grid mainly in the following aspects:

- Large transmission build-out may be reduced and transmission losses can be reduced.
- Improved voltage quality via coordinated reactive power control and constrained active power dispatch.
- Relief of congested networks and devices, for example, during peak loading through selective scheduling of micro source outputs.
- Microgrid can facilitate the use of renewable energy sources and their outputs can be scheduled.
- Enhancement of supply reliability via partial or complete islanding during loss of main grid. When the total number of microgrids reaches a sufficiently high share in LV substations, similar technical benefits can be expected in upstream grids as a consequence of multi-microgrid operation.
- Power generation units are small and are located in close proximity to load.
- The use of both electricity and heat permitted by the close proximity of the generator to the user can increase the overall energy efficiency.

The current level of technical benefits depends mainly on two factors: the optimality of micro source allocation and the degree of coordination among different players. Effective planning of micro source size and location can maximize unit contribution to system performance and random interconnection of oversized micro sources in weak grids, but it can create more technical problems than the benefits it provides. It is also critically important that a real-time, multi-unit coordination platform, in either centralized or decentralized form is needed to maintain targeted microgrid technical performance at all times.

Example 11.2 To illustrate the abovementioned technical benefits, the following calculations were performed to the microgrid illustrated in Figure 11.9 with PSAT:

1. Total losses in the system.
2. The best topology for the system considering loss reduction and a radial configuration.
3. List of the lines whose ends have to change the status.
4. Comparison of the total losses results with those obtained for a mesh network.

---

1. Total losses in the system:

   The total losses and losses by feeder for the given system were calculated running a load flow analysis. The results are indicated in Table 11.2.

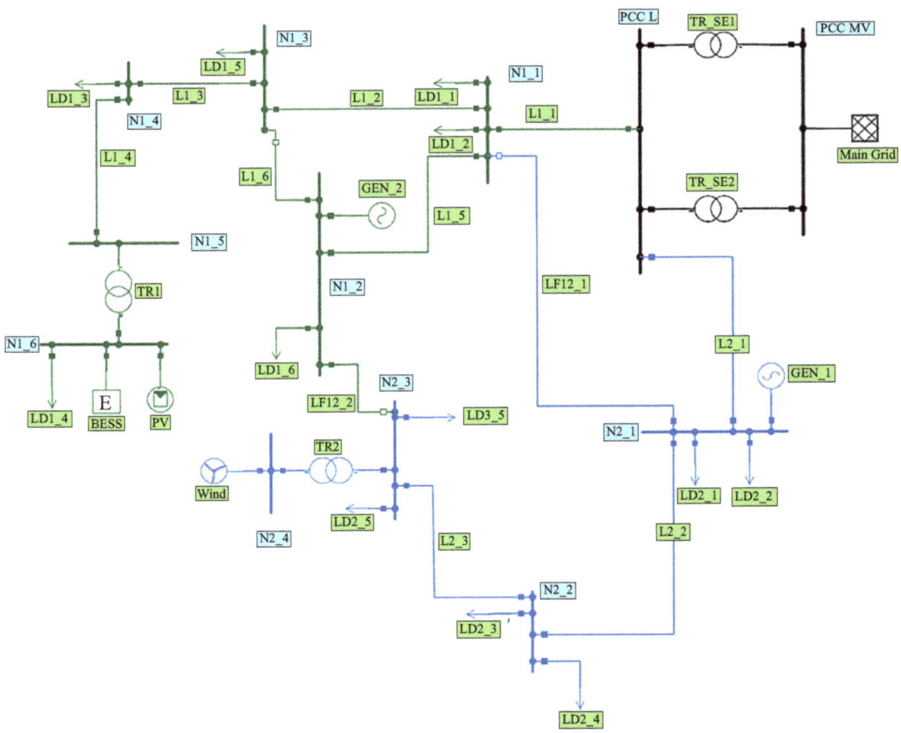

Figure 11.9 Two-feeder microgrid for Example 11.2

Table 11.2 Total losses for the given system in Example 11.2

	P_{Loss} (MW)	Q_{Loss} (MVAR)
Network	0.201	0.343
Feeder_01	0.068	0.077
Feeder_02	0.127	0.125

Table 11.3 Element losses for the given system in Example 11.2

U_n (kV)	$P_{Loss\ Line}$ (MW)	$Q_{Loss\ Line}$ (MVAR)	$P_{Loss\ Transformer}$ (MW)	$Q_{Loss\ Transformer}$ (MVAR)
20	0.193	0.097	0.002	0.106
110	0	0	0.006	0.14

Table 11.3 indicates the total losses for the elements in the given system at the different voltage levels.

2. The best topology for the system considering loss reduction and a radial configuration:

   The best topology was determined with a switching optimization for a system reconfiguration using PSAT. Figure 11.10 illustrates the best topology considering loss reduction and a radial configuration.

   The total losses and losses by feeder for the new topology were calculated running a load flow analysis. The results are indicated in Table 11.4.

   Table 11.5 indicates the total losses of the elements for the reconfigured system at different voltage levels.
3. List of the lines whose ends have to change the status:

   Figure 11.11 illustrates the lines whose ends have to change their status in order to allow the system's reconfiguration for its optimal topology.
4. Comparison of the total losses results with those obtained for a mesh network:

   Table 11.6 indicates the total losses obtained for the given system, the best topology, and for the system in mesh configuration. As can be seen, the mesh configuration allows lower losses in the system.

11.6.3 Environmental and social benefits of a microgrid

Environmental benefits of a microgrid can be expected from two aspects: shift toward renewable or low-emission fuels and adoption of more energy-efficient energy supply solutions.

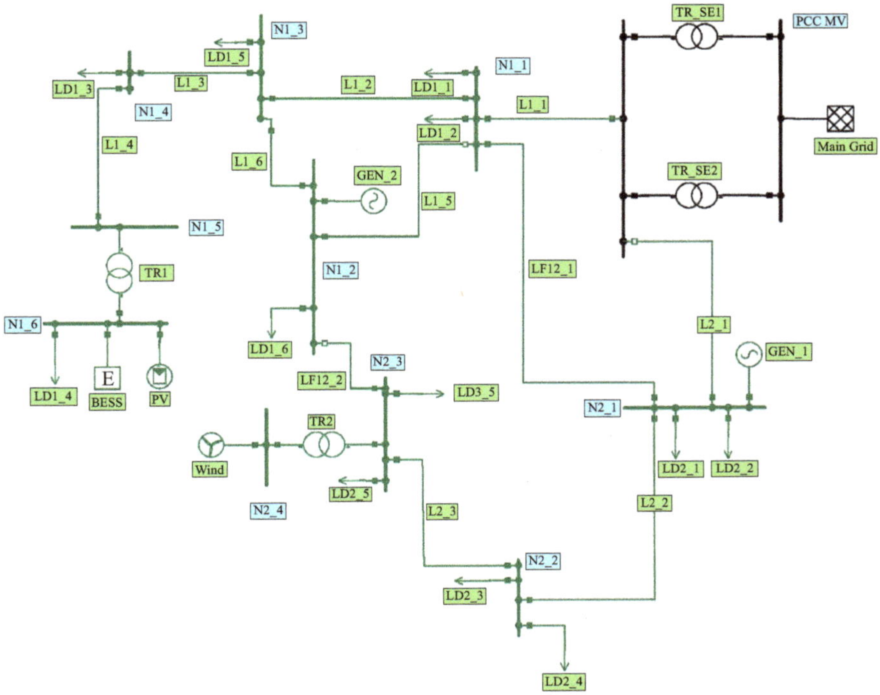

Figure 11.10 Reconfigured microgrid for Example 11.2

Table 11.4 Total losses for the new topology for Example 11.2

From area/zone	P_{Loss} (MW)	Q_{Loss} (MVAR)
Network	0.125	0.318
Feeder_01	0.119	0.177
Feeder_02	0	0

Table 11.5 Line losses for the new topology for Example 11.2

U_n (kV)	$P_{Loss\ Line}$ (MW)	$Q_{Loss\ Line}$ (MVAR)	$P_{Loss\ Transformer}$ (MW)	$Q_{Loss\ Transformer}$ (MVAR)
20	0.117	0.071	0.002	0.106
110	0	0	0.006	0.14

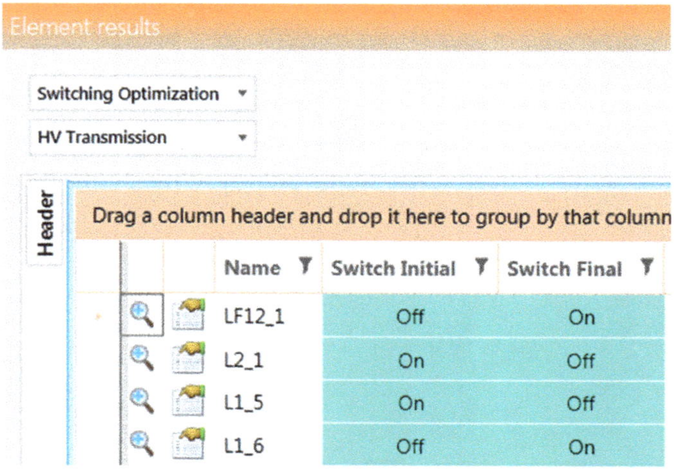

Figure 11.11 Lines whose ends have to change their status

Table 11.6 Total losses comparison for the given system, the new topology, and mesh configuration for Example 11.2

	P_{Loss} (MW)			Q_{Loss} (MVAR)		
	Initial network	**New topology**	**Mesh network**	**Initial network**	**New topology**	**Mesh network**
Network	0.201	0.125	0.089	0.343	0.318	0.292
Feeder_01	0.068	0.119	0.064	0.077	0.177	0.14
Feeder_02	0.127	0	0.019	0.125	0	0.012

With supportive policies for distributed renewable resources, the fuel-switching credit of microgrid is expected to grow as renewable energy source costs decrease over the years.

Social benefits of microgrids include the following:

- Large land use impacts are avoided.
- Raising public awareness and fostering incentives for energy saving and greenhouse gas emission reduction.
- Microgrids can provide electric service to regions and communities that are currently unserved.
- Creation of new research and job opportunities.

11.7 Proposed exercises

1. List the components of a microgrid and give a brief explanation of each one. A section of a typical distribution system is shown in Figure 11.12. Indicate the missing elements required to consider this system as a microgrid.
2. According to the main characteristics of microgrids, give comments on how the power will flow and how the voltages and losses may vary due to its integration into the distribution system and their main reasons.
3. For the microgrid shown in Figure 11.13, determine if the generated energy meets the energy that is being consumed when the microgrid operates disconnected from the main grid. Consider the following:
   (i) Use the load profile for all loads in the microgrid. This profile provides the consumed power in pu.

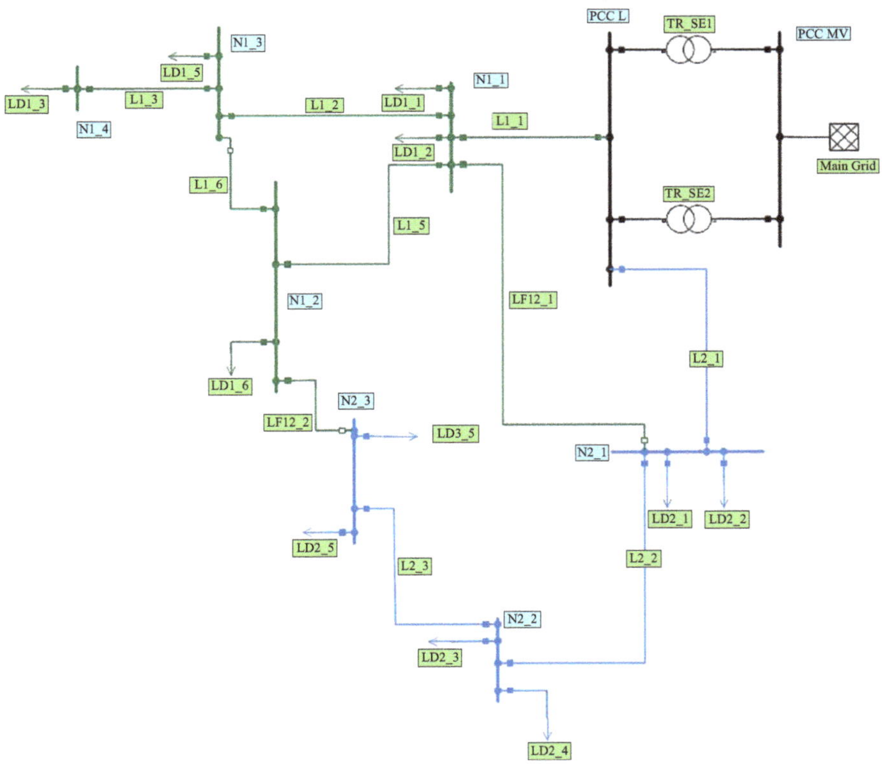

Figure 11.12 Radial distribution system for exercise 1

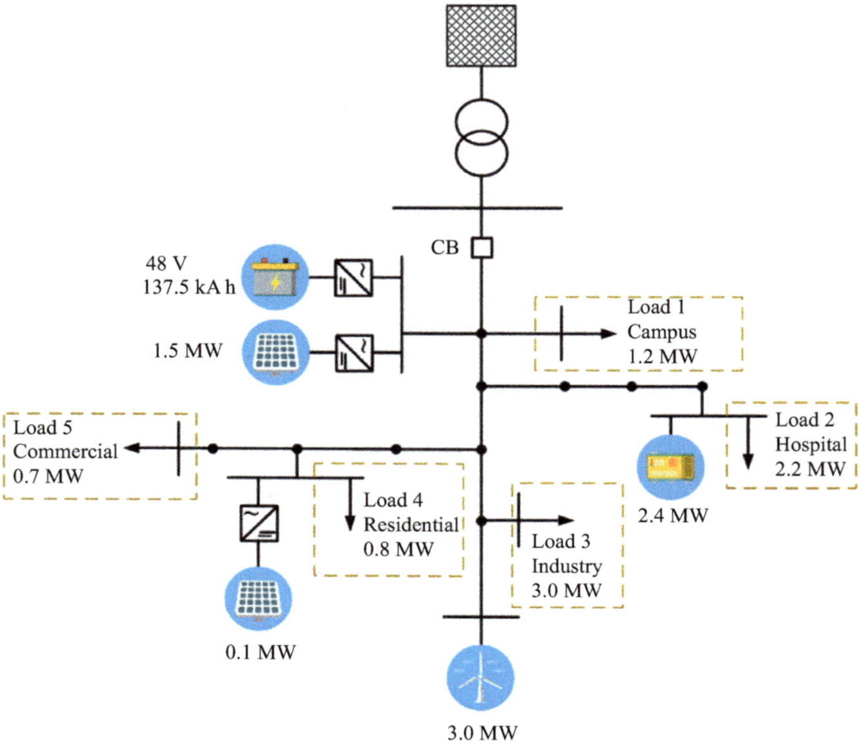

Figure 11.13 Microgrid for exercise 2

(ii) Use the generation profile for the generation units in the microgrid. This profile provides the output power in pu.
(iii) Compute the produced and consumed energy for a period between 6.00 AM and 8.00 PM of one day.
(iv) Calculate the amount of energy that can be provided by the ES system to the unsupplied load.

The reference load and generation profiles are indicated in Figure 11.14 and the values to reproduce the curves above are shown in Table 11.7.

4. Consider the microgrid that is shown in Figure 11.15 with the tie and section already switches located. For each one of the faults F_1, F_2, F_3, and F_4, indicate the operating sequence of the tie and section switches in order to restore service to the highest possible number of customers, as it is requested in Table 11.8.
5. How the protection selectivity is affected after a fault happens and the system is reconfigured in a microgrid, and what actions should be carried out to guarantee the proper selectivity for each scenario?

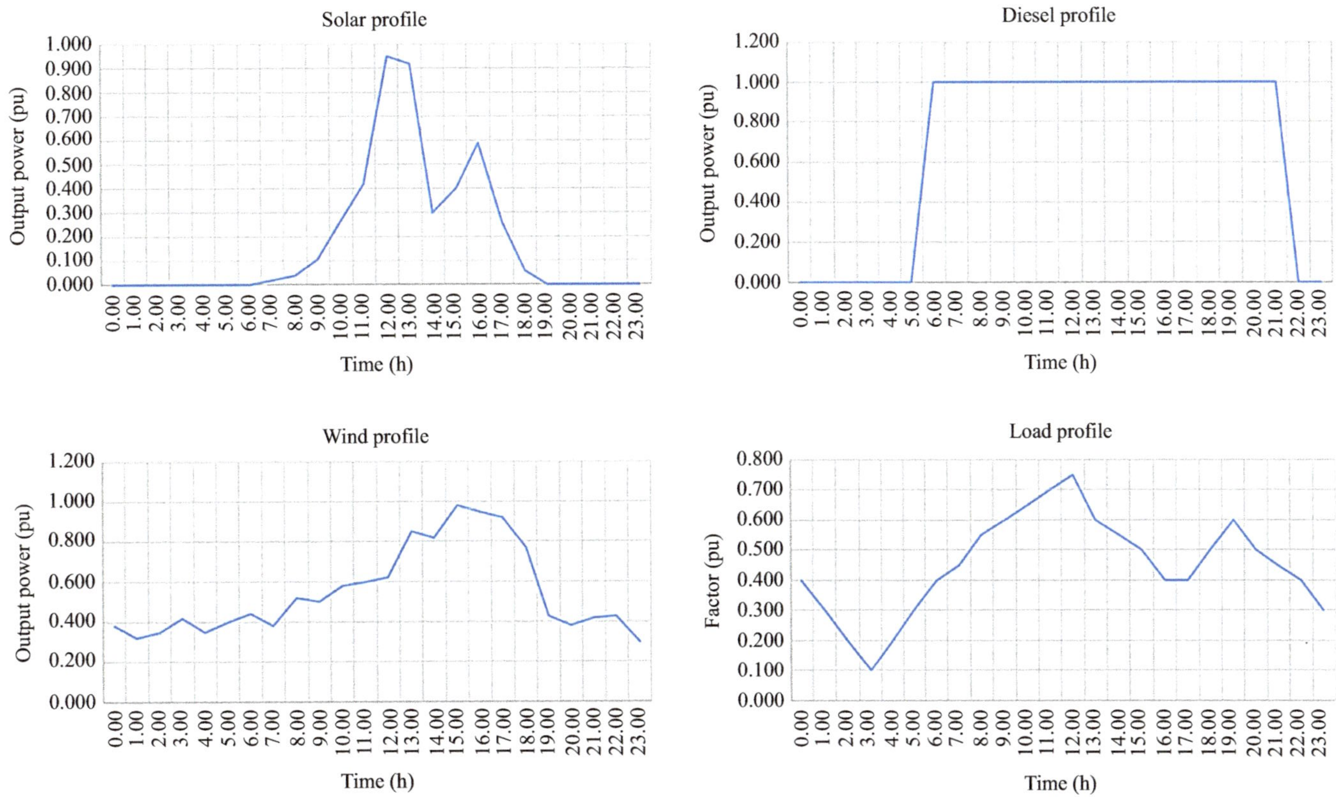

Figure 11.14 Load and generation profiles for exercise 2

Table 11.7 Load and generation profiles data for exercise 3

Time (h)	Factors (pu)				Time (h)	Factors (pu)			
	Solar	Wind	Diesel	Load		Solar	Wind	Diesel	Load
0.00	0.000	0.380	0.000	0.400	12.00	0.950	0.620	1.000	0.750
1.00	0.000	0.320	0.000	0.300	13.00	0.920	0.850	1.000	0.600
2.00	0.000	0.350	0.000	0.200	14.00	0.300	0.820	1.000	0.550
3.00	0.000	0.420	0.000	0.100	15.00	0.400	0.980	1.000	0.500
4.00	0.000	0.350	0.000	0.200	16.00	0.590	0.950	1.000	0.400
5.00	0.000	0.400	0.000	0.300	17.00	0.260	0.920	1.000	0.400
6.00	0.000	0.440	1.000	0.400	18.00	0.060	0.770	1.000	0.500
7.00	0.020	0.380	1.000	0.450	19.00	0.000	0.430	1.000	0.600
8.00	0.040	0.520	1.000	0.550	20.00	0.000	0.380	1.000	0.500
9.00	0.110	0.500	1.000	0.600	21.00	0.000	0.420	1.000	0.450
10.00	0.270	0.580	1.000	0.650	22.00	0.000	0.430	0.000	0.400
11.00	0.420	0.600	1.000	0.700	23.00	0.000	0.300	0.000	0.300

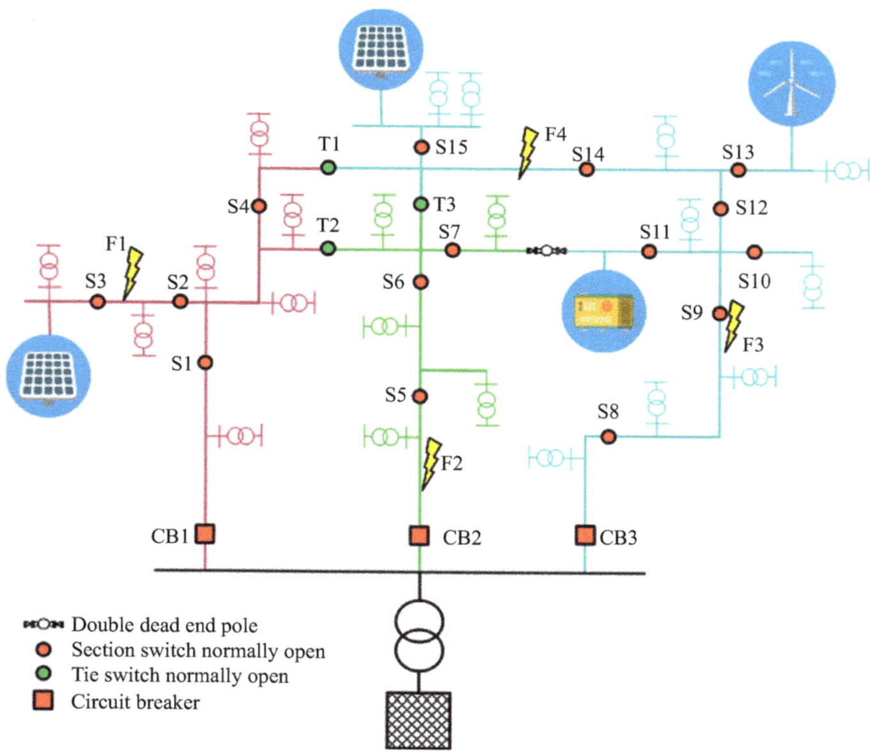

Figure 11.15 Microgrid for exercise 4

Table 11.8 Switches operation sequence for exercise 4

Fault	Open	Close	Total affected customers
F_1			
F_2			
F_3			
F_4			

References

[1] Hatziargyriou, N. D. Microgrids. Architectures and Control. Wiley and Sons, Chichester, UK, 2014.

[2] Office of Electricity Delivery and Energy Reliability Smart Grid R&D Program. DOE Microgrid Workshop Report. San Diego, CA, USA, 2011.

[3] IEEE. IEEE Std. 446-1995. Recommended Practice for Emergency and Standby Power Systems for Industrial and Commercial Applications, vol. 1995. IEEE, New York city, NY, USA, 1996.

[4] IEEE. IEEE Std. 519-2014. Recommended Practices and Requirements for Harmonic Control in Electric Power Systems. IEEE, New York city, NY, USA, 2014.

[5] Institute of Electrical and Electronics Engineers. IEEE Std. 1547-2018. Standard for Interconnection and Interoperability of Distributed Energy Resources with Associated Electric Power Systems Interfaces, *IEEE Std 1547-2018 (Revision of IEEE Std 1547-2003)*. IEEE, New York city, NY, USA, pp. 1–138, 2018.

[6] IEC. "IEC TS 62257-2015. Recommendations for Renewable Energy and Hybrid Systems for Rural Electrification." 2015. [Online]. Available: https://webstore.iec.ch/publication/23502. [Accessed: 25-Sep-2019].

[7] IEC. IEC TS 62898-1:2017. Microgrids – Part 1: Guidelines for Microgrid Projects Planning and Specification. IEC, Geneva, Switzerland, p. 33, 2017.

[8] IEC. IEC TS 62898-2:2018. Microgrids – Part 2: Guidelines for Operation. IEC, Geneva, Switzerland, p. 38, 2018.

[9] Fusheng, L., Ruisheng, L., and Fengquan, Z. Microgrid Technology and Engineering Application. Elsevier, Amsterdam, 2016.

[10] Zambroni de Souza, A. and Castilla, M. Microgrids Design and Implementation. Springer, Cham, Switzerland, 2019.

[11] Hooshyar, A. and Iravani, R. "Microgrid protection," Proceedings of the IEEE, vol. 105, no. 7, pp. 1332–1353, 2017.

[12] Miller, J. Power System Optimization. Smart Grid, Demand Dispatch and Microgrids. National Energy Technology Laboratory. Conference proceedings, 2011.

[13] Almasalma, H., Engels, J., and Deconinck, G. Peer-to-Peer Control of Microgrids. Proc. IEEE YRS 2016. IEEE Benelux PELS/PES/IAS Young Researchers Symposium. Eindhoven, Netherlands, 12-13 May 2016 arXiv:1711.04070v1.

[14] Gers, J. and Holmes, E. Protection of Electricity Distribution Networks, 3rd ed. Institution of Engineering and Technology, IET, Stevenage, 2011.

Chapter 12

Communications in Smart Grids

For years, analogue communication networks have provided channels that allowed long-distance information exchange. However, with the arrival of digital communications, it was possible to employ physical mediums that have their own capacity to transfer large data packets in a reliable and efficient manner.

Digital communication networks and data services have provided solutions that optimize thousands of processes worldwide. In the smart network environments, such solutions have proved successful. A good example of this is the application on protective relaying. It is known that protection functionalities have maintained the theoretical foundations since they were first developed.

However, the operation of protective relaying has been greatly improved in time response due to the availability of much faster relays based on the numerical technology as well as the wonderful development of communication capabilities that have been attained in recent years.

The standardization of protocols brought a convergent platform to implement solutions based on open and flexible solutions. However, since high levels of reliability, availability, and security are required by electrical system networks, the protocols needed to be adapted to offer lower levels of latency and larger broadband.

In the following sections, reference will be made to important topics pertaining to the effect of communications on distribution automation and in general on Smart Grid, including the OSI model, the most popular protocols, and the transmission mediums. Special attention is given at the end of the chapter to the IEC 61850 Standard, considering its huge impact in all aspects of power system automation handling.

12.1 ISO–OSI model

The open system interconnection reference model (the OSI model) was developed in the late 1970s by the International Organization for Standardization (ISO) for open systems interface [1]. It has served as a conceptual model to represent interactions and events necessary for establishing a communication channel between two data terminals.

Figure 12.1 shows the various layers. Each layer constitutes a stage where an activity takes place to store the information in segments, packets, branches, and bits until reaching the physical layer where the analogue wave is transmitted with the corresponding digital information to its destination. The opposite process is used to interpret the message.

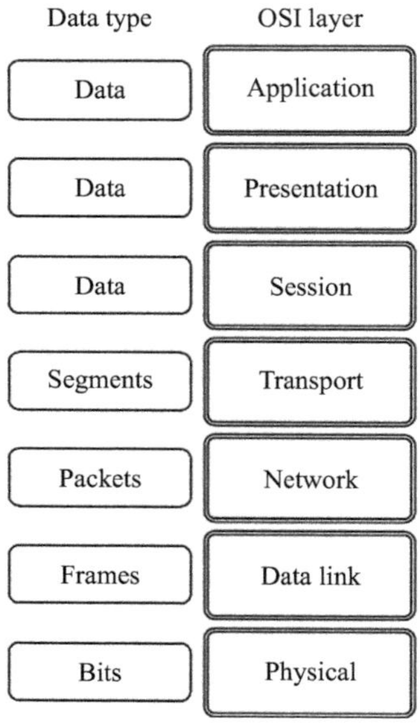

Figure 12.1 OSI model

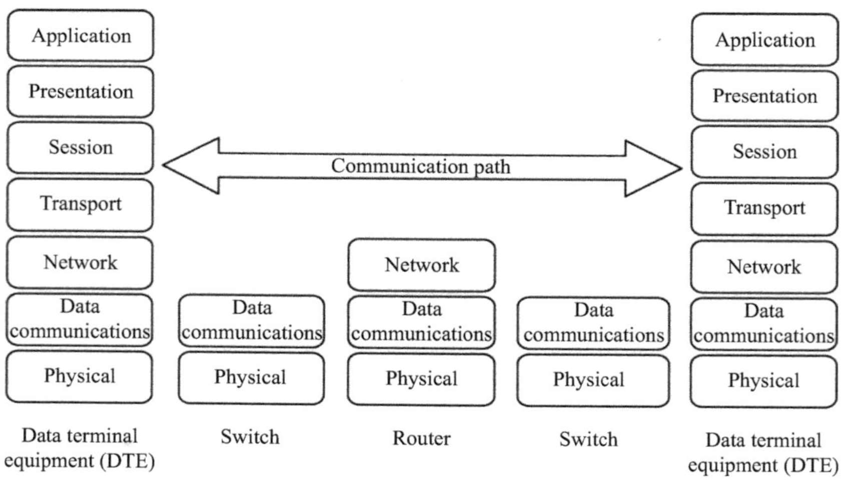

Figure 12.2 TCP/IP link applying OSI model

Figure 12.2 illustrates how the OSI model is applied to an Internet network. It is based on the TCP/IP protocols that are closely related to the OSI model. The intermediate communication elements such as switches and routers are illustrated in the center row. They enable addressing and using efficiently the network elements like switches and routers.

12.2 Communication solutions for the power system world

The spectrum of Smart Grid solutions increases, and each new solution involves additional communication solutions for each particular case. Traditionally, communications, like wired communication between field equipment, auxiliary services, and controllers, serial communication between intelligent electronic devices (IEDs), communication between equipment in different places, and communication through wide area network (WAN) for the integration of substations in a control center station, are considered in the power system environment. Communications have been converted to the convergence of multiple designs appropriate to each domain in an information model offering interoperability between systems and a distributed computing system to reach optimum levels of usage. Consequently, standards such as Common Information Model, specified in IEC 61968 and IEC 61970, were developed [2,3].

Figure 12.3 illustrates sample communication solutions applied to Smart Grid domains. This figure shows a communication infrastructure for Smart Grid that enables utilities to interact with their devices and grid systems as well as with customers and facilities for distributed generation and energy storage. To fully

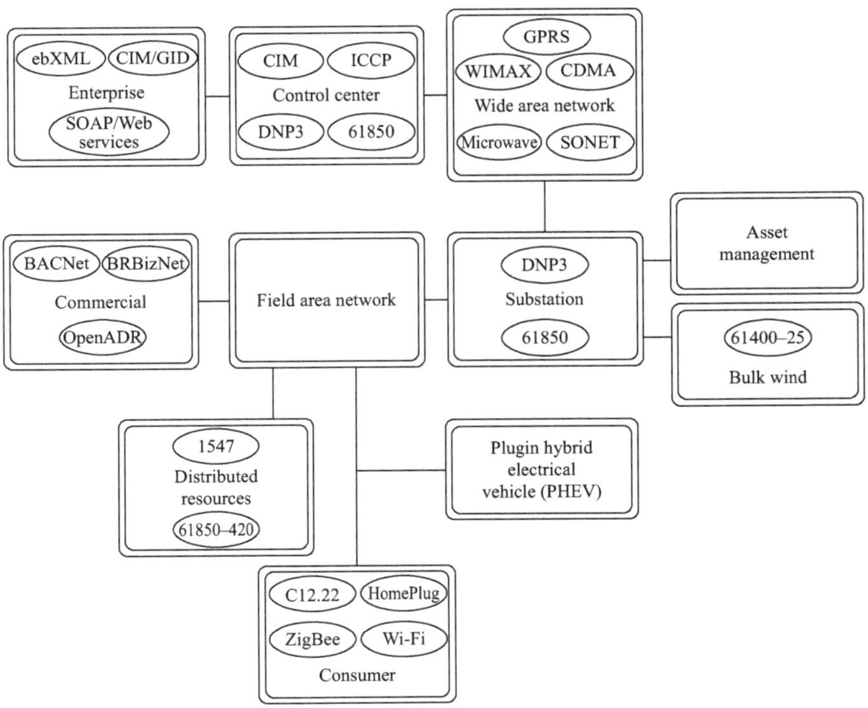

Figure 12.3 Communication options in the Smart Grids

achieve the vision of a Smart Grid, companies need to support multiple communication networks: home area network for energy efficiency on the client side, neighborhood area network for advanced measurement applications, and WAN for distributed automation and Smart Grid backbone.

Communication applications have evolved rapidly in Smart Grid, particularly in two areas: intelligent metering infrastructure and distribution network, and substation communications.

12.2.1 Communication solutions in AMI

Advanced metering infrastructure (AMI) provides multiple solutions like phone lines, GPRS/3G/4G, PLC, fiber optic, and Ethernet, which could employ traditional proprietary protocols and, due to the Smart Grid demands, turn them into DLMS (Device Language Message Specification)/COSEM open standards for data interchange with metering equipment.

Other emerging technologies worth mentioning are those implemented under the Internet of Things (IoT) philosophy, which combine smart meters and cloud computing. They are supported by low-speed wireless communications, which use the existing cellular network or the satellite communications. This provides a complete platform that integrates the collection and analysis of large amount of data in real time using the power of big data and cloud computing. Some of these technologies include LoRaWan, SigFox, NB-IoT, among others. NB-IoT unlike SigFox or LoRaWan is not IP based which allows it to handle more data.

DLMS is an application level protocol integrated in IEC 62056 for device message specification [4]. Similarly, COSEM is defined as companion specification for energy metering.

12.2.2 Distribution network communications

12.2.2.1 IEC 61850

One of the smart network objectives is the interoperability among systems, as is defined by the National Institute of Standards and Technology (NIST) Framework and Roadmap for Smart Grid Interoperability Standards, Release 3.0. Substation automation has been restricted to the solution offered by the various protection and control equipment manufacturers that needed to interface and integrate their units without restrictions. IEC started to develop a common standard for substation communications in 1994. Similarly, IEEE developed a common communication standard known as UCA (Utility Communication Architecture). In 1997, IEEE and IEC agreed to work in conjunction to create a common substation communication standard IEC 61850 [5].

The IEC 61850 Standard in Section 8-1 discusses the critical and noncritical data interchange method in time across local area networks (LANs) employing Abstract Communication Service Interface (ACSI) mapping in multimedia messaging (multimedia messaging service, MMS) over frames ISO/IEC 8802-3. Similarly, in Section IEC 61850-9-2, the specific communication service mapping (SCSM) is defined for the transmission of sampled values according to the IEC 61850-7-2, also using frames ISO/IEC 8802-3 [6].

Figure 12.4 illustrates how critical data like Generic Object Oriented Substation Events (GOOSE) messages and sampled values are directly mapped to lower Ethernet layers, and how metered values or time synchronizations pass through TCP/IP protocols.

IEC 61850 provides the multicast-based generic substation events (GSEs) as a way to quickly transfer event data over an entire substation network. Furthermore, part 9-1 and 9-2 specify a process bus for use with IEDs.

12.2.2.2 DNP3-IEEE Standard 1815

The IEEE 1815-2012 standard defines the DNP3 protocol used for communications media employed in utility automation systems were it describes the structures, functions or applications. The revision of the IEEE 1815-2010 standard includes improved protocols that aid in preventing cyber security concerns particular to the communication media employed in utility automation systems [7].

12.2.2.3 IEC 60870-5 as the standard for remote control

The trend is to adopt IEC 61850 architecture for substation communications and remote control. IEC 60870 has experienced a massive expansion and wide application [8].

IEC 60870-5 is the standard covering telecontrol equipment and systems. Part 5 covers transmission protocols. The standard is well established worldwide.

The IEC 60870-5-104 or IEC 104 protocol is a standard based on IEC 60870-5-101 or IEC 101. It employs the TCP/IP network interface to provide LAN and WAN connectivity. The original IEC 101 itself is also retained for some particular data and services utilized.

The IEC 60870-5-101 and 104 parts are mainly used for exchanging information between substations and control center. Application areas range from primary substations (high and medium levels of voltage) to the secondary ones (medium or low voltage).

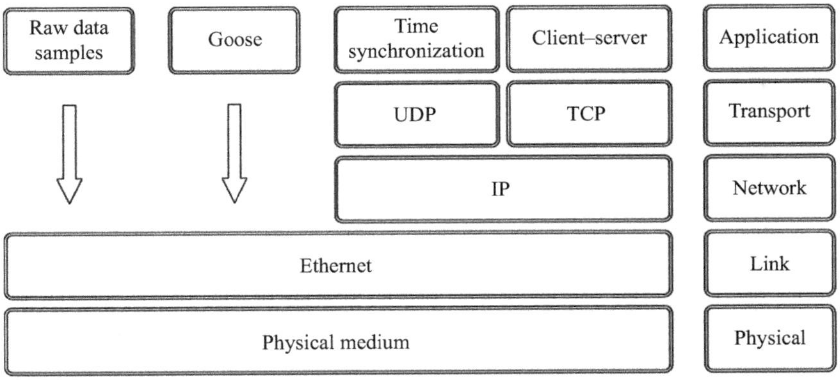

Figure 12.4 Message communication OSI-7 stack

12.3 Transmission mediums

The transmission methods supporting data transfer continue to be similar to what has been used for decades: wired, electric, wireless, and fiber optic. More recently, modulation technologies have come to be used as data highways although they were not designed for this purpose.

12.3.1 Wired and electric mediums

These are the traditional means used for transferring digital data like electricity across a conductor. Wired serial interfaces such as RS-232, RS-485 RS-449, V.35, G.703, RS-530, RS-422, although proven to be robust and reliable, are being replaced by IEC 61850-8-2 where the binary signals are sent across Ethernet cables in GOOSE messages utilizing IEC 61850-8-2 standards. Similarly, metering data of electrical parameters collected and integrated in serial interfaces are being replaced in applications.

This classification includes the power systems networks utilized for data transmission employing BPL (broadband over power lines), PLC (power line carrier), or the power line communication technology patented by ACLARA Technologies where, instead of using wave carrier technology, the data is transmitted around the power wave zero crossing enabling data to flow across the entire distribution network without requiring bridges or jumpers in transformers or capacitor banks.

12.3.2 Wireless mediums

Frequencies from 2 GHz to 16 GHz, usually called as microwaves, constitute a reliable route for data transmission from multiple integrated substations to control centers employing modern digital modulation and multiplication technologies.

Technologies like Wi-Fi, Wimax, and other wireless applications are used in communication systems where data transmission is not time-critical, even though there have been developments in modern digital modulation techniques at fast data transmission speeds. Their application is still limited for reliability considerations, which is influenced by propagation medium.

Wi-Fi has been able to improve computer security in wireless networks using the IEEE 802.11 standard (also known as WPA-2) with the implementation of Advanced Encryption Standard [9].

Technologies like ZigBee and Bluetooth are considered viable for short distance tasks like domotics, control applications in small industries or homes, and non-sensible data collection.

ZigBee works on top of the IEEE 802.15.4 standard [10], in the unlicensed 2.4 GHz or 915/868 MHz bands. An important feature of ZigBee is the possibility of handling mesh-networking, thereby extending the range and making a ZigBee network self-healing.

Bluetooth is a communication protocol especially designed for low consumption devices that require a low emitting range and based on low cost transceivers.

Bluetooth devices are divided into class 1, 2, and 3 in regards to their transmission power, where class 1 devices cover a range of 100 m.

Solutions operating with cellular networks have been accepted by the power systems sector to reach remote locations where using wired methods or even wireless methods have proved challenging. Due to their low reliability and cost, they are mainly used for telemetering or noncritical equipment control.

Satellite communications could eventually offer new remote supervision and control solutions in the transmission and distribution level.

12.3.3 Optical mediums

Traditionally, fiber optics has been employed in the electrical sector for fast data transfer. However, as a result of new developments, fiber optics could replace serial or point-to-point communications within substations implementing IEC 61850.

The use of optical fiber media has dramatically increased in the last 15 years. This is due to multi-vendor availability, much lower purchase and installation costs, dielectric characteristics, increased bandwidth potentials, and communication speed. Optical fibers are available as part of the overhead ground wire, in a self-supported dielectric conductor, as a messenger conductor, or designed for an underground environment. Utilities are using a variety of installation options to build a very reliable communication backbone with minimum common mode failures.

12.4 Information security as the crucial element in smart networks

It is understandable that in an interconnected world, reliability risks increase and security varies according to protocols and Smart Grid domains where they are implemented. Some security standards are applicable to certain protocols, while others are applied to particular profiles.

When extremely sensible and critical data is transferred, security levels should be elevated to meet basic information security like reliability, confidentiality, and integrity.

In the electric field, associations like North American Electric Reliability Corporation (NERC) have developed the NERC 1300 Standard for power system security. These standards are divided into eight specific areas from CIP-002-1 to CIP-009-2 focused on establishing requirements for power system owners, operators, and users to guard critical assets with the best security practices. Also, IEC 62351 or FIPS180-4, 186-3 must be mentioned. For example, IEC 62351 is developed for handling the security of protocols, including IEC 60870-5 series, IEC 60870-6 series, IEC 61850 series, IEC 61970 series, and IEC 61968 series [11].

It must be noted that cybersecurity standards in the application levels are part of applicable models in the computer sector, such as ISO 27002, or when employing the common critical model known as Common Criteria for Information Technology Security Evaluation, ISO 15408 [12].

12.5 Cybersecurity

One important aspect that must be analyzed after information is flowing around the system is the role of the actors and parties that are involved. Cybersecurity in Smart Grids is required in order to guarantee that data around the system and its components is shared with security, integrity, and reliability. Also, aspects like unauthorized access, physical integrity of equipment, user's privacy, and protection of information against cyberattacks must be considered. The standard NISTR 7628-2014 shows relevant aspects to be considered in cyber security [13]. Table 12.1 summarizes the requirements considered in cybersecurity.

Some standards related to cybersecurity are Department of Homeland Security—DHS-2009, "DHS Cyber Security Procurement Language for Control Systems" [14]; IEC 62351-2018, "Power systems management and associated information exchange—Data and communications security"; IEEE C37.240-2014, "Cyber Security Requirements for Substation Automation, Protection and Control Systems" [15]; IEEE 1686-2013 "IEEE Standard for Intelligent Electronic Devices Cyber Security Capabilities" [16]; ISO/IEC TR 27019, "Information technology—Security techniques—Information security management guidelines based on ISO/IEC 27002 for process control systems specific to the energy utility industry" [17]; NIST 1108–2014, "NIST Framework and Roadmap for Smart Grid Interoperability Standards, Release 3.0" [18]; NISTIR 7628-2014, "Guidelines for smart grid cyber security."

12.6 IEC 61850 overview

Substations designed in the past made use of protection and control schemes implemented with single-function, electromechanical or static devices, and hard-wired relay logic. SCADA functions were centralized and limited to the monitoring of circuit loadings, bus voltages, aggregated alarms, control of circuit breakers and tap changers, etc. Disturbance recording and sequence-of-event data if available were centralized and local to the substation.

With the advent of microprocessor-based multifunction IEDs, more functionality into fewer devices was possible, resulting in simpler designs with reduced wiring. In addition, owing to the communication capabilities of the IEDs, more information could be accessed remotely, translating into fewer visits to the substation.

Microprocessor-based protection solutions have been successful because they offered substantial cost savings while fitting very well into preexisting frameworks of relay application. A modern microprocessor-based IED replaces an entire panel of electromechanical relays with external wiring intact, and internal DC wiring replaced by integrated relay logic. Users retained total control over the degree of integration of various functions, while interoperability with the existing environment (instrument transformers, other relays, control switches, etc.) has been maintained using traditional hard-wired connections.

Table 12.1 Requirements for cybersecurity [13]

Requirement	Description	Requirement	Description
Access control	The focus of access control is ensuring that resources are accessed only by the appropriate personnel, and that personnel are correctly identified	Security program management	The security program lays the groundwork for securing the organization's enterprise and Smart Grid information system assets. Security procedures define how an organization implements the security program
Audit and accountability	Periodic audits and logging of the Smart Grid information system need to be implemented to validate that the security mechanisms present validation testing are still installed and operating correctly	Personnel security	The organization may consider implementing a confidentiality or nondisclosure agreement that employees and third party users of facilities must sign before being granted access to the Smart Grid information system
Security assessment and authorization	Security assessments include monitoring and reviewing the performance of Smart Grid information system. Internal checking methods, such as compliance audits and incident investigations, allow the organization to determine the effectiveness of the security program	Risk management and assessment	Risk management planning is a key aspect of ensuring that the processes and technical means of securing Smart Grid information systems have fully addressed the risks and vulnerabilities in the Smart Grid information system
Continuity of operations	The ability for the Smart Grid information system to function after an event is dependent on implementing continuity of operations policies, procedures, training, and resources	Smart Grid information system and services acquisition	Smart Grid information systems and services acquisition covers the contracting and acquiring of system components, software, firmware, and services from employees, contactors, and third parties

(Continues)

Table 12.1 (Continued)

Requirement	Description	Requirement	Description
Identification and authentication	Identification and authentication is the process of verifying the identity of a user, process, or device, as a prerequisite for granting access to resources in a Smart Grid information system	Smart Grid information system and communication protection	Smart Grid information system and communication protection consist of steps taken to protect the Smart Grid information system and the communication links between Smart Grid information system components from cyber intrusions
Smart Grid information system development and mainte-nance	Maintenance activities encompass appropriate policies and procedures for performing routine and preventive maintenance on the components of a Smart Grid information system	Smart Grid information system and information integrity	Maintaining a Smart Grid information system, including information integrity, increases assurance that sensitive data have neither been modified nor deleted in an unauthorized or undetected manner
Media protection	The security requirements under the media protection family provide policy and procedures for limiting access to media to authorized users	Testing and certification of Smart Grid cybersecurity	The testing and certification of the Smart Grid cybersecurity requirements provide assurance that systems and system components are conformant to the requirements selected by the organization
Physical and environmental security	Physical and environmental security encompasses protection of physical assets from damage, misuse, or theft		

In terms of SCADA integration, the first generation of such systems achieved moderate success especially in cases where the end user could lock into a solution from a single vendor. Integrating systems made up of IEDs from multiple vendors invariably led to interoperability issues on the SCADA side. Integration solutions tended to be customized.

Owners of such systems were faced with long-term support and maintenance issues. During this period, two leading protocols emerged: DNP 3.0 and IEC 60870.

Beginning in the early 1990s, initiatives were undertaken to develop a communication architecture that would facilitate the design of systems for protection, control, monitoring, and diagnostics in the substation. The primary goals were to simplify the development of these multi-vendor substation automation systems (SASs) and to achieve higher levels of integration reducing even further the amount of engineering and wiring required.

In 1994, EPRI/IEEE started a work UCA2 with focus on the station bus. In 1996, IEC TC57 (technical committee 57) began work on IEC 61850 to define a station bus. This led to a combined effort in 1997 to define an international standard that would merge the work of both groups that were focused on the development of a standard in which devices from all vendors could be connected together to share data, services, and functions. The result was the international IEC 61850 Standard Ed. 1 ''Communication Networks and Systems in Substation Automation.''

The International IEC 61850 Standard was issued in 2005, and it was developed to control and protect power systems by standardizing the exchange of information between all IEDs within an automated substation and a remote control link. IEC 61850 provides a standardized framework for substation integration that specifies the communication requirements, the functional characteristics, the structure of data in devices, the naming conventions for the data, how applications interact and control the devices, and how conformity to the standard should be tested.

The development of the IEC 61850 Standard is continuing. This work was resulting in Ed. 2 of the standard, which was published during 2010. IEC 61850 was originally defined exclusively for SASs but has since been extended to other application areas—as is reflected in its changed title "IEC 61850 Ed. 2. Communication Networks and Systems for Power Utility Automation."

The IEC 61850 is increasingly being used for the integration of electrical equipment into distributed control systems in process industries. The fact that new application areas such as hydro and wind power are being added is yet another indication of its success.

Some of the benefits of the IEC 61850 Standard are

- reduce dependence on multiple protocols,
- higher degree of integration,
- reduce construction cost by eliminating most copper wiring,
- flexible programmable protection schemes,
- communication networks replacing hard-wired connections,
- advanced management capability,
- high speed peer-to-peer communications,

- improved security/integrity, and
- reduced construction and commissioning time.

Eliminating copper wiring with Ethernet/fiber cables means no more binary inputs and outputs for control and protection functions. The traditional method of tripping a breaker via a contact could be replaced by GOOSE messages sent via Ethernet or fiber optic cables.

12.6.1 Standard documents and features of IEC 61850

The standard is composed of ten standard documents that cover all the requirements that have to be fulfilled by a SAS. It is important to keep in mind that the evolution of this technology is very rapid, and therefore more changes and additions are feasible. The following are the parts of the standard at the moment of writing the book:

- IEC 61850-1: Introduction and overview,
- IEC 61850-2: Glossary,
- IEC 61850-3: General requirements,
- IEC 61850-4: System and project management,
- IEC 61850-5: Communication requirements for functions and device models,
- IEC 61850-6: Configuration language for communication in electrical substations related to IEDs—Ed. 2,
- IEC 61850-7: Basic communication structure for substation and feeder equipment,
- IEC 61850-7-1: Principles and models—Ed. 2,
- IEC 61850-7-2: Abstract communication service interface (ACSI)—Ed. 2,
- IEC 61850-7-3: Common Data Classes—Ed. 2,
- IEC 61850-7-4: Compatible logical node classes and data classes—Ed. 2,
- IEC 61850-8: Specific communication service mapping (SCSM),
- IEC 61850-8-1: Mappings to MMS (ISO/IEC9506-1 and ISO/IEC 9506-2)—Ed. 2,
- IEC 61850-9: Specific communication service mapping (SCSM),
- IEC 61850-9-2: Sampled values over ISO/IEC 8802-3—Ed. 2,
- IEC 61850-10: Conformance testing.

In IEC 61850 Ed. 2, Part 9-1 is deprecated as per conformance table. Ethernet-based sampled value transmission (IEC 61850 9-2) will take over process bus communication. Now all IEC 61850 Specific communication service mappings are using Ethernet technology. Some of the features of IEC 61850 include

1. data modeling;
2. reporting schemes;
3. fast transfer of events, GSE—GOOSE and GSSE;
4. commands;
5. sampled data transfer;
6. setting groups; and
7. data storage—SCL (substation configuration language).

1. *Data modeling*: In IEC 61850, series functionalities that the real devices comprise are decomposed into the smallest entities, which are used to exchange information between different devices. These entities are called logical nodes. Complete functionality of the substation is modeled into different standard logical nodes. The logical nodes are the virtual representation of the real functionalities. The intent is that all data that could originate in the substation can be assigned to one of these logical nodes. Several logical nodes from different real devices build up a logical device.

   Logical devices, logical nodes, and data objects are all virtual terms. They represent real data, which is used for communication. One device (e.g., a control unit) only communicates with the logical nodes or its data objects of another device (e.g., an IED). The real data, which the logical nodes represent, are hidden, and they are not accessed directly. This approach has the advantage that communication and information modeling is not dependent on operating systems, storage systems, and programming languages. The concept of virtualization is shown in Figure 12.5, where the real device on the right hand side is modeled as a virtual model in the middle of the figure. The logical nodes (e.g., XCBR and circuit breaker) defined in the logical device (Bay) correspond to well-known functions in the real devices. In this example, the logical node XCBR represents a specific circuit breaker of the bay to the right.

   Based on their functionality, a logical node contains a list of data (e.g., position) with dedicated data attributes. The data have a structure and a predefined semantic. The information represented by the data and their attributes are exchanged by the communication services according to the well-defined rules and the requested performances.

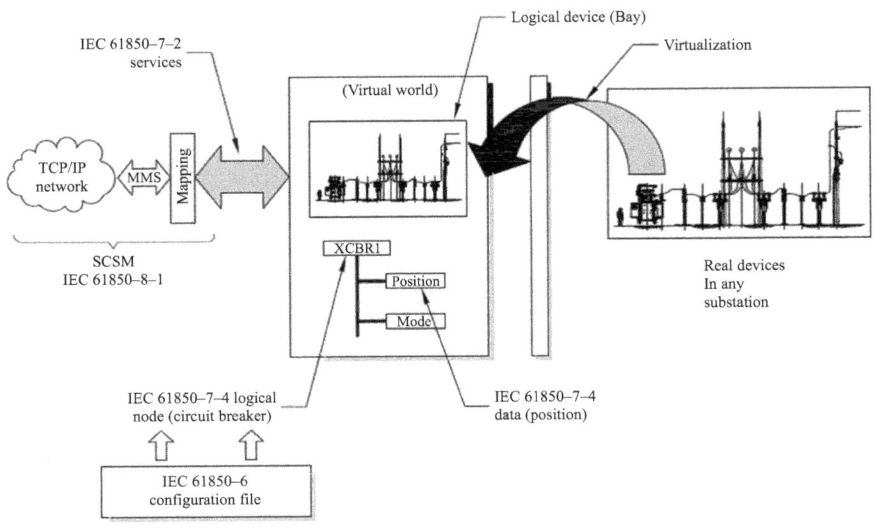

Figure 12.5 Virtual and real world

To illustrate more clearly how the logical devices, logical nodes, classes, and data concepts map to the real world, imagine an IED that is a container as shown in Figure 12.6.

The container is the physical device, which is containing one or more logical devices. Each logical device contains one or more logical nodes, each of which contains a predefined set of data classes. Every data class contains many data attributes (status value, quality, etc.).

IEC 61850-7-4 Ed. 2 defines a list of logical node groups as shown in Table 12.2.

In Ed. 2 LN list has been extended from 92 to 208 based on new requirements. New system logical nodes are introduced to indicate the status of GOOSE and sampled value subscription. Grouping of logical node is also extended to include relevant areas.

IEC 61850-5 Ed. 2 defines the protection logical nodes which are shown in Table 12.3.

2. *Reporting schemes*: There are various reporting schemes (BRCB and URCB) for reporting data from server through a server–client relationship that can be triggered based on predefined trigger conditions.
3. *Fast transfer of events*: GSE is defined for fast transfer of event data for a peer-to-peer communication mode. This is subdivided into GOOSE and GSSE (generic substation status event), which provide backward compatibility with the UCA GOOSE.

   In IEC 61850 Ed. 2 GSSE is deprecated and moved to Annex of 7 2. This will cause older less flexible GSSE-based systems move out from IEC 61850 scope. Hence, GOOSE will have added importance in inter-bay communication.

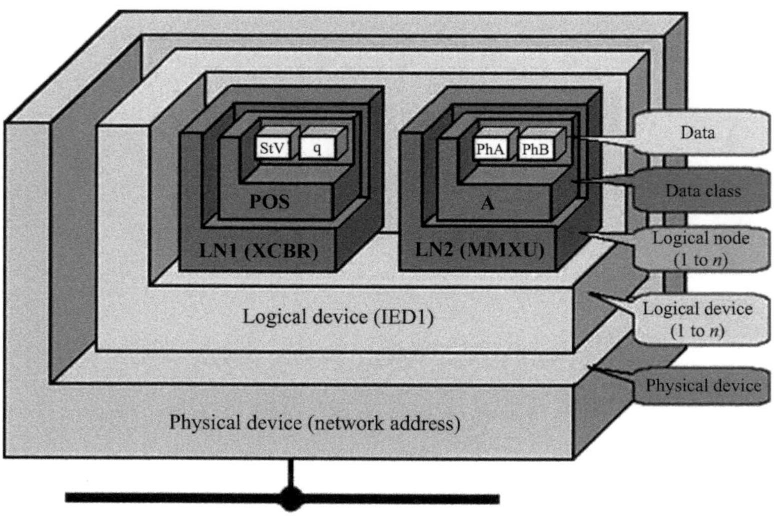

Figure 12.6 Physical and logical device

Table 12.2 Protection logical nodes defined by IEC 61850-7-4 Ed. 2

Group indicator	Logical node groups
A	Automatic control
B	Reserved
C	Supervisory control
D	Distributed energy resources
E	Reserved
F	Functional blocks
G	Generic function references
H	Hydro power
I	Interfacing and archiving
J	Reserved
K	Mechanical and nonelectrical primary equipment
L	System logical nodes
M	Metering and measurement
N	Reserved
O	Reserved
P	Protection functions
Q	Power quality events detection related
R	Protection-related functions
S[a]	Supervision and monitoring
T[a]	Instrument transformer and sensors
U	Reserved
V	Reserved
W	Wind power
X[a]	Switchgear
Y[a]	Power transformer and related functions
Z[a]	Further (power system) equipment

[a]LNs of this group exist in dedicated IEDs if a process bus is used. Without a process bus. LNs of this group are the I/Os in the hardwired IED one level higher (e.g., in a bay unit), representing the external device by its inputs and outputs (process image).

4. *Setting groups*: The setting group control blocks are defined to handle the setting groups so that user can switch to any active group according to the requirement.
5. *Sampled data transfer*: Schemes are also defined to handle transfer of sampled values using sampled value control blocks.
6. *Commands*: Various command types are also supported by IEC 61850 that include direct and select before operate commands with normal and enhanced securities.
7. *Data storage*: SCL is defined for complete storage of configured data of the substation in a specific format. SCL originally was the abbreviation for substation configuration language. Now that the use of IEC 61850 beyond substations has become reality, the name changed to system configuration language.

Table 12.3 Protection logical nodes defined by IEC 61850-5 Ed. 2

Functionality allocated to LN	IEC	IEEE	LN function	LN class	LN class naming
Transient earth fault protection			PTEF	PTEF	Transient earth fault
Sensitive directional earth fault		(37) (67N)	PSDE	PSDE	Sensitive directional earth fault
Thyristor protection			PTHF	PTHF	Thyristor protection
Protection trip conditioning			PTRC	PTRC	Protection trip conditioning
Checking or interlocking relay		3	CILO	CILO	Interlocking
Over speed protection	w>	12	POVS		
Zero speed and under speed protection	w<	14	PZSU	PZSU	Zero speed or underspeed
Distance protection	Z<	21	PDIS	PDIS PSCH	Distance protection Protection scheme
Volt per Hz protection		24	PVPH	PVPH	Volts per Hz
Synchronism check		25	RSYN	RSYN	Synchronism check
Over temperature protection	>	26	PTTR	PTTR	Thermal overload
(Time) undervoltage protection	U<	27	PTUV	PTUV	Undervoltage
Directional power/reverse power protection	P >	32	PDPR	PDOP PDUP	Directional over power Directional under power
Undercurrent/Underpower protection	P<	37	PUCP	PTUC PDUP	Undercurrent Directional under power
Loss of field/Underexcitation protection		40	PUEX	PDUP PDIS	Directional under power (distance)
Reverse phase or phase balance current protection, negative sequence current relay	$I_2>$	46	PPBR	PTOC	Time overcurrent
Phase sequence or phase balance voltage protection, negative sequence voltage relay	$U_2>$	47	PPBV	PTOV	Overvoltage protection
Motor start-up protection		48, 49, 51LR66	PMSU	PMRI PMSS	Motor restart inhibition Motor starting time supervision
Thermal overload protection	$\theta>$	49	PTTR	PTTR	Thermal overload
		49R	PROL	PTTR	Thermal overload
Rotor thermal overload protection					
Rotor protection		49R 64R (40) 50 51	PROT	PTTR PHIZ PDUP PDIS	Thermal overload Time overcurrent Ground detector Distance (impedance)
Stator thermal overload protection		49S	PSOL	PTTR	Thermal overload
Instantaneous overcurrent or rate of rise protection	$I \gg$	50	PIOC	PIOC	Instantaneous overcurrent
AC time overcurrent protection	$I>, t$	50TD 51	PTOC	PTOC	Time overcurrent
Voltage controlled/dependent time overcurrent protection		51V	PVOC	PVOC	Voltage controlled time overcurrent

Table 12.3 (Continued)

Functionality allocated to LN	IEC	IEEE	LN function	LN class	LN class naming
Power factor protection	cos φ>	55	PPFR	POPF	Over power factor
	cos φ<			PUPF	Under power factor
(Time) Overvoltage protection	U>	59	PTOV	PTOV	Overvoltage
DC-overvoltage protection		59DC	PDOV	PTOV	Overvoltage
Voltage or current balance protection		60	PVCB	PTOV	Overvoltage
				PTOC	Time overcurrent
Earth fault protection, ground detection	I_E>	64	PHIZ	PTOC	Time overcurrent
				PHIZ	Ground detector
Rotor earth fault protection		64R	PREF	PTOC	Time overcurrent
				PHIZ	Ground detector
Stator earth fault protection		64S	PSEF	PTOC	Time overcurrent
				PHIZ	Ground detector
Interturn fault protection		64W	PITF	PTOC	Time overcurrent
AC directional overcurrent protection	I>	67	PDOC	PTOC	Time overcurrent
Directional protection		87B	PDIR	PDIR	Direction comparison
Directional earth fault protection	I_E>	67N	PDEF	PTOC	Time overcurrent
DC time overcurrent protection		76	PDCO	PTOC	Time overcurrent
Phase angle or out-of-step protection	ϕ>	78	PPAM	PPAM	Phase angle measuring
Frequency protection		81	PFRQ	PTOF	Over frequency
		7		PTUF	Under-frequency
				PFRC	Rate of change of frequency
Differential protection		87	PDIF	PDIF	Differential (Impedance)
Busbar protection		87B	PBDF	PDIF	Differential
				PDIR	Direction comparison
Generator differential protection		87G	PGDF	PDIF	Differential
Differential line protection		87L	PLDF	PDIF	Differential
Motor differential protection		87M	PMDF	PDIF	Differential
Restricted earth fault protection		87N	PNDF	PDIF	Differential
Phase comparison protection		87P	PPDF	PDIF	Differential
Differential transformer protection		87T	PTDF	PDIF	Differential
Harmonic restraint			PHAR	PHAR	Harmonic restraint

12.6.2 System configuration language (SCL)

The fact that IEC61850 is a worldwide standard makes it also possible that relays from different manufacturers are able to exchange information as long as the relays conform to the IEC61850 standard.

To guarantee interoperability and enhance the configuration phase, the IEC61850-6 introduced a common language that can be used to exchange information between different manufactures using SCL.

Each proprietary tool must have a function that allows the export of the IED's description into this common, XML-based language. The ICD (IED Capability Description) file contains all information about the IED, which allows the user now to configure a GOOSE message.

The development process of a project based on IEC 61850 depends on the availability of software tools that make use of the SCL language. The SCL specifies a common file format for describing IED capabilities, a system-specific caption that can be viewed in terms of a single line diagram, and a SAS description. IEC 61850-6 introduces four types of common files. These files are the ICD, Configured IED Description (CID), Substation Configuration Description (SCD), and System Specification Description files. IEC 61850-6 Ed. 2 introduced the new "Instantiated IED Description" file. Figure 12.7 describes the complete engineering process using SCL language.

The configuration can be performed by a manufacturer-independent tool, the so-called IEC61850 System Configurator. Some manufacturer advanced there proprietary tools in a way that they can be used as IEC61850 System Configurator; however, there are also some third party tools available. All ICD files get imported into the IEC61850 System Configurator, and the GOOSE messages can be programmed by specifying the sender (publisher) and the receiver (subscriber) of a message. On the end, the whole description of the system, including the description of the GOOSE messages, get stored in the SCD file. Each proprietary tool must be able to import this SCD file and extract the information needed for the IED. The information out of the SCD file is typically the list of the GOOSE messages.

SCL language does not cover all features of today's IEDs. In fact it was not the intention of the editors of IEC 61850 standard to standardize all aspects of IEDs. This is because of the wide variety of functionality provided by each manufacturer.

It is important to know that in addition to the configuration information specified by part 6 of IEC 61850 the complete configuration of an IED should be completed with a proprietary IED configuration tool provided by the manufacturer in order to configure the device parameters and gain access to all of the internal functions supported.

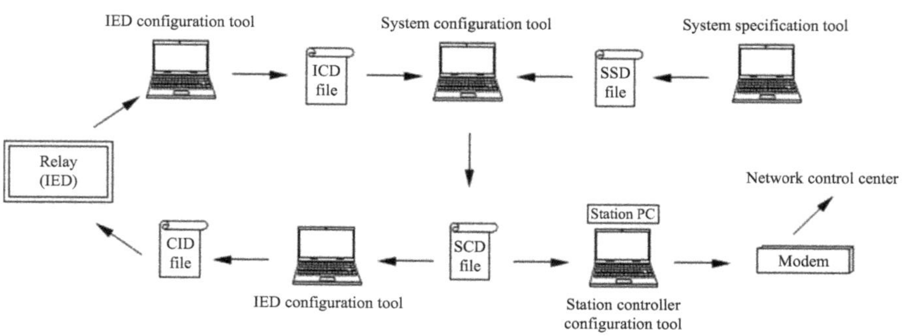

Figure 12.7 Substation engineering process using SCL language

The following list shows examples of features that can only be set-up with a vendor's tool:

- logic and trips equations,
- graphical display on an IED's HMI,
- internal mappings, and
- non-IEC 61850 and vendor specific parameters.

Although the present practice of connecting a notebook computer directly to the front port of the IED will likely be eventually phased out in favor of remote access solutions via the LAN, it is assumed that some activities will continue to need to be performed at the relay, particularly during initial commissioning. It is also assumed that each individual vendor's IED configuration methodology will continue to be unique to each device or product family. Continuous innovation in IED design and competitive market forces would tend to preclude standardization of individual IED configuration software, though some activity is underway in this regard.

The final configuration done by the proprietary tool is individualized per IED and can have either a proprietary format or a standard CID format. Nowadays many vendors have decided to have a CID file using a proprietary format, while a few vendors create the CID file in a similar way as the ICD file.

12.6.3 Configuration and verification of GOOSE messages

This section will describe how to configure and verify GOOSE messages for an automated IEC 61850 substation. The type of scheme being configured to illustrate the process is a communications-based breaker failure scheme. Figure 12.8 shows the one line diagram of the system.

The system has four IEDs from different vendors which are shown as relays A, B, C, and D. Relays B, C, and D are the main feeder protection, and relay A is the backup protection.

Not shown are the IEDs in the breaker cabinet that take the GOOSE messages from each relay and convert them to a physical output to energize the breaker coil. In this example, a fault has been placed at location A, marked in the one line diagram. The instantaneous overcurrent element in relay B will transmit a trip GOOSE to breaker B. Simultaneously, relay A will receive the trip GOOSE from relay B as the breaker failure initiate signal. Relay A will receive this GOOSE and it will initiate the breaker failure timer, which will expire in ten cycles. If for any reason the breaker failure initiate signal is not received and breaker B does not open, relay A will trip on a definite time overcurrent in 15 cycles. This section will only show the configuration between relays A and B. This same process is repeated for the other IEDs on the system. Figure 12.9 shows the logic of the breaker failure scheme.

The programming of this scheme can be achieved in two parts. The first part is the programming of the individual IEDs in order to obtain the SCL file required for the configuration of the system. These files will contain the published GOOSE messages for each IED. With this information, the full system configuration can be accomplished. Before any configuration can occur, it is good practice to create a

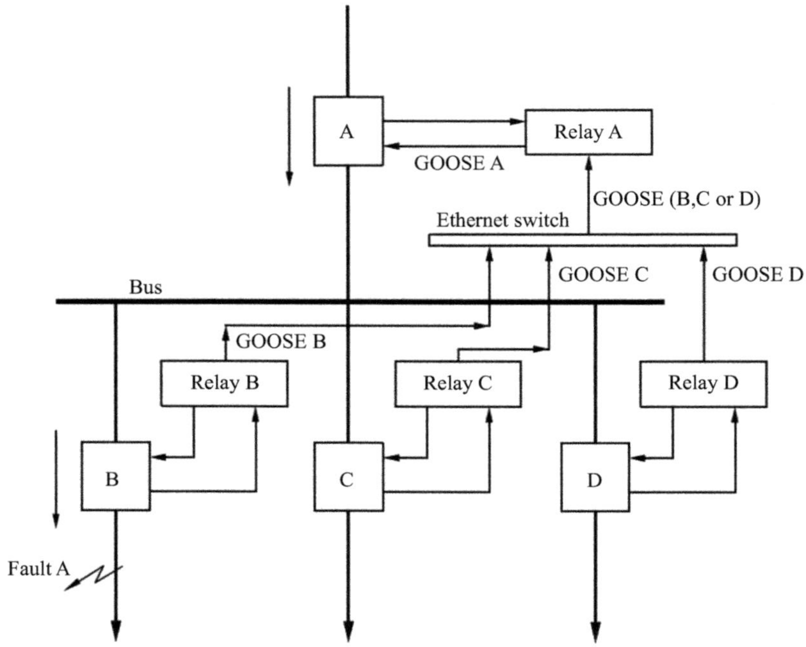

Figure 12.8 System single line diagram

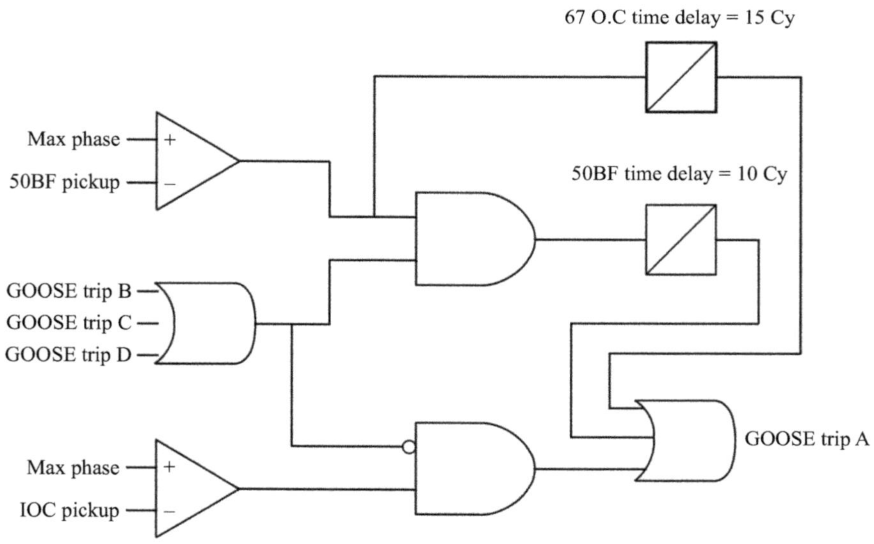

Figure 12.9 Logic of breaker failure scheme

virtual wiring map. This map will show what messages are going to be required in order to configure the scheme of interest. The map provides such information as to what GOOSE messages each IED will publish and subscribe.

The IED is configured via its own proprietary configuration tool. One of the biggest challenges in configuring an IEC 61850–based protection scheme is becoming familiar with the different IED configuration software packages. It is important to get familiar with how each configuration tool works. This will save time in the long run, especially when trying to configure the system scheme. The primary goal of configuring the IED is to obtain the SCL file types necessary for the system configuration. Figure 12.11 shows examples of the configuration tools used to configure the individual IEDs.

The SCL files provide the overall information as to how the relay is configured but most importantly, what GOOSE messages the particular IED will publish or subscribe. The configuration process would begin by configuring the IEDs that will only be publishing and subscribing to the least number of GOOSE messages.

When the IED is configured, an SCL file is exported. This same process is repeated until all IEDs of interest have been configured and an SCL file is generated.

If desired, a simple test can be performed to verify that the published GOOSE messages are correct. This can be done by connecting some form of network analyzer and capturing the GOOSE messages on the network. An illustration of proprietary configuration tools used for the configuration of IEDs is shown in Figure 12.10.

Some of the details of the GOOSE configuration were left out due to differences in the configuration of the IEDs. For example, in some IEDs the GOOSE messages are going to be published through GGIO (generic GOOSE I/O), while

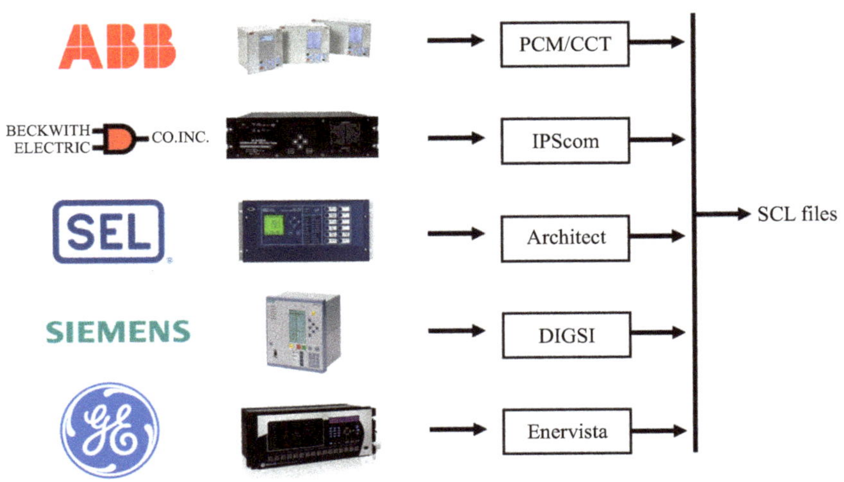

Figure 12.10 Proprietary configuration tools used for the configuration of IEDs

others will be published through the protection node such as PDIS or PTOC. This difference can sometimes lead to the GOOSE message being delayed by some fixed time set in the IED. To avoid these problems, it is recommended to read the IED's manual and see where the IED will publish its high-speed GOOSE.

12.6.4 Configuration of the system

In order to configure the system, it is necessary to have all ICD files available. These files are going to be used by a system configuration tool or the IED's individual configuration tool. The system configuration can be done by a substation configuration tool. This type of tool imports all the ICD files required for the system configuration. The substation configuration tool will generate an SCD file. Another method is by opening up the configuration tool of the IEDs that will subscribe to a GOOSE message.

For example, the configuration of relay A is done by importing all the ICD files into its configuration tool. The ICD file is intended to be a template for an IED type, which could be instantiated multiple times—e.g., per feeder—to build a system.

When the system is configured, a system verification test should be performed. This test will help identify any problems with the system configuration. The first time a system is configured, some configuration problems may occur. Many of these problems are due to the differences in the configuration of the IEDs. Some of the IEDs require a more manual approach, while others are a bit more automatic.

12.6.5 System verification test

The system verification test will identify any configuration conflicts in the system. The tools required are going to be a network analyzer (software) and a modern test set that is IEC 61850 compliant. The test set must be able to receive and send GOOSE messages via the substation LAN. This would require the test system to interrogate the network, acquire the right GOOSE message, and stop a timer with minimal effect on time.

Figure 12.11 presents the test connections used for standalone IEC 61850 and illustrates how the trip-GOOSE messages are sent from a numerical relay to an Ethernet switch and from this to a protective relay test system.

Modern test systems must be able to receive and send GOOSE messages via the substation LAN. This would require the test system to be able to interrogate the network, acquire the right GOOSE message, and stop injections or timer in less than 2 ms. Also the test system would have to be able to read SCL files and map inputs to the various GOOSE messages available in the SCL file. If an SCL file is not available, the test system would have to be able to interrogate the network and display all available GOOSE messages on the network, to allow the user to be able to map these messages to binary inputs on the test system.

12.6.6 Substation IT network

As communications in the substation take on more critical roles in the protection and control task of the utility, it is important for the protection engineer to

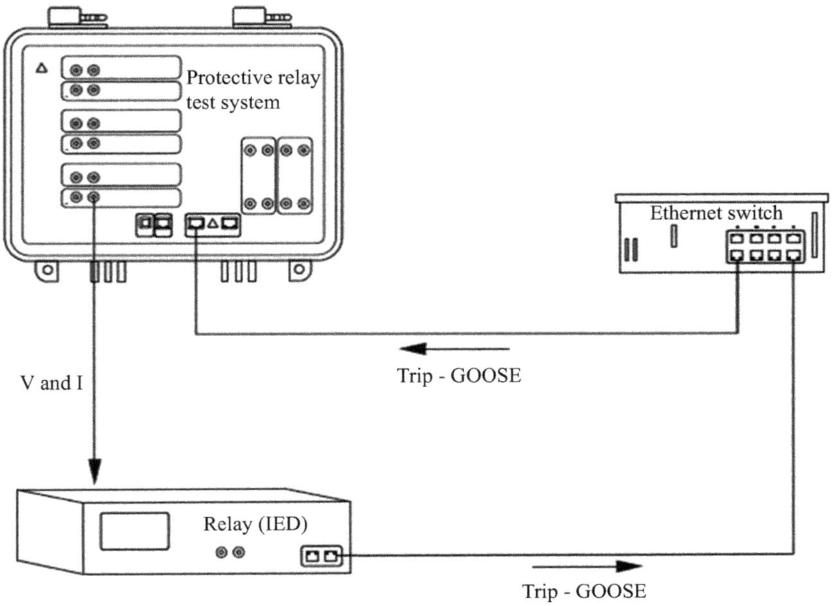

Figure 12.11 Test connections used for standalone IEC 61850–based IED

understand the basics of the IT network. The protection engineer must also understand the behavior and characteristics of components like Ethernet switches, Ethernet ports, and router, as well as being familiar with terminology such as LAN, VLAN, Mac address, network topology, latency, priority tag, firewalls, RSTP, HSR/PRP, etc.

Many experienced protection engineers find discussion of IT network issues to be dense and perhaps intimidating, because until now they have not faced the need to understand the behavior and performance characteristics of IT networks. In the modern substations, Ethernet switches are as important to understand as protective relays in order to achieve availability, dependability, security, and maintainability goals of the substation.

12.6.7 Process bus

Process bus is defined by IEC 61850 Ed. 2 Part 9-2. Process bus is the digitalization of all analogue signals in the substation. This is achieved by connecting all current transformers, potential transformers, and control cables to merging units. These units convert the analogue signals to binary signals and send the information via the process bus to all the devices that subscribe for that information. There are some pilot projects installed around the world with this technology reporting successful results. Figure 12.12 shows a full implementation of IEC 61850—process bus and station bus.

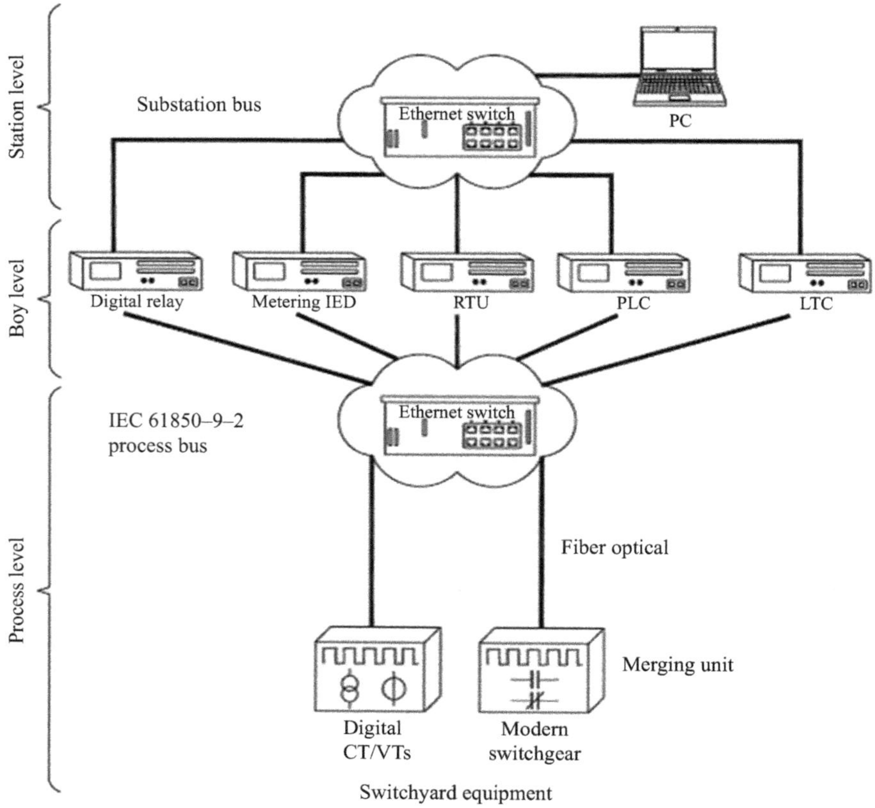

Figure 12.12 Substation network—process bus and station bus

12.6.8 Communications redundancy networks IEC 618590

It is shown how GOOSE messages replace the wired logic between IED protection relays by implementing an Ethernet network designed for substation communications. The relevance of communications engineering in substation network is defined based on the criticality of packets circulating between the different automation elements from either an IED to the SCADA system or another IED within the same network. When an event is detected, IEDs use a multicast transmission to notify the event to any device in the network that uses IEC61850. This requires that sending of messages are done as fast as possible, and the data is transmitted in a reliable way.

This gives the requirement for communication to be redundant, which allows ensuring the survival of data in the case of a failure and providing routes of alternative data when a communication path between devices fails. In data networks, there are two basic redundancy techniques: active redundancy, in which both connection links are active at the same time, and passive redundancy, where one link is active and the other is in standby mode. Choosing one or the other will depend on the criticality of the network and the type of application. For instance, passive redundancy would not be suitable for high availability environments, such as substations, metal refineries, chemical plants, and other applications.

Critical GOOSE messages and control commands must be received by subscribing devices within a specific time window to ensure the reliability of protection and control performance. IEC 61850 has adopted the recommendations of IEC 62439-3 and specifies the active redundancy protocols in its implementation. This corresponds to the Parallel Redundancy Protocol and the Seamless High Availability Redundancy Protocol. Both protocols duplicate the data transmitted through the network and allow an immediate switching time when a fault occurs in a cable or switch.

In order to ensure a correct, secure and scalable implementation of the systems, compliance of reliability, cybersecurity, and other communications requirements is required.

References

[1] ISO/IEC 7498. Information Technology – Open Systems Interconnection – Basic Reference Model, 1996.
[2] IEC 61968. Common Information Model (CIM)/Distribution Management, 2005.
[3] IEC 61970. Common Information Model (CIM)/Energy Management, 2012.
[4] IEC 62056. Electricity Metering Data Exchange, 2002.
[5] IEC 61850 Ed. 2. Communication Networks and Systems for Power Utility Automation Scheduled for 2010, 2013.
[6] IEC 8802. Information Technology, 2013.
[7] IEEE. IEEE Std. 1815-2012 – IEEE Standard for Electric Power Systems Communications-Distributed Network Protocol (DNP3), 2012.
[8] IEC 60870. Telecontrol Equipment and Systems.
[9] IEEE. IEEE Std. 802.11-2016 – IEEE Standard for Information Technology—Telecommunications and Information Exchange Between Systems Local and Metropolitan Area Networks—Specific Requirements Part 11: Wireless LAN Medium Access Control (MAC) and Physical Layer (PHY) Specifications. 2016.
[10] IEEE. IEEE P802.15.4w/D5, Sep 2019 – IEEE Draft Standard for Low-Rate Wireless Networks Amendment for a Low Power Wide Area Network (LPWAN) Extension to the Low Energy Critical Infrastructure Monitoring (LECIM) Physical layer (PHY). 2019.
[11] IEC 62351. Power Systems Management and Associated Information Exchange, 2007.
[12] ISO 15408. Information Technology — Security Techniques, 2014.
[13] National Institute of Standards and Technology. NISTIR – 7628 – Guidelines for Smart Grid Cybersecurity. 2014.
[14] Department of Homeland Security – DHS-2009. DHS Cyber Security Procurement Language for Control Systems, 2009.

[15] IEEE C37.240-2014. Cyber Security Requirements for Substation Automation, Protection and Control Systems, 2014.
[16] IEEE 1686-2013. IEEE Standard for Intelligent Electronic Devices Cyber Security Capabilities, 2013.
[17] ISO/IEC TR 27019. Information Technology – Security Techniques – Information Security Management Guidelines Based on ISO/IEC 27002 for Process Control Systems Specific to the Energy Utility Industry.
[18] National Institute of Standards and Technology. NIST Framework and Roadmap for Smart Grid Interoperability Standards, Release 3.0. USA, NIST Special Publication 1108. 2010.

Further reading

[1] Aguilar, R. and Ariza, J. "Testing and Configuration of IEC 61850 Multivendor Protection Schemes," in IEEE Transmission and Distribution Conference and Exposition, IEEE, New Orleans, LA, USA, 2010.
[2] Aravinthan, V., Karimi, B., Namboodiri, V., and Jewell, W. "Wireless Communication for Smart Grid Applications at Distribution Level — Feasibility and Requirements," in Power and Energy Society General Meeting, 2011 IEEE, 24–29 July 2011.
[3] Chen, A. C. M. "Power/energy: automated power distribution: increasingly diverse and complex power operation and distribution systems will mean a larger role for microprocessor and communications technologies," *IEEE Spectrum*, vol. 19, no. 4, pp. 55–60, 1982.
[4] Choi, D., Kim, H., Won, D., and Kim, S. "Advanced key-management architecture for secure SCADA communications," *IEEE Transactions on Power Delivery*, vol. 24, no. 3, pp. 1154–1163, 2009.
[5] Goraj, M. and Herrmann, J. Experiences in IEC 61850 and Possible Improvements of SCL Languages, *Praxis Profiline*. p. 60, April 2007.
[6] Holbach, J. and Dufaure, T. "Comparison of IEC61850 GOOSE Messages and Control Wiring Between Protection Relays," in 62th Annual Georgia Tech Protective Relaying Conference, 2008.
[7] Kasztenny, B., Whatley, J., Urden, E. A., Burger, J., Finney, D., and Adamiak, M. "IEC 61850: A Practical Application Primer for Protection Engineer," in 60th Annual Georgia Tech Protective Relaying Conference, 2006.
[8] National Institute of Standards and Technology. "IEC 61850 Objects/DNP3 Mapping," NIST, 2009.
[9] National Institute of Standards and Technology. NIST Smart Grid Standards Roadmap. National Coordinator for Smart Grid Interoperability, USA, 2009 [Online]. Available: https://www.nist.gov/fusion-search?s=smart+grid+roadmap.
[10] Parikh, P. P., Kanabar, M. G., and Sidhu, T. S. "Opportunities and Challenges of Wireless Communication Technologies for Smart Grid Applications," in Power and Energy Society General Meeting, 2010 IEEE, 25–29 July 2010.

[11] Power System Relaying Committee Working Group H6. Application Consideration of IEC 61850/UCA2 for Substation Ethernet Local Area Network Communication for Protection and Control, 2005.
[12] Shreyas, A. Analysis of Communication Protocols for Neighborhood Area Network for Smart Grid. Master thesis, California State University, Sacramento, CA, USA, 2010.
[13] Tosic, V. and Djordjevic-Kajan, S. "The Common Information Model (CIM) Standard – An Analysis of Features and Open Issues," in Proceedings of TELSIKS '99 – 4th International Conference on Telecommunications in Modern Satellite, Cable, and Broadcasting Services, Nis, Yugoslavia, October 1999, 1999, vol. 2, pp. 677–680.

Chapter 13

Interoperability concepts in power electric systems

A complete integration of all the components of the power system from generation to the end user is required to achieve the objectives of a Smart Grid. For instance, the integration of renewable energy sources will affect existing electricity grid infrastructure, operations, and the functioning of the electricity market itself. This can be tackled by electricity grids operating smartly and cost-efficiently. They should relate to each other in a transparent way and independently of the applied technologies.

Utilities business needs to integrate control center applications (Supervisory Control and Data Acquisition (SCADA), Alarm Processing, History, etc.) and the new and legacy systems in Transmission and Distribution (energy management system (EMS), distribution management system (DMS), Asses Management, Outage Managing, Geographic Information Systems (GIS), Planning Systems, etc.). Also, it is expected to include the control center into an enterprise-wide integration strategy (business integration) and the exchange model and status information with other utilities/sites for security and open access. In order to achieve this, it is required to implement a common power system model, a common information exchange format, and an integration framework based on standards.

There are many limitations on communication among the components of a power system, because of incompatibilities between protocols and technologies. Additionally, it is also required to combine a large number of autonomous IT systems into a homogeneous IT landscape. However, conventional network control systems could only be integrated with considerable effort. Interfaces solve this problem but they have the disadvantage of limiting the information system's growth. Therefore, it is required to standardize the representation of the power system's information, in such a way that each component accesses the data independently of the source. A seamless and efficient information exchange is necessary at various stages, between an increasing number of companies (transmission system operators (TSOs), distribution system operators (DSOs), generators, etc.). A solution for this is achieved when power systems fulfill the conditions of interoperability.

IEEE Std. 2030-2011 [1] defines interoperability as the capability of two or more networks, systems, devices, applications, or components to externally exchange and readily use information securely and effectively. The same standard defines a methodology to guarantee the interoperability in power systems from the perspectives of three systems: the power system itself, the information system, and the communication system. The process has to be done in such a way as to guarantee long life cycle permitting upgrading.

This concept applied to Smart Grid networks ensures efficient communication whether the information systems are used on different types of infrastructure or even at a distance and protecting the investment costs. Such information exchanges have become indispensable in network planning, power system operation (real-time information on the generation output, balancing control, etc.), market (generation schedules, trades, balancing resource management, etc.). Interoperability must be vertical and horizontal of the involved products and systems and a common definition of data model must be done.

The concept of ontology from computer science helps one to understand this idea. Ontology is the philosophical study of the nature of the basic categories and their relationships. Ontology-based strategies applied to interoperability constitute a framework for organizing information. They are used in the representation of components of power systems.

13.1 Elements required for interoperability

Entities such as the Electric Power Research Institute (EPRI) and National Institute of Standards and Technology (NIST) have defined the concept of interoperability by using architectural principles [2]. Table 13.1 enumerates the elements that are required to define the concept of interoperability.

The various functions in IT have the responsibility of distributing safely and reliably information to all points of the network where this is required for monitoring or decision-making. Thanks to these information technologies, Smart Grid networks will become much more dynamic in their configuration and will approach operating conditions that give greater opportunities for real-time analysis and optimization technologies.

The utilities must process large amounts of operational and not operational information, both in real-time and stored, in either static or dynamic form. They also receive large amounts of information remotely, directly from the analysis done at fault sites and events, and advanced metering infrastructure (AMI) deployments along the system, thanks to the installation of intelligent electronic devices and other Smart Grid devices.

13.2 Information exchange processes

Interoperability in the exchange of information ought to allow two applications to exist whenever each one can use the interchange information to develop its own results. This exchange should be carried out in a dataflow sequence along different data paths. The applications will use a link data series (or communication systems technology) that employs sufficient protocols to bear the dataflow requests. In addition, the structure and the meaning of the information exchange should be understood by both systems. This implies that they share an understanding of the unit of measure, the context of use, indices of information, and validity. A scheme of information exchange can be seen in Figure 13.1.

Table 13.1 Architecture of the principles of interoperability [2]

Principle	Description
Standardization	The elements of the infrastructure and the ways in which they interrelate are clearly defined, published, useful, open, and stable over time
Openness	The infrastructure is based on technology that is available to all qualified stakeholders on a nondiscriminatory basis. Providers of the technology have the ability to modify access to technology through time to ensure continuous openness and standardization
Interoperability	The standardization of interfaces within the infrastructure is organized such that (a) the system can be easily customized for particular geographical, application-specific, or business circumstances, but (b) customization does not prevent necessary communications between elements of the infrastructure
Security	The infrastructure is protected against unauthorized access and interference with normal operation. The utility continuously implements information privacy and other security policies as required
Extensibility	The infrastructure is not designed with built-in constraints against extending its capabilities as new applications are discovered and developed. Toward this goal, (a) its data is defined and structured according to a CIM. (b) It separates the definition of data from the methods used to deliver it. (c) Its components can announce and describe themselves to other components
Scalability	The infrastructure can be expanded with no inherent limitations on its size
Manageability	The components of the infrastructure can have their configuration assessed and managed. Faults can be identified and isolated. The components are remotely manageable
Upgradeability	The configuration, software, algorithms, and security credentials of the infrastructure can be upgraded safely and securely with minimal remote-site visits. This also touches on manageability
Shareability	The infrastructure uses shared resources that offer economies of scale, minimize duplicative efforts, and, if appropriately organized, encourage the introduction of competing innovative solutions
Ubiquity	Authorized users of the Smart Grid can readily take advantage of the infrastructure and what it provides regardless of geographic or other barriers
Integrity	The infrastructure operates at a high level of availability, performance, and reliability. It reroutes communications automatically
Ease of use	There are logical, consistent, and, preferably, intuitive rules and procedures, for the use and management of the infrastructure. The system maximizes the information and choices available to users of the Smart Grid, while minimizing the actions they must take to participate if they choose to do so

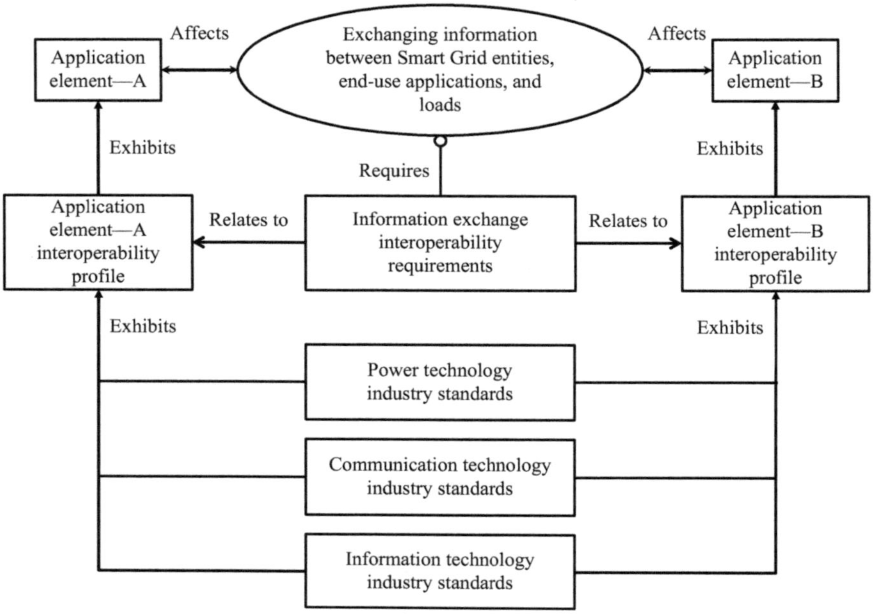

Figure 13.1 Information exchange between two application elements [1]

13.3 Data models and international standards

A model of well-defined data elements allows not only the simple direct exchange of information but also assists in fulfilling requests from development and security. The existence of a data model for the architecture is almost always a good indicator of a well-governed process of facilitating the use of the data by multiple applications because of the establishment of a common semantics.

Two global standards exist for the integration of businesses in the electricity sector. One of them is MultiSpeak®, sponsored by National Rural Electric Cooperative Association (NRECA) [3], and the other is the Common Information Model (CIM), an international standard maintained by Technical Committee 57 (TC 57) of the International Electrotechnical Commission (IEC) [4].

The CIM data model describes the electrical network, the connected electrical components, the additional elements, and the data needed for network operation as well as the relations between these elements.

Working Group 14 of TC 57 that deals with work processes to distribution utilities has recently defined the Network DataSet message that specifies how the CIM should be used to exchange detailed models for distribution engineering analysis [5]. CIM-compatible Network DataSet model exchanges are in operation at several utilities.

Figure 13.2(a) depicts a single-line diagram of a one single busbar system switchgear that has a connection to a grid via a 132-kV line. The system has an

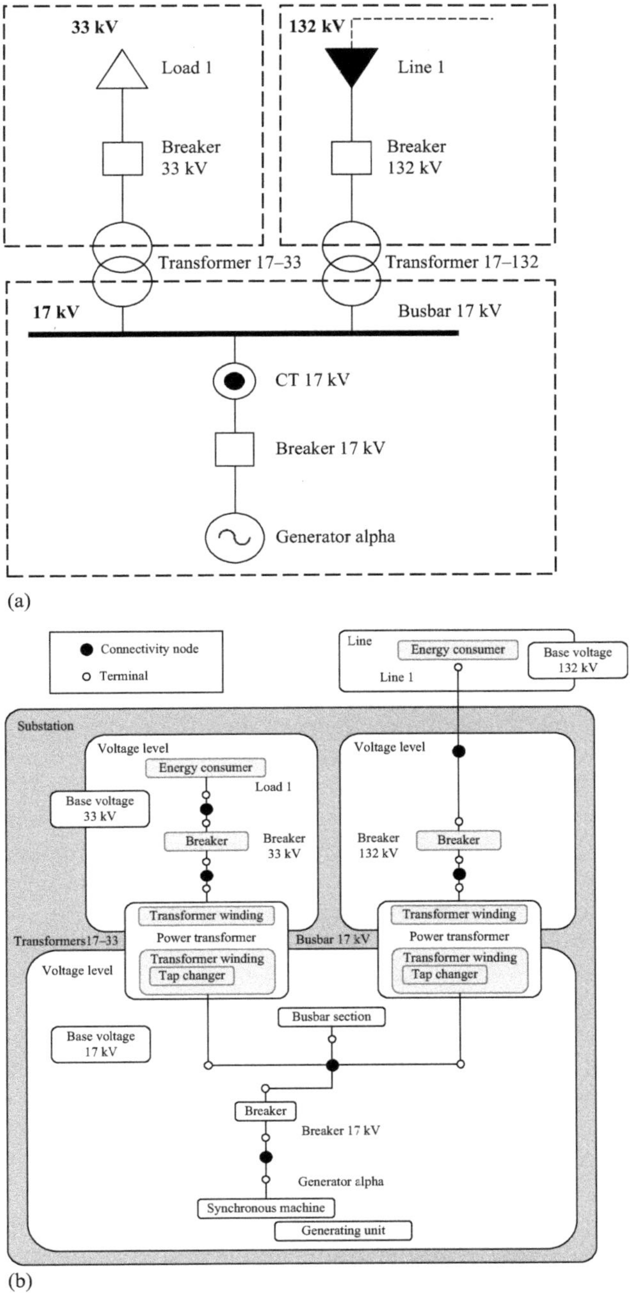

Figure 13.2 Example circuit (a) as a line diagram; (b) with CIM mappings

internal generator of 17-kV and a 17/138-kV step-up transformer. The load is fed at 33 kV through a 33-kV step-down transformer. Figure 13.2(b) represents all the elements of the power system shown in Figure 13.2(a) converted to CIM objects, which have different attributes in a common base for each element.

Further work may provide service descriptions and protocol specifications for electricity metering. Good data models will enhance the return on investment of the Smart Grid by enabling more applications to use the data, and improving its value for newly developed analytics that may examine the information in novel ways.

Ontology-based strategies are commonly used with success in creating and manipulating data models since they provide easy export or translation to Extensible Markup Language (XML) or Unified Modeling Language (UML). Data without a clear and intuitive meaning cannot be used. Within the context of a single-user interface application, developers strive to make the meaning clear, but when data is transferred to another system, the meaning may be lost.

The objects of the real world are represented by means of classes, which conform to the basic unit that encompasses all their information. Thanks to these classes, the elements of power systems (transformers, sectionalizers, circuit breakers generators, etc.) can be modeled. UML is used to represent the different relations among the classes that the model uses. These relations among classes are known as association. For each class, the attributes can be observed, i.e., the information or data that represent the object.

The UML, a standardized, object-oriented method that is supported by various software tools, is used as the descriptive language. CIM is used primarily to define a common language for exchanging information via direct interfaces or an integration bus and for accessing data from various sources.

The CIM model is subdivided into packages such as basic elements, topology, generation, load model, measurement values, and protection. The sole purpose of these packages is to make the model more transparent. Relations between specific types of objects being modeled may extend beyond the boundaries of packages.

The electrically conductive connections between the elements are defined via terminals and nodes (connectivity nodes). Every conductive element has one or more terminals. A terminal connects the element, such as a generator, or one side of, for example, a circuit-breaker, to a node. A node can hold any number of terminals and provides an impedance-free connection linking all elements connected to it. A topology processor can determine the current network topology via these relations and with the current states of the circuit breakers. This topology model can also be used to describe gas, water, district heating, and other networks for tasks such as modeling interconnected control centers.

Figure 13.3 is an example of a diagram of the classes of a transformer, represented in UML. The relationship of the different classes can be seen for some of the subclasses like the tap changer, transformer winding, and the conducting equipment. In order to be simpler, the attributes of each class were hidden but they could be visible if required.

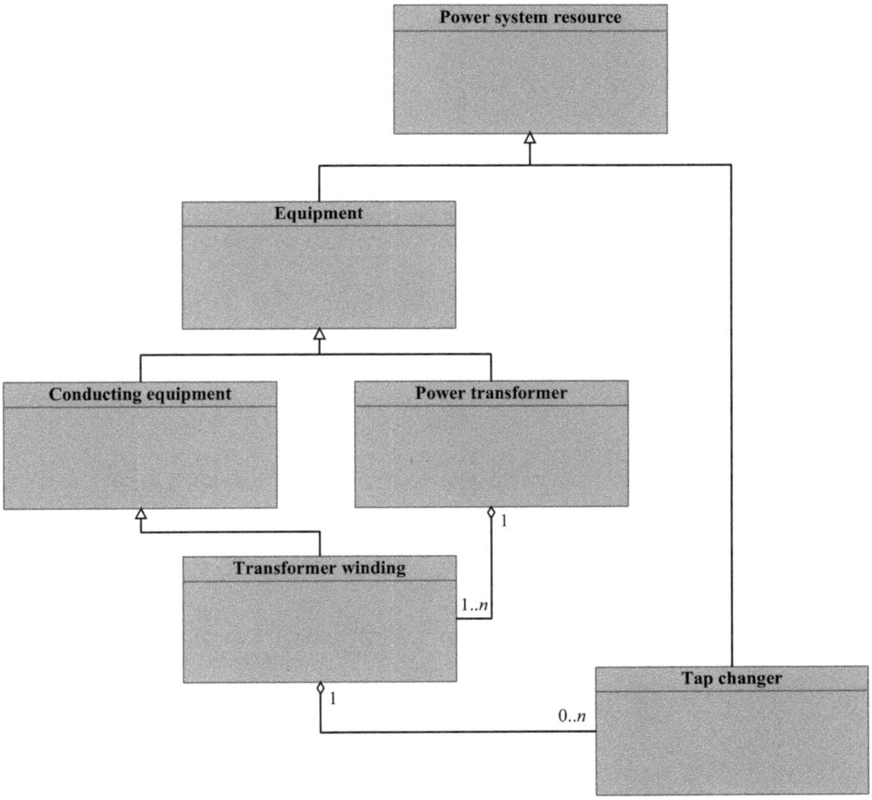

Figure 13.3 Transformer class diagram

When the elements to be modeled have been already defined, the XML is utilized to create instances, which constitute the physical model of the elements of the real world power network. Figure 13.4 shows the XML representation of some elements of a power system modeled with NEPLAN [6]. In this case, it is clear that the modeled elements have been already defined, and it is possible to create the instances in the physical model of the power system. The illustration here represents a 13.8 kV Bus 9 in XML. The rectangle in red of part (a) of the figure indicates the node that will be represented in XML. The rectangle in red of part (b) is a portion of the code generated in XML to illustrate how the information is represented with this format.

A modern network control system provides a service-oriented architecture (SOA) with standardized process, interface, and communication specifications based on standards IEC 61968 and IEC 61970 [7,8]. They form the basis for integrating the network control system in the enterprise service environment of the utility.

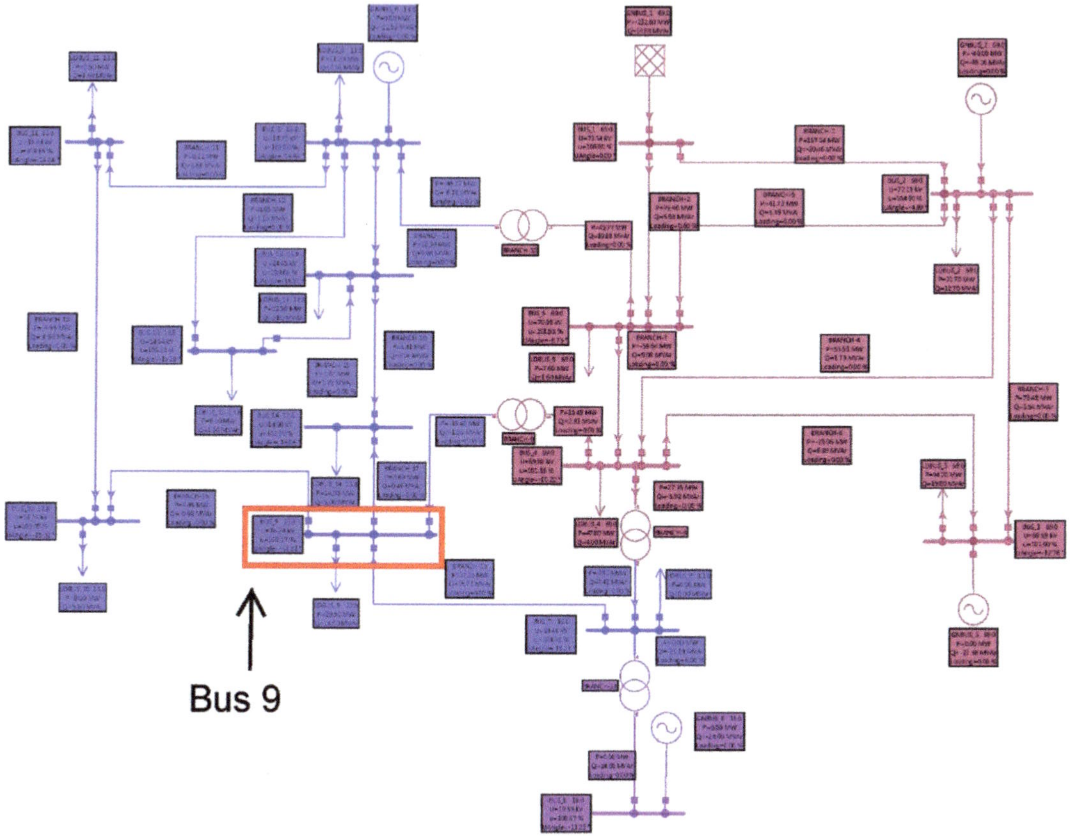

Figure 13.4 Example of (a) power system representation and (b) CIM/XML representation [6]

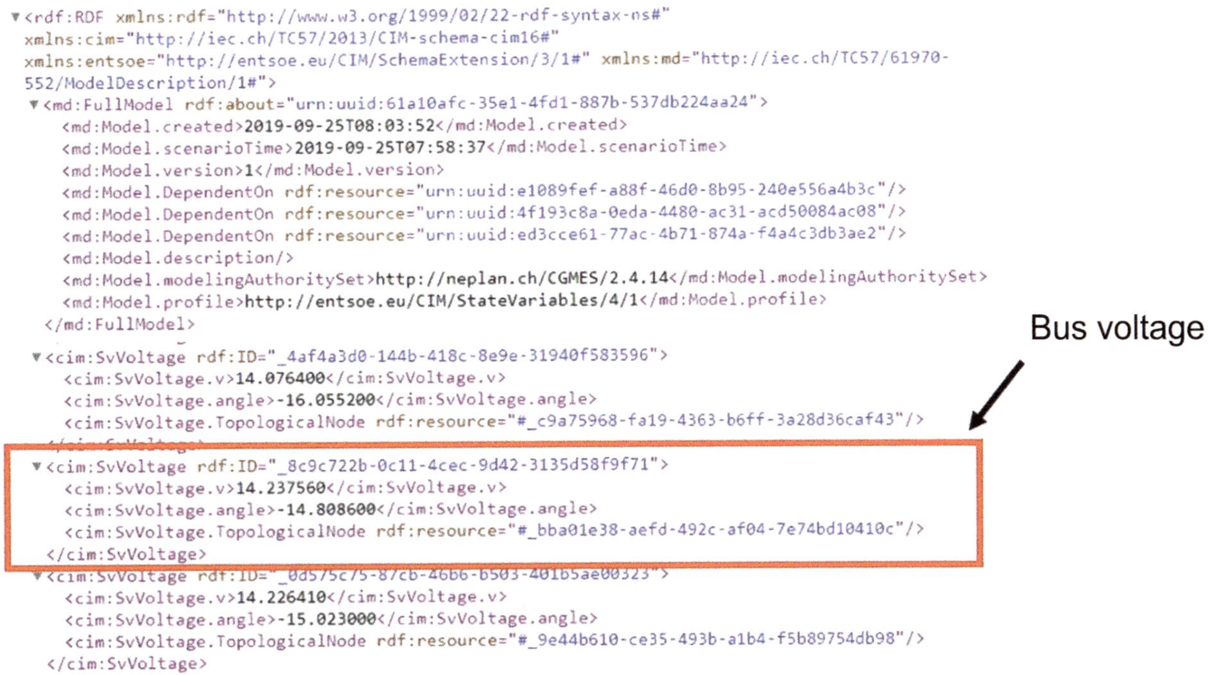

Figure 13.4 (*Continued*)

13.4 Implementation of common information models

In order to implement an interoperable model of information as in the case of the CIM model, it is required to identify all the elements of the physical infrastructure into which all the applications and functions will be integrated.

It should be verified that the utility has a data bus that permits it to exchange messages among the different functions and services and that facilitate the integration of the system. Thus, it is of vital importance to implement at least one business bus. Usually, these buses utilize the SOA, which is an assembly of guidelines designed to support user-defined functional interfaces. These interfaces take care of the marketing activities, the connection of the service, and the connection of new clients. Normally, for the messages exchanged among services, the XML is utilized.

SOA only establishes the structure, architecture, media, and formats of messages. Nevertheless, a message sent between applications can contain any sequence of information in any order, structure or hierarchy. Semantic SOA (SSOA) is the application in the SOA service that provides a model of information and a common language to facilitate the integration of applications into the established infrastructure, so that the messages sent between applications follow a plan, or common semantic model.

In Figure 13.5, one can see that the implementation of an Enterprise Service Bus together with the SSOA allows the integration of all the functions within the CIM model, significantly reducing the number of interfaces to implement. In this figure, the elements for implementing interoperability in information systems are represented by Power System Analysis Tools, GIS, SCADA, Event Log, Asset Management, and an Electromagnetic Transient Programme.

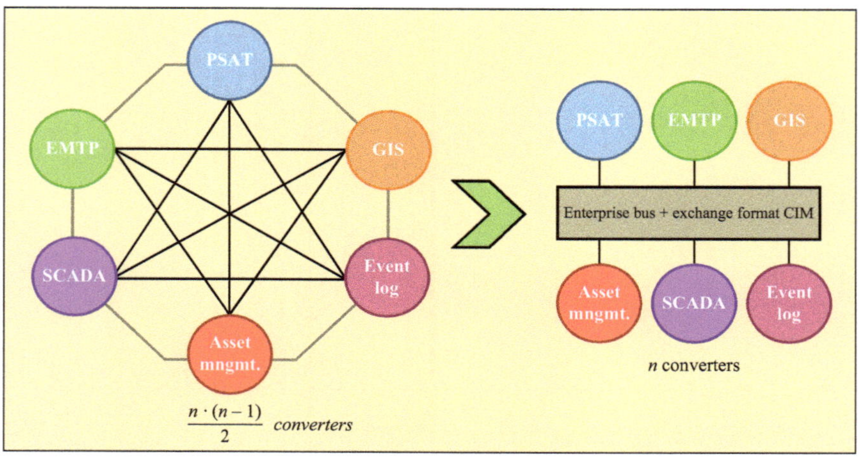

Figure 13.5 Elements for implementing interoperability in information systems

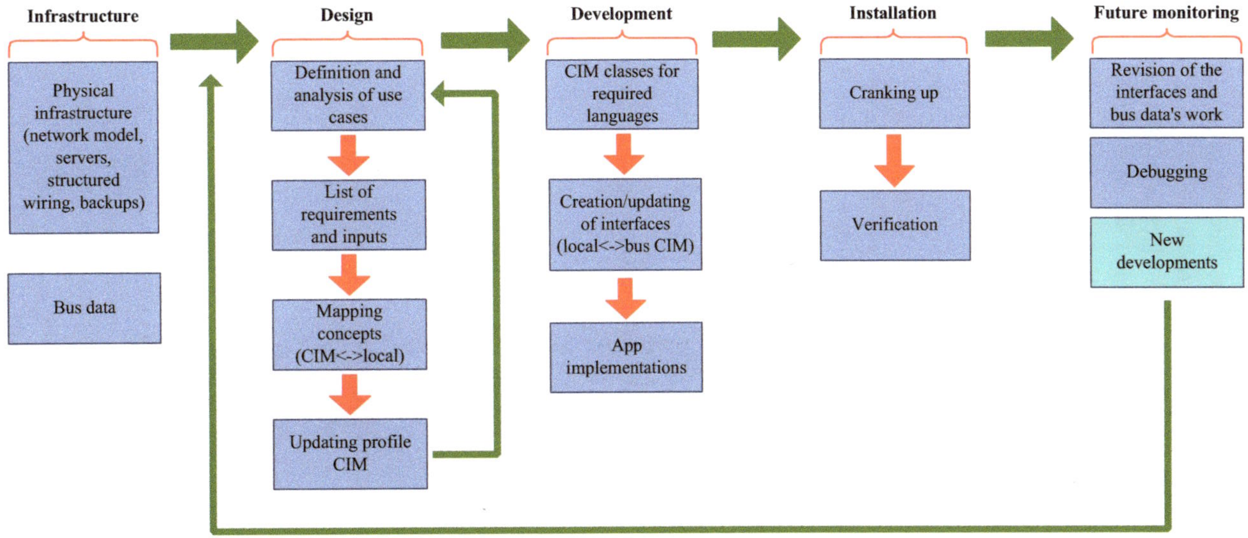

Figure 13.6 Stages in a CIM implementation

After having characterized all the existing resources, the required function implementations are designed. It is of vital importance that the purpose of the system be completely documented and defined through a case of use. For example, to guarantee integration within a system that generates work orders to a team within a GIS, it is necessary to have well-defined reasons for this integration. In this case, it is necessary to generate automatic work orders to the local personnel. This helps in defining the data that will be interchanged, for example, the name of the worker, the type of event, the place of the event, the status, and availability.

With these elements identified, the CIM mapping is performed. The relationships among the characteristics that will be exchanged among the applications in the model are expressed by a good approximation of their condition/value in the system. This procedure is followed for each case of use, that is to say, for each one of the elements that have to be integrated. The integration and the characteristics shared with the other systems should be well documented.

After all the aspects of the design are defined, the phase of development starts, where the classes that will be utilized in order to create the CIM profile are determined. This profile is nothing more than the representation of the function or system in terms of the model. In it, all the elements that are to be integrated fit perfectly in this reference that is common to them all. With this profile the applications are installed. As in all software development, it is necessary to verify that the interfaces function adequately and that they are revised constantly to correct errors that are found later.

The stages required in a CIM implementation are shown in Figure 13.6.

References

[1] IEEE. IEEE Std. 2030 – IEEE Guide for Smart Grid Interoperability of Energy Technology and Information Technology Operation With the Electric Power System (EPS), End-Use Applications, and Loads. IEEE, New York, NY, USA, 2011.

[2] Electric Power Research Institute (EPRI). Report to NIST on the Smart Grid Interoperability Standards Roadmap. National Coordinator for Smart Grid Interoperability, USA, 2009 [Online]. Available: https://www.energy.gov/oe/downloads/report-nist-smart-grid-interoperability-standards-roadmap.

[3] National Rural Electric Cooperative Association (NRECA). MultiSpeak and the NIST Smart Grid. National Rural Electric Cooperative Association (NRECA), 2014.

[4] IEC Technical Committees & Subcommittees. TC 57, Power Systems Management and Associated Information Exchange [Online]. Available: https://www.iec.ch/dyn/www/f?p=103:7:0::::FSP_ORG_ID:1273.

[5] IEC Subcommittees and Working Groups. TC 57 – WG14, System Interfaces for Distribution Management (SIDM) [Online]. Available: https://www.iec.ch/dyn/www/f?p=103:14:4780461338828::::FSP_ORG_ID,FSP_LANG_ID:2393,25.

[6] NEPLAN. IEEE14 – AC Steady State Load Flow. NEPLAN, 2018.
[7] IEC 61968. Common Information Model (CIM)/Distribution Management. 2ed., International standard, TC57, 08-03-2019.
[8] IEC 61970. Common Information Model (CIM)/Energy Management. International standard, TC57, 2005.

Further reading

[1] De Vos, A., Widergren, S. E., and Zhu, J. "XML for CIM Model Exchange," PICA 2001. Innovative Computing for Power - Electric Energy Meets the Market. 22nd IEEE Power Engineering Society. International Conference on Power Industry Computer Applications (Cat. No.01CH37195), Sydney, NSW, Australia, pp. 31–37, 2001.
[2] Office of the National Coordinator for Smart Grid Interoperability. NIST Framework and Roadmap for Smart Grid Interoperability Standards, Release 3.0. NIST Special Publication 1108r3, 2014.
[3] Pradeep, Y., Seshuraju, P., Khaparde, S. A., and Joshi, R. K. "CIM-based connectivity model for bus-branch topology extraction and exchange," IEEE Transactions on Smart Grid, vol. 2, no. 2, pp. 244–253, 2011.
[4] Tosic, V. and Djordjevic-Kajan, S. "The Common Information Model (CIM) Standard – An Analysis of Features and Open Issues," in Proceedings of TELSIKS'99 – 4th International Conference on Telecommunications in Modern Satellite, Cable, and Broadcasting Services, Nis, Yugoslavia, October 1999, vol. 2, pp. 677–680.

Index

www.ingramcontent.com/pod-product-compliance
Lightning Source LLC
LaVergne TN
LVHW051048210926
848486LV00001B/47